Encyclopedia of Petroleum: Biotechnology and its Environmental Applications

Volume I

Encyclopedia of Petroleum: Biotechnology and its Environmental Applications

Volume I

Edited by **Andy Margo and**
Michael Dedini

NY RESEARCH PRESS
New York

Published by NY Research Press,
23 West, 55th Street, Suite 816,
New York, NY 10019, USA
www.nyresearchpress.com

Encyclopedia of Petroleum: Biotechnology and its Environmental Applications
Volume I
Edited by Andy Margo and Michael Dedini

International Standard Book Number: 978-1-63238-154-5 (Hardback)

Printed in China.

Contents

Preface

The changing world requires newer and better methods of pollution free use of resources. There are emerging fields that work towards such a goal, and environmental technology is one such field which has roles in advancing fields of industry. Environmental Biotechnology is the system of engineering and scientific knowledge that uses microorganisms and their related products to prevent any pollution of the environment through biological treatment of solid, liquid and gaseous wastes as well as biological remediation of polluted environments and monitoring environmental processes. The main reason that biotechnology is being considered as a lead in waste treatment is because it has an economically and technically reasonable rate of biodegradability for high volume waste. Any type of biotechnological treatment of waste can be accelerated using optimum environmental factors. It is mainly used in the biodegradation of organic matter in wastewater and biodegradation of hazardous substances in water. Biotechnology is also useful in the field of petroleum. Since most of the world's crude oil is trapped in reservoirs, it is often impossible to recover it by conventional methods. The search for an environmentally friendly and economical solution has led researchers to apply biotechnology in the process of oil recovery. The use of microbes to enhance oil recovery has become an alternative method of recovery that uses microorganisms and their products. As such, the fields of petroleum and biotechnology are fast becoming intertwined.

This book aims to compile research done in both fields and study the data provided. I am thankful to those whose hard work has resulted in the successful completion of this book.

Editor

Adsorption and Desorption Behavior of Chlorotriazine Herbicides in the Agricultural Soils

Anil Kumar Singh and Swaranjit Singh Cameotra*

Institute of Microbial Technology, Sector 39 a, Chandigarh 160036, India

Abstract

The chlorotriazine herbicides have been applied worldwide for controlling of broadleaf weeds. These herbicides are known carcinogen and thus, their occurrence in surface and ground water is a cause of concern. In the present study, adsorption–desorption behavior of chlorotriazine herbicides namely simazine and atrazine was investigated. For the study, two different types of agricultural soil A (loam, 13.7% organic matter content) and soil B (clay-loam, 4.8% organic matter content) was collected from Punjab, India. The adsorption–desorption study was carried out by batch equilibration procedure. The data fitted well into the Freundlich equation. The adsorption K_f values indicated that the simazine has more soil adsorption ability than atrazine. The extent of herbicides adsorbed to the soil was more under low pH conditions and decreased with the increase in pH value. Increase in temperature decreased herbicides adsorption ability for both the soil types. The Gibb's free energy (ΔG) values were found less negative with the increase in temperature. Atrazine exhibited better desorption behavior than simazine. Soil A with high organic matter content exhibited sorption–desorption hysteresis. The results indicate that the soil organic matter content and aqueous solubility play an important role in the adsorption–desorption behavior of chlorotriazine herbicides. The present understanding of chlorotriazine herbicides sorption–desorption process will help to determine the herbicides fate and availability in soil, biodegradation, runoff and leaching.

Keywords: Simazine; Atrazine; Organic matter; Clay; Sorption–desorption hysteresis

Introduction

The chlorotriazine herbicides namely atrazine and simazine have been widely used since 1950s for controlling of broadleaf weeds and annual grasses in agricultural and non-crop fields [1]. A very small fraction of applied herbicides reaches the target weeds, while the large fraction enters the environment, causing soil and water pollutions [2]. Consequently, the chlorotriazine herbicides have been detected in the surface and groundwater of the United States, Europe and Australia [3,4].

United States Environmental Protection Agency (EPA) has reported atrazine and simazine as potential carcinogenic herbicides. Owing to their carcinogenic potentials, the presence of these herbicides in water is a cause of concern for public and regulatory agencies [5,6]. Atrazine and simazine has been reported to cause tumors and cancers, including breast, ovarian, and uterine, leukemia and lymphoma [1,7-9]. Atrazine is an endocrine disrupting chemical, interrupting regular hormone function, causing birth defects, reproductive tumors and weight loss particularly in amphibians as well as humans [7-9].

Literatures suggest that adsorption of herbicide to the soil is the key process that affects their ecotoxicological impact, environmental mobility and the rate of degradation [10,11]. The desorption process of herbicides is also important since it determines the release rate and the potential mobility of herbicides in the soil [10]. The adsorption-desorption process of herbicide is influenced by several factors like organic matter content, soil texture, pH, temperature, etc. [10-13]. The risk of herbicide entering the surface water and groundwater mainly comes from the adsorbed herbicide present in the soil. To protect surface and groundwater from pesticide contamination and evaluate their impact, extensive knowledge concerning degradation and sorption–desorption processes in the environment is required [13,14].

The use of herbicides in India had gradually increased in recent years especially in the state of Punjab [15]. In 1999-2000 Punjab used nearly 923g/ha of pesticides which was highest among all the Indian states [15]. According to the Directorate of Plant Protection, Quarantine and Storage, Government of India, Punjab is among the top pesticide consuming state in India, thus having high risk of soil and water pollution. In Punjab chlorotriazine herbicides are mainly used for non-selective weed control on waste land and selective weed control in crops such as sorghum, maize, sugarcane, etc [11].

The aim of this work was to determine adsorption-desorption behavior of chlorotriazine herbicides namely atrazine and simazine in the typical agricultural soils of Punjab, India. Kinetics and isotherm adsorption experiments were carried out to have a better insight into the adsorption process. Factors affecting the adsorption efficiency of herbicides such as initial pH of the solution and temperature were also studied.

Material and Methods

Chemical and physical characterization of agriculture soil

The two representative soils used in the present study were collected from two different agricultural sites namely soil A (Rajpura; 30.48°N, 76.6°E) and soil B (Kharar; 30.74°N, 76.64°E) from Punjab, India. The soils were collected from the surface layer (0–20 cm depth), air-

***Corresponding author:** Swaranjit Singh Cameotra, Institute of Microbial Technology, Sector 39 A, Chandigarh 160036, India, Tel: 0172-6665233; Fax: 0172-2690632/585

dried, and sieved to pass through 2.0 mm mesh. Chemical and physical characterization of agriculture soil was performed following previously described methods [16,17].

Chlorotriazine adsorption to the soil

The atrazine and simazine herbicides adsorption isotherms of both the soils were performed using the batch equilibration procedure. To minimize changes in the ionic strength and avoid dispersion, 5mM $CaCl_2$ solution was used as a background. First, herbicides were dissolved in methanol. Then the methanol solution was diluted by 5mM $CaCl_2$ solution to obtain the working solutions of different concentrations (5, 10, 20, 50 and 100 mg/l). The concentration of methanol in the working solution was kept constant at 0.1%. Aliquots (8ml) of aqueous herbicide solution were added to 2g of soil. Soil slurries were shaken on a horizontal shaker at 200rpm for a period of 24 h at 25 ± 1°C. Aliquots (600µl) were sampled from the supernatant at different time interval of 0, 0.5, 1, 6, 12 and 24 h. The samples were mixed with methanol 1:1 v/v, centrifuged (4000 rpm for 5 min) for soil separation and then quantified by High-Performance Liquid Chromatography (HPLC). The amount of pesticide adsorbed was calculated as the average of the three replicates. A blank sample was also kept to assess the effects of chlorotriazine herbicide adsorption onto the tubes and the following possible degradation during the process.

Chlorotriazine desorption from the soil

Desorption experiments were performed instantly after the adsorption experiments. Supernatant was decanted and the herbicides residue present in the soil was analyzed. Then equal volume of the fresh 5 mm $CaCl_2$ solution was added. The mixtures were shaken on a horizontal shaker at 200rpm for 8 h at 25 ± 1°C. These steps were repeated three times consecutively.

Chlorotriazine herbicides adsorption enthalpy

Chlorotriazine herbicides adsorption enthalpy on both the soils was determined using the batch experiments as described above. The adsorption process was performed at two different temperatures (25°C and 35°C). Herbicides concentrations used for the study were 5, 10, 20 and 50 mg/l, respectively.

Effect of pH on chlorotriazine herbicides adsorption

The pH values of 5mM $CaCl_2$ solutions were adjusted to 5.0, 7.0 and 9.0 with the help of 5N HCl or 5 N NaOH. The chlorotriazine herbicides were then added to the solution so as to have concentrations of 5, 10, 20 and 50 mg/l. Further, the experiment was performed in the same way as done for the kinetic study described above.

Chlorotriazine quantification

Chlorotriazine herbicides used in the present study were quantified using HPLC. Chromatographic analysis was performed with a Prominence HPLC system (Shimadzu, Japan) equipped with a UV-VIS detector. The column was Phenomax- C18 (5µ, 250 mm x 4.6 mm) while mobile phase was methanol: water (40/60, v/v) at a flow rate of 0.5mg/ml. The samples were injected with constant injection volume of 20 µl and detected at 254 nm absorbance wavelength. The chlorotriazine herbicides were identified and quantified by comparing the retention peaks of samples and authentic standard procured from Sigma-Aldrich, USA. External calibration curves with standard solutions between 1 and 50 mg/l were used for the calculations. The chemical structure of atrazine and simazine is shown in Figure 1.

Data analysis

Adsorption - desorption data were fitted on the Freundlich model: $S = K_f \bullet C^{1/n}$, where S is the amount of herbicides sorbed per mass of soil (mg/ kg soil), C is the concentration of the herbicide in the liquid phase (mg/l), K_f is the Freundlich sorption coefficient (l/kg), and

n is a dimensionless parameter. Sorption coefficients K_f and n were determined by plotting the logarithmic form of the Freundlich equation.

The adsorption coefficient (K_f) was also calculated as a function of the Organic Carbon (OC) content and Organic Matter (OM) of the soil following equations $K_{OC} = K_f/\%OC \times 100$ and $K_{OM} = K_f/\%OM \times 100$, respectively. The free energy change (ΔG, cal/ mol) in the adsorption process was calculated by the following equation $\Delta G = -RT \ln K_{OM}$, where R is the gas constant (2.0 cal/K/ mol), and T is Kelvin temperature.

Hysteresis coefficient, H, is calculated for the adsorption–desorption isotherms according to the following equation H = (n desorption)/ (n adsorption), where n desorption and n adsorption are the Freundlich constants obtained for desorption and adsorption isotherms, respectively.

Results and Discussion

Chemical and physical characterization of soil

Table 1 shows chemical and physical features of agricultural soil samples used in the present study. Soil A showed higher organic matter content (13.7%) than soil B (4.8%). The chlorotriazine herbicides used in the present study were not detected in the soil samples, thus suggesting absence of atrazine and simazine pre-contamination.

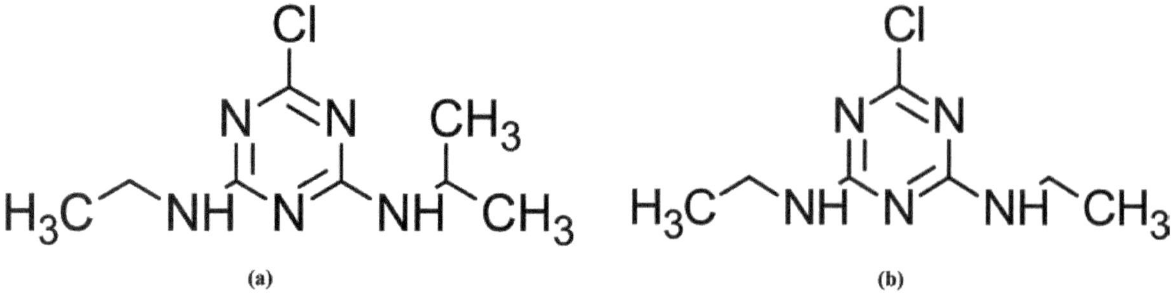

Figure 1: Chemical structure of **(a)** atrazine [1-Chloro-3-ethylamino-5-isopropylamino-2, 4, 6-triazine] and **(b)** simazine [6-chloro-N, N'-diethyl-1, 3, 5-triazine-2, 4-diamine]. The solubility of atrazine and simazine in water is 5 mg/l and 30mg/l (pH 7, 25°C).

Chlorotriazine adsorption kinetics

A preliminary adsorption experiment was performed to determine the contact time required for attaining adsorption equilibrium. As evident from Figure 2 the adsorption kinetics exhibited two distinct stages in both the soil types. The initial step was a very rapid adsorption followed by a slow adsorption. This observation is due to the fact that in initial stage a large number of vacant surface sites were available for adsorption, but once the vacant surfaces were occupied, the repulsive forces between the solute molecules of the solid and bulk phases came into play, thus delaying the adsorption process. Similar observation has been reported for few other organic herbicides [18,19].

Herbicides adsorption equilibrium was reached within 5h of incubation in both the soil types. The maximum chlorotriazine herbicides adsorption was observed in soil A. This may be due to high organic matter content in the soil [2]. As compared to atrazine, simazine exhibited better adsorption on both the soil. This may be attributed to the different water solubility of the herbicides [19]. Simazine with very low water solubility has greater tendency to get adsorbed to the soil.

Chlorotriazine adsorption isotherms

In the present study, no detectable amount of chlorotriazine degradation was found in the supernatant during the adsorption process. Thus, the reduction in the herbicide concentration in the solution was considered to be due to soil adsorption. The chlorotriazine herbicides exhibited non-linear sorption isotherms on soil A and B (Figure 3). Table 2 shows the adsorption parameters of chlorotriazine in soils A and B. The sorption of chlorotriazine herbicides in both the soil types fitted well within the Freundlich model. According to Calvet shape of the adsorption isotherm provides information regarding the adsorption mechanisms [19]. The value of the exponent 1/n gives an indication of the favorability and the capacity of the adsorbent-adsorbate system. Values $n > 1$ represent favorable adsorption conditions and $1 < n < 10$ shows beneficial adsorption. The soils A and B showed non-liner isotherm ($n \neq 1$), which indicated that chlorotriazine could adsorb to the clay fraction of soil as well as organic matter present in the soil. Similar adsorption isotherms have been previously reported for simazine [13,20] and atrazine adsorption [21,22]. Apart from soil organic matter, clays can also be an important sorbents for the adsorption of organic pesticides in the soil [23]. Simazine can sorb on hydrophobic microsites of smectites [24], and montmorillonite clay minerals [23]. Celis et al. [25] has reported that atrazine and simazine initially sorbed as neutral species on hydrophobic microsites of clay (montmorillonite).

Soil	pH	Organic matter (%)	Organic carbon (%)	Conductivity (µS/cm)	Texture	Particles Size		
						Clay	Silt	Sand
A	6.4	13.7	8.6	186.2	Loam	25	30	45
B	8.0	4.8	2.1	178.1	Clay Loam	37	36	27

Table 1: Physical and chemicals properties of the agricultural soils collected from Punjab, India.

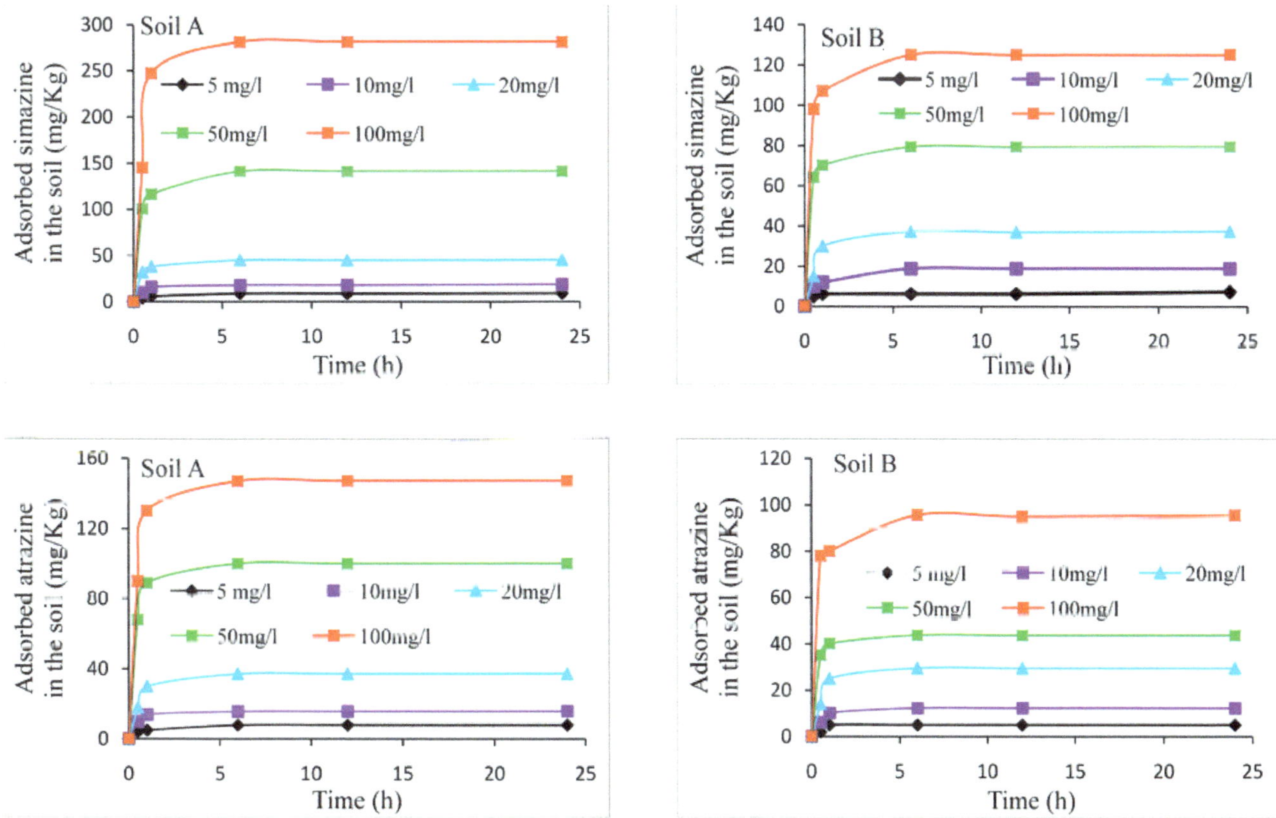

Figure 2: Adsorption kinetics of simazine and atrazine on agricultural soils A and B with different initial herbicide concentrations. The values represent an average value of three independent experiments.

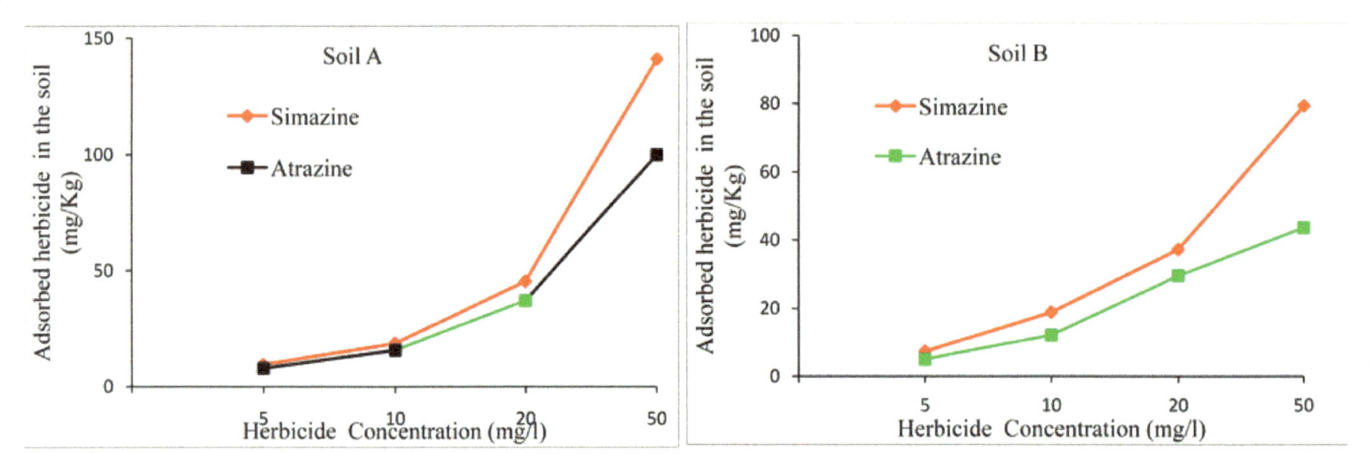

Figure 3: Adsorption isotherms of simazine and atrazine in soils A and B.

Herbicides	Temperature (K)	Soil	*Kf* (mg/Kg)	1/n	R^2	*Koc* (mg/kg)	*Kom* (mg/kg)
Simazine	298	A	1.1581	0.1496	0.9978	13.4662	8.4532
		B	0.9138	0.3251	0.9868	43.5142	19.0357
	308	A	0.8559	0.3293	0.9933	17.8312	6.2474
		B	0.6039	0.4747	0.9925	28.7571	12.5812
Atrazine	298	A	1.0113	0.2121	0.9889	11.7593	7.3817
		B	0.8395	0.2865	0.9811	39.9762	17.4896
	308	A	0.8009	0.3598	0.9846	9.3128	5.8459
		B	0.5898	0.3876	0.9942	28.0857	12.2875

Table 2: The adsorption parameters of simazine and atrazine in agricultural soils at different temperature. The values represent an average value of three independent experiments.

Soil organic matter plays a crucial role in the adsorption of organic pesticides [2]. In the present study soil A with high organic matter content exhibited better herbicide sorption ability. In the soil A 47.5-70.5% and 40.0-50.0% of available simazine and atrazine was adsorbed, respectively. The soil B, with less organic matter content exhibited 36.4-47.1% and 25.0 -36.9% of adsorption for simazine and atrazine, respectively. The sorption of s-triazines on organic matter is governed by H bonds and proton transfer between s-triazines and acidic groups of humic substances [10]. The K_{oc} values observed in the present study suggest that mineral constituents as well as soil organic matter contribute towards sorption of herbicides. Literatures suggest that the adsorption of chlorotriazine is more hysteretic for humic acid than for the clay [13]. Herbicide desorbs relatively easily from clay than from the organic matter, suggesting a stronger interaction between chlorotriazine and organic matter than between the chlorotriazine herbicides and clay. Thus, soil B that has more clay and less organic matter exhibited comparative low herbicide sorption.

Chlorotriazine adsorption enthalpy

Temperature play crucial role in determining the extent of herbicide that gets adsorbed on the soil. To determine chlorotriazine herbicides adsorption enthalpy in the agricultural soils, adsorption experiments were conducted at 25°C and 35°C. Table 2 shows Freundlich constants at 25°C and 35°C. At higher temperature, K_f values declined for both the herbicides used in the present study. The values of the Freundlich exponent 'n' also decreased when the temperature increased from 298 K to 308 K. In the soils A the ΔG values for simazine changed from -1272 kJ/mol to – 1128 kJ/mol when the temperature was increased from 298K to 308 K, thus indicating the feasibility and spontaneous nature of adsorption process at lower temperature. A similar observation was also made in soil B where ΔG values increased from -1756 kJ/mol to -1559 kJ/mol with change in the temperature. In the soils A the ΔG values for atrazine was found to change from -1191 kJ/mol to – 1087 kJ/mol when the temperature was increased from 298K to 308 K. A similar observation was also made in soil B where ΔG values changed from -1705 kJ/mol to -1545 kJ/mol. Decrease in ΔG value suggest the feasibility and spontaneous nature of the adsorption process [13].

The chlorotriazine herbicides adsorption increased at lower temperature, indicating that the process is exothermic. The exothermic adsorption of chlorotriazine on soil may be explained by the fact that the adsorbate-adsorbent bonds were weakened with increasing temperature. Temperature induced change in aqueous solubility of herbicides may also influence the adsorption process [26]. Higher temperatures increase herbicides solubility and lower the tendency to get sorbed on the soils. During summer in Punjab temperature generally remain higher than 25°C. Thus, the probability of surface and ground water contamination with chlorotriazine herbicides is more during summer rains than in the winter. Chlorotriazine herbicides applied during winter (when temperature remains below 25°C) have more tendency to remain adhered to the soil.

Effect of pH on chlorotriazine adsorption

The adsorption capacity of herbicides is significantly influenced by the initial pH of the solution. In the present study adsorption of chlorotriazine herbicides in the soils were more at low pH and decreased with the increase in pH of the solution (Table 3). Thus, indicating that the availability of the adsorption sites to chlorotriazine molecules reduced as the pH value increased. This observation is in accordance with the general trend observed for many other pesticides [13,27,28]. The number of protonated surface on the adsorbent increases with decreasing pH and thus the coulombic attractive forces between more positively charged surfaces and herbicide increases at lower pH values [13]. In soil A, the K_f value for simazine was 1.6235 mg/kg at pH 5, which is 3.2 times higher than those of corresponding value at pH 9. For soils B the K_f value for simazine at pH 5 was 1.1610 mg/kg which is 2.3 times higher than the corresponding value at pH 9. Similarly in soil A, the K_f value for atrazine was 1.4938 mg/kg at pH 5, which was 2.7 times higher than those of corresponding value at pH 9. In soils B, the K_f value at pH 5 and pH 9 changed by ~ 1.5 times only. The differences may be due to the different concentration of organic matter. Soil A with high organic matter content has a better buffer capacity for pH change [13]. Organic matter present in the soil is known to act as a pH buffering agent in nature [29].

Desorption behavior of chlorotriazine on soils

Looking into the desorption process of herbicides is vital since it determines the release rate and the potential mobility of herbicides in the soil as well as the treatment strategy for the contaminated soils. The herbicides with a lower desorption rate may possess higher risk to the successive crops. Thus, desorption kinetic studies were conducted to assess the desorption potential of adsorbed chlorotriazine herbicides and the results are shown in Table 4. The rate of desorption was comparatively fast in soil B as compared to the soil A. The organic matter content may play an important role in the desorption process [13]. Soil B with little organic matter content exhibited the highest desorption K_f for both the herbicides (Table 4). The K_f values of the desorption process suggested that atrazine has better desorption ability as compared to simazine. This observation may be attributed to the different solubility of atrazine and simazine. Atrazine with a comparatively better water solubility has higher tendency to enter the aqueous phase and get desorbed from the soil.

Hysteresis in adsorption and desorption

Occurrence of hysteresis in adsorption-desorption reaction in the soils suggests that adsorption of pesticides occurs with a limited degree of reversibility depending upon both the physico-chemical properties of the molecules and the soils involved in the process [13,30]. Generally, a value of hysteresis index close to 1 means that desorption and sorption has similar rate, therefore, hysteresis is absent. On the other hand, a value of hysteresis index less than 1 indicates that the rate of desorption is lower than that of sorption, thus hysteresis takes place [31]. In the soil A, the hysteresis index was less than 1 for both the herbicides (Table 4). However, in the soil B the hysteresis index for simazine and atrazine was less than 1 and nearly equal to 1, respectively. The soil B with low organic matter content and atrazine with a higher water solubility supported desorption process in the present study. Results suggest that organic matter and aqueous solubility of herbicides significantly influence the adsorption–desorption hysteresis. The entrapping of herbicide molecules within the condensed soil organic matter makes a significant contribution to adsorption–desorption hysteresis. In soils with high organic matter content, the adsorption–desorption hysteresis is mainly controlled by organic matter. However, involvement of the soil mineral composition in desorption behavior of herbicides cannot be ignored [14]. The hysteresis index values observed in the present study, suggest that the adverse effect of simazine and atrazine on the succession crop should be noticed, especially when chlorotriazine were applied on soils with high organic matter content, for example, soil A. While applying atrazine to soil with low organic matter content like soil B, attention should be paid to the risk of groundwater and surface water contamination from the herbicide.

Conclusions

This study contributes to the understanding of simazine and atrazine sorption-desorption process in the two types of agricultural soil of Panjab, India. Organic matter content of soil has significant influence on the adsorption of simazine and atrazine. Soil with high organic matter content has better herbicide's adsorption ability. Soils of Punjab with higher organic matter content exhibited enhanced simazine and atrazine sorption capacity. The increase in temperature and pH of the solution had a negative effect on the adsorption process. Organic matter content also contributed significantly towards simazine and atrazine desorption process. The results from the present study would help in designing of effective herbicide management strategies in Punjab, India. This also highlights the feasibility of adverse effect of soil adsorbed chlorotriazine on the succession crops.

Herbicides	pH	Soil	*Kf* (mg/Kg)	1/n	R^2	*Koc* (mg/kg)	*Kom* (mg/kg)
Simazine	5	A	1.6235	0.0376	0.9928	18.8779	11.8503
		B	1.1610	0.2881	0.9926	55.2857	24.1875
	7	A	1.1581	0.1496	0.9978	13.4662	8.4532
		B	0.9138	0.3251	0.9868	43.5142	19.0357
	9	A	0.5071	0.2296	0.9894	5.8965	3.7014
		B	0.4991	0.3005	0.9874	23.7809	10.4041
Atrazine	5	A	1.4938	0.0024	0.9794	17.3697	10.9036
		B	1.0956	0.2132	0.9878	52.1714	22.825
	7	A	1.0113	0.2121	0.9889	11.7593	7.3017
		B	0.8395	0.2865	0.9811	39.9762	17.4896
	9	A	0.5459	0.3301	0.9744	6.3476	3.9846
		B	0.6968	0.3231	0.9719	33.1809	14.5166

Table 3: The adsorption parameters of simazine and atrazine in agricultural soils at different pH. The values represent an average value of three independent experiments.

Herbicides	Soil	*Kf* (mg/Kg)	1/n	R^2	*Koc* (mg/kg)	*Kom* (mg/kg)	Hysteresis index
Simazine	A	0.5717	0.4738	0.9957	6.6476	4.1729	0.315
	B	0.4979	0.4940	0.9814	23.7095	10.3729	0.658
Atrazine	A	0.5238	0.4043	0.9936	6.090	3.8233	0.525
	B	0.6780	0.3015	0.9774	32.285	14.125	0.950

Table 4: The desorption parameters of simazine and atrazine in the soils.

Acknowledgements

Anil K Singh thanks the UGC, Govt. of India for his research fellowship. The authors thank the Director, IMTECH for providing the facilities for this work.

References

1. Kettles MK, Browning SR, Prince TS, Horstman SW (1997) Triazine herbicide exposure and breast cancer incidence: an ecologic study of Kentucky counties. Environ Health Perspect 105: 1222-1227.
2. Arias-Estévez M, López-Periago E, Martínez-Carballo E, Simal-Gándara J, Mejuto JC, et al. (2008) The mobility and degradation of pesticides in soils and the pollution of groundwater resources. Agr Ecosyst Environ 123: 247-260.
3. Cox L, Celis R, Hermosín MC, Cornejo J (2000) Natural soil colloids To retard simazine and 2,4-D leaching in soil. J Agric Food Chem 48: 93-99.
4. Cox C (1992) A New List of Carcinogenic Pesticides Used on Food. Journal of Pesticide Reform.
5. Eldridge JC, Tennant MK, Wetzel LT, Breckenridge CB, Stevens JT (1994) Factors affecting mammary tumor incidence in chlorotriazine-treated female rats: hormonal properties, dosage, and animal strain. Environ Health Perspect 11: 29-36.
6. Wetzel LT, Luempert LG 3rd, Breckenridge CB, Tisdel MO, Stevens JT, et al. (1994) Chronic effects of atrazine on estrus and mammary tumor formation in female Sprague-Dawley and Fischer 344 rats. J Toxicol Environ Health 43: 169-182.
7. Fan W, Yanase T, Morinaga H, Gondo S, Okabe T, et al. (2007) Atrazine-induced aromatase expression is SF-1 dependent: implications for endocrine disruption in wildlife and reproductive cancers in humans. Environ Health Perspect 115: 720-727.
8. Hayes TB, Khoury V, Narayan A, Nazir M, Park A, et al. (2010) Atrazine induces complete feminization and chemical castration in male African clawed frogs (Xenopus laevis). P Natl Acad Sci USA 107: 4612-4617.
9. Weber GJ, Sepúlveda MS, Peterson SM, Lewis SS, Freeman JL (2013) Transcriptome alterations following developmental atrazine exposure in zebrafish are associated with disruption of neuroendocrine and reproductive system function, cell cycle, and carcinogenesis. Toxicol Sci 132: 458-466.
10. Liu Y, Xu Z, Wu X, Gui W, Zhu G (2010) Adsorption and desorption behavior of herbicide diuron on various Chinese cultivated soils. J Hazard Mater 178: 462-468.
11. Singh P, Suri CR, Cameotra SS (2004) Isolation of a member of Acinetobacter species involved in atrazine degradation. Biochem Biophys Res Commun 317: 697-702.
12. Siripattanakul S, Wirojanagud W, McEvoy J, Limpiyakorn T, Khan E (2009) Atrazine degradation by stable mixed cultures enriched from agricultural soil and their characterization. J Appl Microbiol 106: 986-992.
13. Flores C, Morgante V, González M, Navia R, Seeger M (2009) Adsorption studies of the herbicide simazine in agricultural soils of the Aconcagua valley, central Chile. Chemosphere 74: 1544-1549.
14. Wu C, Zhang S, Nie G, Zhang Z, Wang J (2011) Adsorption and desorption of herbicide monosulfuron-ester in Chinese soils. J Environ Sci (China) 23: 1524-1532.
15. Gill MS (2003) Punjab society : perspectives and challenges. Concept Pub. Co New Delhi India.
16. Singh AK, Cameotra SS (2013) Efficiency of lipopeptide biosurfactants in removal of petroleum hydrocarbons and heavy metals from contaminated soil. Environ Sci Pollut Res Int.
17. Nelson DW, Sommers LE (1996) Total carbon, organic carbon, and organic matter. In: Methods of Soil Analysis, Part 3, 4th edn Bigham, JM. ASA and SSSA, Madison, WI, 961-1010.
18. Rama Krishna K, Philip L (2008) Adsorption and desorption characteristics of lindane, carbofuran and methyl parathion on various Indian soils. J Hazard Mater 160: 559-567.
19. Calvet R (1989) Adsorption of organic chemicals in soils. Environ Health Perspect 83: 145-177.
20. Albarrán A, Celis R, Hermosín MC, López-Piñeiro A, Cornejo J (2004) Behaviour of simazine in soil amended with the final residue of the olive-oil extraction process. Chemosphere 54: 717-724.
21. Dehghani M, Nasseri S, Amin S, Nadafi K, Taghavi M, et al. (2005) Atrazine adsorption-desorption behavior in Darehasaluie Kavar corn field soil in Fars Province of Iran. Iran J Environ Health Sci Eng 2: 221-228.
22. Krutz LJ, Senseman SA, McInnes KJ, Zuberer DA, Tierney DP (2003) Adsorption and desorption of atrazine, desethylatrazine, deisopropylatrazine, and hydroxyatrazine in vegetated filter strip and cultivated soil. J Agric Food Chem 51: 7379-7384.
23. Rafael CG, Juan C, Hermosin Gavino MC, Koskinen WC (1997) Sorption-desorption of atrazine and simazine by model soil colloidal components. Soil Sci Soc Am J 61: 436-443.
24. Lucía CM, Rafael CG, Hermosin Gavino MC, Juan C, Zsolnay A, et al. (2000) Effect of organic amendments on herbicide sorption as related to the nature of the dissolved organic matter. Environ Sci Technol 34: 4600-4605.
25. Celis R, Cornejo J, Hermosín MC, Koskinen WC (1998) Sorption of atrazine and simazine by model associations of soil colloids. Soil Sci Soc Am J 62: 165-171.
26. Gunasekara AS, Troiano J, Goh KS, Tjeerdema RS (2007) Chemistry and fate of simazine. Rev Environ Contam Toxicol 189: 1-23.
27. Boivin A, Cherrier R, Schiavon M (2005) A comparison of five pesticides adsorption and desorption processes in thirteen contrasting field soils. Chemosphere 61: 668-676.
28. Ertli T, Marton A, Földényi R (2004) Effect of pH and the role of organic matter in the adsorption of isoproturon on soils. Chemosphere 57: 771-779.
29. Helling CS, Chesters G, Corey RB (1964) Contribution of Organic Matter and Clay to Soil Cation-Exchange Capacity as Affected by the pH of the Saturating Solution. Soil Sci Soc Am J 28: 517-520.
30. Kovaios ID, Paraskeva CA, Koutsoukos PG, Payatakes ACh (2006) Adsorption of atrazine on soils: model study. J Colloid Interface Sci 299: 88-94.
31. Tang Z, Zhang W, Chen Y (2009) Adsorption and desorption characteristics of monosulfuron in Chinese soils. J Hazard Mater 166: 1351-1356.

Algae Fuels as an Alternative to Petroleum

Evan Stephens[1,2]*, Rocky de Nys[3,4], Ian L Ross[1,2] and Ben Hankamer[1,2]

[1]The University of Queensland, Solar Biofuels Research Centre, Queensland, Australia
[2]The University of Queensland, Institute for Molecular Bioscience, Queensland, Australia
[3]James Cook University, School of Marine and Tropical Biology, Queensland, Australia
[4]James Cook University, Centre for Sustainable Tropical Fisheries and Aquaculture, Queensland, Australia

Abstract

Here we examine the scale of petroleum consumption and the current lack of scalable petroleum alternatives. We highlight the contribution that macroalgae and microalgae can collectively make as feedstocks in the future energy mix, discuss recent advancements and current development pathways, and consider the potential and the limitations of these production systems for economic and environmental sustainability moving towards a scale of 10-20% of global petroleum consumption.

Keywords: Macroalgae; Microalgae; Fossil fuel; Biofuels; Bioproducts; Feedstock

Introduction

The scale of global fossil fuel consumption is massive. In 2011 almost 11,000 million tons of oil equivalent (Mtoe) were consumed in the form of oil, gas, and coal-based fossil fuels [1]. While fossil fuel consumption continues to increase (~32% increase from 2000 to 2011) to sustain our growing population and the advancement of developing nations (e.g. China and India), most of this increase in consumption comes from coal (~57% increase 2000-2011) and natural gas (~34% increase 2000-2011). Coal and gas production rates are currently increasing faster than consumption rates (figure 1) [1]. For petroleum oil however, consumption (~14% increase 2000-2011) has grown faster than oil production (~10% increase 2000-2011) [1] in the same period largely due to the plateau in production of conventional oil; a harbinger of some major challenges and changes to the future energy mix.

The petroleum industry can exploit a range of feedstocks for the production, processing and transformation of liquid hydrocarbons, of which conventional oil has, until recently, been the cheapest and most readily accessible. Currently, we are witnessing a necessary transition to a more diverse mix of feedstocks. A significant factor in the choice of future feedstocks will be the impact on global CO_2 emissions [2] for which targets have been set by many governments suggesting a trend that is likely to increase. If effective, these targets would impose a market premium increasingly favouring CO_2 neutral feedstocks, including fuels derived from algae. The technology of producing biocrude and more specific oils from algae has recently made significant progress in addressing economic obstacles and the scalability of photoautotrophic algae production is arguably high (up to ~10-20% of global fuel consumption), when water and nutrient recycling strategies are employed. Here we discuss how algae can contribute to the renewable production of biofuels and bioproducts relative to other sources.

Global Fuel Consumption

Hydrocarbon based fuels represent ~80% of our global energy consumption [1,3]. In comparison, biofuel technologies promoted as a means to address fuel security and climate change concerns, currently represent only ~0.6% of global fuel consumption [1]. Supply of first generation biofuels such as corn ethanol and rapeseed derived biodiesel have increased substantially, but this has been accompanied by a *food versus fuel* debate [4,5] which has now manifested as real world concerns, as biofuel production volumes from such sources are already threatening global food security [6,7]. Petroleum derived oil represents ~56% of global fuel consumption and the scale of this enormous consumption (~4.1×10^9 T in 2011) [1] presents an issue of scalability to advocates of biofuel and bioenergy technologies. Global production of the main food products in 2010 totalled ~7.6×10^9 T [8] yet the embodied energy of our global food harvest is dwarfed by the energy we consume from petroleum (Figure 2) which is around three times greater. Combined with the energy losses associated with processing biomass into energy dense fuels, any attempt to generate a globally significant contribution to petroleum alternatives with food crops or with systems that divert energy from food production, will incur significant economic and social impacts that must be properly evaluated and planned for.

The extraction of petroleum oil is becoming more costly and problematic [9,10] as easily extracted reservoirs are depleted. Ironically, the unprecedented prospect of complete arctic ice loss in the summer [11] is opening up new opportunities for oil exploitation, as well as other deep sea prospects (now ~70% of new oil discoveries are in the deep sea [10]). The technical challenges of liberating unconventional petroleum resources and the prospect of increasing oil prices (predicted rises to ~US\$120-140 bbl^{-1} by 2035 [12,13]) are indeed stimulating technology development in response. It remains controversial as to whether conventional oil production has peaked (largely depending on how "conventional oil" is defined). Predictions range from optimistic estimates of continued increases in production (e.g. [12,13]) through to more polarised debates such as those summarised by Kerr [14-17]. At some point, it is assumed that climate change related effects, particularly temperature rise and precipitation changes (with corresponding effects on food production), must heighten concerns not only over the effects of climate change but its real economic cost

***Corresponding author:** Evan Stephens, The University of Queensland, Solar Biofuels Research Centre, Queensland, Australia

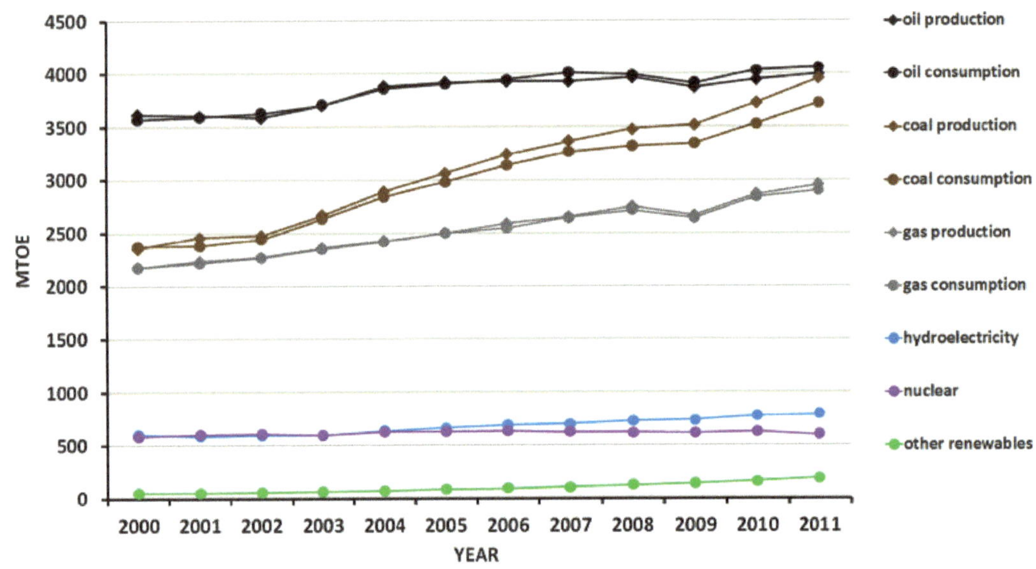

Figure 1: Energy production/consumption since 2000. Data from BP Statistical Review of World Energy 2012 [1]. It should be noted that production values for hydroelectricity, nuclear, and other renewables (which are generally direct electricity generating technologies) are represented by BP data as Mtoe values through the application of a 38% thermal efficiency conversion factor.

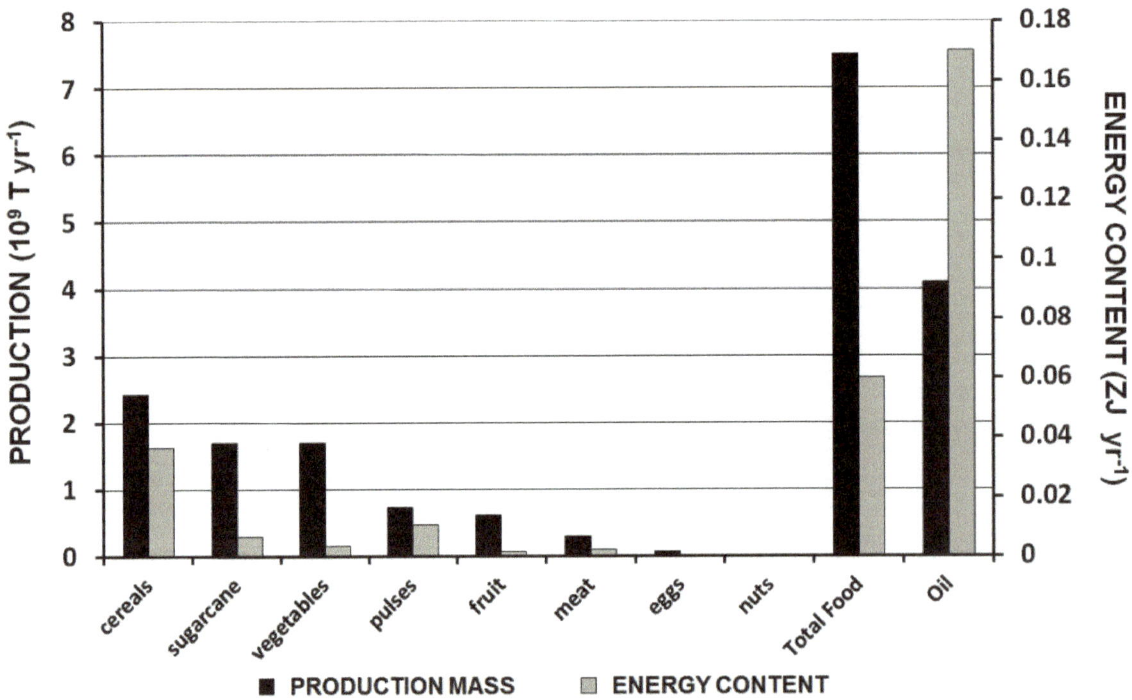

Figure 2: Global Energy Production from Petroleum and Food Products. The conversion of crop biomass to energy dense oils or fuels results in a loss of ≥ 50% of embodied energy. Food crops cannot therefore replace a globally significant proportion of energy as oil, without dramatically increasing pressure upon food security. The data presented for oil is sourced from BP Statistical Review of World Energy 2012 [1] and for food data is sourced from FAOSTAT [8], with general food calorific values calculated as previously by Stephens et al. [43].

e.g. costs of reduced crop yields, damage from extreme weather events and adaptation of communities [18,19]. Subsequently, an obvious social response will be an increasingly vocal promotion of legislative action to limit CO_2 emissions, in part by shifting to renewable fuels and renewable electricity generation. The two non-renewable options for alleviating this pressure are effective and economical CO_2 capture and sequestration technologies, and a massive expansion of nuclear power, neither of which appears to be making dramatic progress at present. In the absence of such progress, claims that the majority of fossil fuels must be left in the ground rather than combusted [20] are indeed sobering, and alternative carbon-neutral production sources for energy and fuel are necessary.

Renewable Energy Systems

Hydroelectric schemes are among the most successful renewable energy technologies and their global energy output has increased at an average and relatively consistent rate of ~3% yr^{-1} for the past 50 years [1]. Other renewable technologies such as solar photovoltaic/thermal and wind energy, have only recently been effectively deployed (almost 30-fold and 15-fold increase respectively from 2000 to 2010 [1]). Nonetheless, they are yet to reach globally significant levels (currently ~0.1% and 0.8% respectively, of current global energy consumption [1]). Though a welcome trend, most renewable energy systems target electricity, which only represents ~20% of our current energy consumption, and the need for energy dense, transportable fuels still remains. Even if increased electrification of our energy consumption can be achieved (increasingly likely with the emergence of new battery technologies such as the lithium-air batteries in development by IBM and collaborators [21]), transition from our current fuel consumption rates will take considerable time, effort, and cost.

Regardless of the future energy mix, there is an inextricable link between the transport of goods and people that requires high-energy density liquid fuels, at least into the foreseeable future. Second generation or advanced biofuel technologies (i.e. those that do not compete for arable land or with food crops) are being considered globally as the successor to fossil derived liquid fuels. However some biofuel technologies (heterotrophic oil production from carbohydrate feedstocks, and farnesene from sugar) are essentially carbon conversion methods and while they may boast impressive technology systems and can produce valuable demand-driven bioproducts, they draw from the existing energy and carbon pools of our primary production (i.e. first generation feedstocks). These heterotrophic systems do not directly capture solar energy at high efficiency and their scalability and ultimate global significance for energy solutions is therefore limited [3] just as first generation biofuel systems have been (e.g. corn ethanol). If such conversion technologies utilize second generation feedstocks or waste streams then they would be much more sustainable and scalable.

Next generation biofuel systems that have less need for arable land and freshwater (and hence greater scalability) are in development (e.g. algae production and lignocellulosic processing of more abundant biomass sources) but while the technical challenge of 'spinning straw into gold' appears surmountable, the expense of these technologies has so far placed them outside of affordable reality (2 to 10 times the current cost of production for petroleum oil). Further, issues of sustainability may be mitigated - but not avoided entirely. It is clear that to ultimately be both environmentally sustainable and scalable to globally significant levels, algae biomass systems for large scale fuel production must be photoautotrophic (or employ waste streams that would otherwise generate CO_2), and must not compete with production of food crops.

Algae Production

The production of macroalgae and microalgae (Figure 3 and Table 1) utilise established technologies which have been practiced commercially for decades, primarily for waste water treatment and the production of food products for direct consumption, phycocolloids, nutraceuticals and other high value products (HVPs). The production of macroalgae in open ponds and coastal seawaters, and microalgae in high rate ponds (HRPs) and photobioreactors (PBRs) can achieve much higher yields than terrestrial crops [3,22,23], but the high capital cost of establishment has traditionally prevented the production of algal biomass at the low costs required for economically sustainable commodity fuel production.

Table 1 does not capture all relevant issues; for example algal systems are much more expensive to establish, operate and harvest than agricultural sugar cane, though more productive (see below) and more efficient in nutrient (due to prevention of leachate and nutrient

Figure 3: Microalgae and Macroalgae. (A) microalgae high rate pond cultivations of *Spirulina* (Earthrise farms in California USA) [65]; (B) microalgae cultivated in photobioreactors (example depicted is LGem tubular system at Solar Biofuels Research Centre, Australia) can have higher yields for specialty bioproducts[66]; (C) freshly harvested microalgal biomass[66]; (D) macroalgae coastal cultivations of *Kappaphycus* [67]; (E) macroalgae cultivation of Ulva in open ponds can be scaled to large ponds on land [68]; (F) freshly harvested macroalgal biomass [67].

	Macroalgae	Microalgae	Sugarcane
Production Yield	~50-110 T ha^{-1} yr^{-1} dry weight [22, 23,69,70]	~70-100 T ha^{-1} yr^{-1} dry weight [38]	~71 T ha^{-1} yr^{-1} wet weight [8]
Oil content % w/w	up to 10% [61,71]	up to 30% (or more) [72]	N/A
Standard Calorific Value	~5-20 MJ kg^{-1} [27,73]	~20-25 MJ kg^{-1} [33,74]	~3.9 MJ kg^{-1}
Calculated Captured Solar Energy (CCSE)	~0.2-2.2 TJ ha^{-1} yr^{-1}	~1.4-2.5 TJ ha^{-1} yr^{-1}	~0.3 TJ ha^{-1} yr^{-1}
Photosynthetic Conversion Efficiency (PCE)	up to 3.0%	up to 3.4%	~0.4%
Oil/Fuel Potential	HTL to biocrude 125-1100 GJ ha^{-1} yr^{-1}	HTL to biocrude 700-1250 GJ ha^{-1} yr^{-1}	Ferment/distill to ethanol ~100 GJ ha^{-1} yr^{-1}

Table 1: Comparative Assessment of Macroalgae and Microalgae. HTL = Hydrothermal liquefaction; PCE calculated as CCSE/20 MJ m^{-2} d^{-1} @ 365 d yr^{-1} solar energy (in regions where similar yields are obtained with lower average solar energy <20 MJ m^{-2} d^{-1} then PCE rates might be reasonably up to 30% higher, but much below 15 MJ m^{-2} d^{-1} and output would be expected to be adversely affected); HTL conversion efficiency for assumed at 50%; ethanol energy output calculated as 71 T ha^{-1} yr^{-1} harvest x ~80% for 'burn & crop' output x 1.7 GJ ethanol per tonne = ~100 GJ ha^{-1} yr^{-1}.

recycling) and freshwater usage (for saline and wastewater systems freshwater consumption can be ~10-fold lower than conventional crops) [24]. Though macroalgae potential fuel/energy output is lower than that for microalgae, the capital and operating costs (especially harvesting [25,26]) associated with mass cultivation can be considerably less costly in some production models.

Obstacles to commercialization of algae fuels

The dream of algae fuels is a good one; massive algae farms absorbing CO_2 from the atmosphere and from our industrial emissions, to produce biomass at high efficiency. This biomass could be used for thermochemical conversion [27,28] or for biochemical derivation to produce liquid fuels [24]. The carbonaceous biomass residues can either be digested or gasified to CH_4 or H_2 [29,30] which can then be used to power the process or to generate nitrogenous fertilisers to make the process more sustainable. Furthermore, nutrient recycling is more efficient than in conventional agriculture allowing the potential for production of other bioproducts, and applications such as wastewater treatment and bioremediation.

The reality, as usual, is a lot tougher than the dream, despite the projections of over-sellers and the cheerleaders [31] who claim overestimated productivities and deliver premature promises of commercial success. While energy return on investment (EROI) can be sufficiently high in well designed production models (greater than that of tar sands oil extraction for example), economic sustainability remains the primary challenge [27,32].

The high capital cost of establishing algae farms, both for growth systems and for downstream processing equipment, is a primary economic factor in the cost of biocrude/oil production. The reduction of initial CAPEX and the improvement of biomass productivity are both primary effectors in sensitivity analysis of the economics [33]. The growth of algae in land-based aquatic systems necessitates high rates of water transfer and the high water content of algae biomass can also present challenges in achieving economical dewatering with low energy inputs. Water sustainability (due to evaporation losses) has been highlighted as an issue for mass cultivation of algae [34], although water sustainability is similarly a concern for other primary production systems [35-37] and with the use of saline and waste water resources algae biomass can be produced with around 10 times less freshwater for microalgae than many conventional crops [38] and even less freshwater required for macroalgae cultivation. Nevertheless, for the large scale of production required to contribute ~10% or more of global fuel use, good water management practices must be engaged and access to sufficient sources of suitable water must be ensured [39,40]. Importantly, the vast majority of the world's 15 million tonnes per annum production of macroalgae, valued at US$8 billion (at costs of ~$350-3,500 T^{-1} [41]), is produced in-sea with natural flow delivering the requisite supply of nutrients and CO_2. The scalable production here is substantial and utilises well developed automated technologies in use across temperate and tropical climates [26,42]. At the large scales of production required to achieve significant global impact, the mass cultivation of microalgae can be limited by the availability of sources of enriched CO_2 and by the availability of major nutrients (nitrogen and especially phosphorous) [43]. One advantage of macroalgae is that it is less constrained by these limitations, as the vast majority (currently >95%) of macroalgae production is produced in-situ in the oceans with no use of added nutrients or freshwater.

It should be emphasised that the potential carbon neutrality of advanced biofuel production is not currently attributed a value, and thus price comparisons against fossil fuel resources are skewed. A balanced techno-economic and life-cycle comparison would need to impose carbon capture and storage (CCS) costs upon fossil fuels, such that the carbon balance is similar to carbon neutral fuel production systems, for a true 'apples to apples' assessment. Furthermore for a true comparison fossil fuel subsidisation should also be noted. The global cost of subsidising gasoline, diesel and kerosene exceeds US$500 billion per year [44]. To put this into context, this is considerably more than Greece's national debt ($331 billion) [45].

Current Advancement in Expanding Global Algae Cultivation

Algae production technologies continue to advance and old paradigms are being challenged, with new production strategies emerging for both microalgae [3] and macroalgae [46,47]. Both macroalgae and microalgae show considerable potential for integration with other systems [33,48-52] and this is arguably an attractive path towards early commercialisation, especially given the potential for mitigation of waste from other production outputs (e.g. CO_2 emissions from industry and waste water, and nutrient waste loads from agriculture, aquaculture, and other industries). However, while the ambition to develop dedicated fuel production systems from algae is more economically challenging, it is also being pursued and is a primary path towards globally significant scales of production (as energy production will most likely saturate markets of protein feedstocks, HVPs, and other applications). For microalgae, harvesting has received considerable attention [25,53], and the ongoing discovery of endogenous bioactives (e.g. dinoflagellate biotoxins) [54] as well as recent development of engineered HVPs including pharmaceuticals [55,56] demonstrates the potential of natural algae biodiversity and genetic engineering for valuable algae bioproducts.

Hydrothermal liquefaction (HTL) of whole algal biomass [27,28] to biocrude is being adopted as an option for improving the economics and energetics over conventional production models. HTL is a thermochemical processing of biomass at high pressure and temperature and importantly, can be operated with wet biomass slurries of ≥15%, minimising the need for intensive dewatering. But while this strategy is promising it is not without its technical challenges. Although traditional and expensive dewatering and extraction processes for oil rich algae are avoided in this HTL approach through the processing of whole biomass, the oil content and general composition of the biomass still impacts on biocrude quality (e.g. energy density). However, this process is the conversion method of choice for macroalgae which have a lower oil yield, (and therefore energy density) but higher productivity per unit area. Though the energy density of HTL algae biocrude of ~32-35 MJ kg^{-1} [27,28] is not quite as high as specific extracted algal oils (which can even marginally exceed petroleum oil energy density of ~42 MJ kg^{-1}), conversion efficiencies of ~40 wt% currently to perhaps ~50 wt% as a future target show potential for this method. Nutrient recycling is also possible with this method, but not all nutrients can be retained and the minimisation of nitrogen in biocrude output is an important consideration to reduce NO_x emissions from algae fuels [28], as is reduction of sulphur for SO_x. The provision of biomass low in nitrogen and sulphur enhances biocrude quality while the use of catalysts can improve the quality of biocrude output [57] however, the economics and resource sustainability of this approach need to be properly examined. The removal of valuable nitrogenous components of the biomass in fact theoretically improves efficiencies in the biocrude process while adding to the total value of the biomass. Hydrothermal carbonisation and hydrothermal gasification technologies also exist but utilising these production strategies for algal biomass faces strong competition from the continued abundance of cheap coal and natural gas, though it is of note that it is estimated that of the 2850 billion tonnes of the world's total indicated fossil fuel reserves, only ~600 billion tonnes (~21%) can be extracted to stay within this 2°C safe limit [20].

Other methods of processing that can utilise wet biomass such as microwave pyrolysis are also worth investigation [58], while in microalgae, naturally occurring and genetically engineered oil secretion systems are also being studied for their potential [59,60]. The extraction of secreted oils from semi-continuous culture has many benefits but the 'milking' of extracellular lipids can require additional costs also (e.g. more complex bioreactors and/or agents to prevent degradation of secreted oils), and the economics of this strategy are yet to be properly examined.

Costs of algae fuel feedstocks are still higher than conventional petroleum but advancement in the last decade has been strong, and well configured production models for algal fuels can potentially achieve positive EROI (potentially up to 10x) and carbon neutrality. Although at full scale an algal fuels industry might not be capable of providing more than 10-20% of global fuel due to resource constraints (CO_2, water, and nutrients), this is nonetheless a highly desirable outcome.

Future potential

For an expanding algae industry there is potential for further advancement in biology (strain development, engineering, crop protection and population management), engineering (improved growth systems, downstream processing, and technology integration for improved efficiency and lower costs), business and economics (markets for new products, contributing to stabilisation of energy security), and social aspects (anticipated higher valuation of renewable production systems).

Most algae production industries have traditionally produced only a small number of species commercially (dominant share from 6 microalgal species and 10 macroalgal species, Table 2) but capacity to reliably cultivate a wider range of species is increasing (e.g. large scale cultivation of *Scenedesmus* by Sapphire Energy (www.sapphireenergy.com) and *Tetraselmis* by Muradel (www.muradel.com)). Total algae biodiversity is very large (over 8,000 species of macroalgae and over 30,000 species of microalgae) relative to terrestrial crops utilised in conventional agriculture, and this remains an untapped and relatively unexplored bioresource [29,61,62].

Algae present a promising avenue for continued development of high efficiency systems to convert solar energy to hydrocarbon fuels, including the generation of more complex non-fuel bioproducts and other environmental applications. Solazyme has been a pioneer in the production of a range of specialty HVPs from algae, and while generally this and the production of mid value products (MVPs) like omega-3 oils for nutraceuticals markets (e.g. from *Schizochytrium* and *Cryphecodinium*) has utilised heterotrophic production, photoautotrophic production systems (e.g. omega-3 EPA from *Nannochloropsis* [38]) are approaching commercial reality. Algal potential for production of bulk protein and amino acids is being investigated for livestock feed markets, and also for renewable chemicals [46,63,64].

Conclusion

Macro- and microalgal fuel alternatives have advanced rapidly, but they still require further development before price parity with conventional oil can be achieved. The key challenge remains biomass productivity, as feedstock value is the key driver in all production models. Hydrothermal processing of whole biomass, advanced production strategies for specific oils, and integration of bioproduct

Products	Macroalgae	Microalgae
Whole food products	*Porphyra* (nori) *Undaria* (wakame) *Saccharina* (kombu)	*Spirulina*, *Chlorella*
Nutraceuticals	*Undaria* for fucoidans *Sargassum* for fucoidans *Ulva* for ulvans	*Dunaliella* for β-carotene *Haematococcus* for astaxanthin *Schizochytrium* for DHA *Crypthecodinium* for DHA
Aquaculture feeds	*Ascophyllum*	*Isochrysis*, *Chaetoceros*, *Tetraselmis*, *Thalassiosira*, *Pavlova*
Phycocolloids	*Saccharina* for alginates *Kappaphycus* for carrageenans *Eucheuma* for carrageenans *Gracilaria* for agar	-
Agriculture	*Ascophyllum* for animal supplements *Ascophyllum* for fertilisers *Ecklonia* for fertilisers	-

Table 2: Conventional examples of commercial production systems for macro- and micro-algae.

outputs in biorefinery models can then each assist in development of sustainable systems. With a low level of investment into this early technology set (~US$1-2 billion to date), relative to fossil fuels (US$ trillions per year), the production of renewable fuel from algae is largely unexplored from an economic perspective. Given the challenge of minimising CO_2 emissions, increasing fuel security through distributed systems while providing a more sustainable fuel source to support economic development, serious questions should not only be asked at the scientific level, but also at the level of policy.

Acknowledgements

The authors would like to thank Dr Liam Wagner for advice on energy related economics issues. Ben Hankamer gratefully acknowledges the financial support of the Australian Research Council DP110101699 & DP130100346 as well as funding from a National and International Research Alliance Project (NIRAP) grant supported by Queensland State Government and project partners (KBR, Neste Oil, Siemens, Cement Australia, The University of Queensland, University of Bielefeld and Karlsruhe Institute of Technology). Rocky de Nys acknowledges the support of the Australian Renewable Energy Agency (ARENA), the Advanced Manufacturing Cooperative Research Centre (AMCRC) funded through the Australian Government's Cooperative Research Centre Scheme, and MBD Energy for supporting the Macroalgal Biofuels and Bioproducts - High Energy Algal Fuels Program.

References

1. BP (2012) Statistical Review of World Energy June 2012.
2. Stocker TF (2013) Climate change. The closing door of climate targets. Science 339: 280-282.
3. Stephens E, Ross IL, Hankamer B (2013) Expanding the microalgal industry - continuing controversy or compelling case? Curr Opin Chem Biol 17: 444-452.
4. Pimentel D, Marklein A, Toth MA, Karpoff MN, Paul GS, et al. (2009) Food Versus Biofuels: Environmental and Economic Costs. Hum Ecol 37: 1-12.
5. Gomiero T, Paoletti M, Pimentel D (2010) Biofuels: Efficiency, Ethics, and Limits to Human Appropriation of Ecosystem Services. J Agric Environ Ethics 23: 403-434.
6. Graziano da Silva J (2012) The US must take biofuel action to prevent a food crisis. Financial Times; August 9, 2012.
7. Lewis B, Kambas M (2012) EU Commission to cap food-based biofuels in major shift. Reuters.
8. Seiden A, Hawley NL, Schulz D, Raifman S, McGarvey ST (2012) Long-term trends in food availability, food prices, and obesity in Samoa. Am J Hum Biol 24: 286-295.
9. Dittrick P, Izundu U (2008) OTC speakers highlight offshore industry's future. Oil Gas J 106: 20-25.
10. Tippee B, Poruban S, Dittrick P (2010) OTC speakers highlight offshore industry's future. Oil Gas J 108: 18-22.
11. Duarte CM, Lenton TM, Wadhams P, Wassman P (2012) Abrupt climate change in the Arctic. Nature Climate Change 2: 60-62.
12. International Energy Agency (2011) World Energy Outlook 2011. International Energy Agency.
13. Energy Information Administration (2011) International Energy Outlook 2011.
14. Kerr RA (2011) Energy supplies. Peak oil production may already be here. Science 331: 1510-1511.
15. Kerr RA (2012) Oil resources. An oil gusher in the offing, but will it be enough? Science 338: 1139.
16. Kerr RA (2012) Oil resources. Technology is turning U.S. oil around but not the worlds. Science 335: 522-523.
17. Kerr RA (2012) Oil resources. Are world oil's prospects not declining all that fast? Science 337: 633.
18. Stern N (2006) The Economics of Climate Change. HM Treasury.
19. Metz B, Davidson OR, Bosch PR, Dave R, Meyer LA (eds) (2007) IPCC Fourth Assessment Report of the of the Intergovernmental Panel on Climate Change 2007. Cambridge University Press, Cambridge, New York, USA.
20. Climate Commision (2013) The Critical Decade 2013: Climate Change Science, Risks and Responses.
21. IBM (2013) The Battery 500 Project.
22. Capo TR, Jaramillo JC, Boyd AE, Lapointe BE, Serafy JE (1999) Sustained high yields of *Gracilaria* (Rhodophyta) grown in intensive large-scale culture. J Appl Phycol 11: 143-147.
23. Mata L, Schuenhoff A, Santos R (2010) A direct comparison of the performance of the seaweed biofilters, *Asparagopsis armata* and *Ulva rigida*. J Appl Phycol 22: 639-644.
24. Frank E, Wang M, Han J, Elgowainy A, Palou-Rivera I (2011) Life-Cycle Analysis of Algal-Based Fuels with the GREET Model. Energy Systems Division, Argonne Laboratory, San Francisco, USA.
25. Uduman N, Qi Y, Danquah MK, Forde GM, Hoadley A (2010) Dewatering of microalgal cultures: A major bottleneck to algae-based fuels. Journal of Renewable and Sustainable Energy 2: 1-15.
26. Paul NA, Tseng CK (2012) Chapter 13: Seaweed and Microalgae. In Aquaculture: Farming Aquatic Animals and Plants, J.S. Lucas and P.C. Southgate, Editors, Wiley-Blackwell Publishing Ltd: Oxford.
27. Rowbotham JS, Dyer PW, Greenwell HC, Theodorou MK (2012) Thermochemical processing of macroalgae: a late bloomer in the development of third-generation biofuels? Biofuels 3: 441-461.
28. Biller P, Ross AB (2012) Hydrothermal processing of algal biomass for the production of biofuels and chemicals. Biofuels 3: 603-623.
29. Sialve B, Bernet N, Bernard O (2009) Anaerobic digestion of microalgae as a necessary step to make microalgal biodiesel sustainable. Biotechnol Adv 27: 409-416.
30. Onwudili JA, Lea-Langton AR, Ross AB, Williams PT (2013) Catalytic hydrothermal gasification of algae for hydrogen production: composition of reaction products and potential for nutrient recycling. Bioresour Technol 127: 72-80.
31. Waltz E (2009) Biotech's green gold? Nat Biotechnol 27: 15-18.
32. Sun A, Davis R, Starbuck M, Ben-Amotz A, Pate R, et al. (2011) Comparative cost analysis of algal oil production for biofuels. Energy 36: 5169-5179.
33. Stephens E, Ross IL, King Z, Mussgnug JH, Kruse O, et al. (2010) An economic and technical evaluation of microalgal biofuels. Nat Biotechnol 28: 126-128.
34. National Research Council (2012) Sustainable Development of Algal Biofuels. National Academies Press.
35. National Research Council (2008) Water Implications of Biofuels Production in the United States. National Academies Press.
36. de Fraiture C, Giordano M, Liao Y (2008) Biofuels and implications for agricultural water use: blue impacts of green energy. Water Policy 10: 67-81.
37. Cominelli E, Galbiati M, Tonelli C, Bowler C (2009) Water: the invisible problem. Access to fresh water is considered to be a universal and free human right, but dwindling resources and a burgeoning population are increasing its economic value. EMBO Rep 10: 671-676.
38. Caspari M (2012) Algae-the optimal crop & growth industry for Western Australia, In: 2nd Algae World Australia: Perth, WA, Australia.
39. Murphy CF, Allen DT (2011) Energy-water nexus for mass cultivation of algae. Environ Sci Technol 45: 5861-5868.
40. Venteris ER, Skaggs RL, Coleman AM, Wigmosta MS (2013) A GIS cost model to assess the availability of freshwater, seawater, and saline groundwater for algal biofuel production in the United States. Environ Sci Technol 47: 4840-4849.
41. Bixler HJ, Porse H (2011) A decade of change in the seaweed hydrocolloids industry. J Appl Phycol 23: 321-335.
42. Chopin T, Sawhney M (2010) Seaweeds and their Mariculture. Marine Agronomy, Academic Press: London. p. 216-225.
43. Stephens E, Wagner LD, Ross I, Hankamer B (2012) Microalgal production systems: Global impact of industry scale up. In Microalgal Biotechnology, C. Posten, Editor, De Gruyter: Berlin, USA.
44. Pascual C, Elkind J (2010) Energy security: Economics, Politics, Strategies, and Implications: Brookings Institution Press.

45. World Bank (2011) Annual Report.

46. Jung KA, Lim SR, Kim Y, Park JM (2013) Potentials of macroalgae as feedstocks for biorefinery. Bioresour Technol 135: 182-190.

47. Kraan S (2013) Mass-cultivation of carbohydrate rich macroalgae, a possible solution for sustainable biofuel production. Mitigation and Adaptation Strategies for Global Change 18: 27-46.

48. Mulbry W, Kondrad S, Pizarro C, Kebede-Westhead E (2008) Treatment of dairy manure effluent using freshwater algae: algal productivity and recovery of manure nutrients using pilot-scale algal turf scrubbers. Bioresour Technol 99: 8137-8142.

49. Lundquist TJ, Woertz IC, Quinn NWT, Benemann JR (2010) A Realistic Technology and Engineering Assessment of Algae Biofuel Production. Energy Biosciences Institute, University of California, California, USA.

50. de Paula Silva PH, de Nys R, Paul NA (2012) Seasonal growth dynamics and resilience of the green tide alga cladophora coelothrix in high-nutrient tropical aquaculture. Aquaculture Environment Interactions 2: 253-266.

51. Saunders RJ, Paul NA, Hu Y, de Nys R (2012) Sustainable sources of biomass for bioremediation of heavy metals in waste water derived from coal-fired power generation. PLoS One 7: e36470.

52. Cole AJ, Mata L, Paul NA, de Nys R (2013) Using CO_2 to Enhance Carbon Capture and Biomass Applications of Freshwater Macroalgae. Global Change Biology Bioenergy.

53. Pragya N, Pandey KK, Sahoo PK (2013) A review on harvesting, oil extraction and biofuels production technologies from microalgae. Renew Sust Energ Rev 24: 159-171.

54. Gallardo-Rodríguez J, Sánchez-Mirón A, García-Camacho F, López-Rosales L, Chisti Y, et al. (2012) Bioactives from microalgal dinoflagellates. Biotechnol Adv 30: 1673-1684.

55. Jones CS, Luong T, Hannon M, Tran M, Gregory JA, et al. (2013) Heterologous expression of the C-terminal antigenic domain of the malaria vaccine candidate Pfs48/45 in the green algae Chlamydomonas reinhardtii. Appl Microbiol Biotechnol 97: 1987-1995.

56. Tran M, Van C, Barrera DJ, Pettersson PL, Peinado CD, et al. (2013) Production of unique immunotoxin cancer therapeutics in algal chloroplasts. Proc Natl Acad Sci U S A 110: E15-22.

57. Biller P, Riley R, Ross AB (2011) Catalytic hydrothermal processing of microalgae: decomposition and upgrading of lipids. Bioresour Technol 102: 4841-4848.

58. Budarin V, Ross AB, Biller P, Riley R, Clark JH, et al. (2012) Microalgae biorefinery concept based on hydrothermal microwave pyrolysis. Green Chem 14: 3251-3254.

59. Roessler PG, Chen Y, Liu B, Dodge CN (2009) Secretion of Fatty Acids by Photosynthetic Microorganisms. World Patent Application (WO2009/076559A1).

60. Vasudevan V, Stratton RW, Pearlson MN, Jersey GR, Beyene AG, et al. (2012) Environmental performance of algal biofuel technology options. Environ Sci Technol 46: 2451-2459.

61. Gosch BJ, Magnusson M, Paul NA, de Nys R (2012) Total lipid and fatty acid composition of seaweeds for the selection of species for oil-based biofuel and bioproducts. Global Change Biology Bioenergy 4: 919-930.

62. Georgianna DR, Mayfield SP (2012) Exploiting diversity and synthetic biology for the production of algal biofuels. Nature 488: 329-335.

63. Chang HN, Kim NJ, Kang J, Jeong CM (2010) Biomass-derived volatile fatty acid platform for fuels and chemicals. Biotechnology and Bioprocess Engineering 15: 1-10.

64. Lammens TM, Franssen MCR, Scott EL, Sanders JPM (2012) Availability of protein-derived amino acids as feedstock for the production of bio-based chemicals. Biomass Bioenerg 44: 168-181.

65. Google Inc. (2013) Google Earth v7.1.1.1580.

66. Photo courtesy of Solar Biofuels Research Centre University of Queensland (2013).

67. Magnusson, M, Mata, L, de Nys, R and Paul, NA (2013) Biomass, Lipid and Fatty Acid Production in Large-Scale Cultures of the Marine Macroalga Derbesia tenuissima in review.

68. Photo courtesy of Centre for Sustainable Tropical Fisheries and Aquaculture, JCU.

69. Neori A, Msuya FE, Shauli L, Schuenhoff A, Kopel F, et al. (2003) A novel three-stage seaweed (Ulva lactuca) biofilter design for integrated mariculture. J Appl Phycol 15: 543-553.

70. Langdon C, Evans F, Demetropoulos C (2004) An environmentally-sustainable, integrated, co-culture system for dulse and abalone production. Aquacult Eng 32: 43-56.

71. Kumari P, Bijo AJ, Mantri VA, Reddy CR, Jha B (2013) Fatty acid profiling of tropical marine macroalgae: an analysis from chemotaxonomic and nutritional perspectives. Phytochemistry 86: 44-56.

72. Chisti Y (2007) Biodiesel from microalgae. Biotechnol Adv 25: 294-306.

73. Lawton RJ, de Nys R, Paul NA (2013) Selecting Reliable and Robust Freshwater Macroalgae for Biomass Applications. PLOS One 8: e64168.

74. Illman AM, Scragg AH, Shales SW (2000) Increase in Chlorella strains calorific values when grown in low nitrogen medium. Enzyme Microb Tech 27: 631-635.

Alkaline Protease Production Using Proteinaceous Tannery Solid Waste

Javed Ahmad* and Tauseef A Ansari

Department of Biochemical Engineering and Food Technology, Harcourt Butler Technological Institute, Nawabganj, Kanpur 208002, India

Abstract

Leather industry has commonly been associated with high pollution due to the bad smell, organic wastes and high water consumption caused during traditional manufacturing processes. Different forms of waste in quality and quantity, which emerge during the transformation of hides and skins into leathers in thousands of leather factories, from primitive to modern all around the world, have negative impacts on the environment. In leather factories solid waste containing protein that constitute more than 60% of raw hide weight are disposed to the environment without turning them into good use at industrial level. A biological approach such as alkaline protease production by different microorganism is prominent way to utilization of tannery proteinaceous solid waste.

In this research work four different species *Bacillus licheniformis, Bacillus coagulans, Bacillus polymxa* and *Bacillus amyloliquefaciens* were isolated from soil sample and identified by biochemical tests. These species efficiently produce alkaline using animal fleshing (ANFL) which is an untanned tannery solid waste as prominent source of protein .The efficiency of protease production was measured on different temperature, pressure and incubation time. *Bacillus licheniformis* gave highest protease production. Due to efficient production of alkaline protease by same bacillus species, it can be use at industrial level.

Keywords: Leather industries; Environment; Solid waste; ANFL; Biological process; *Bacillus species*; Alkaline proteases

Introduction

The leather industry generally uses hides and skins as raw materials, which are the by-products of meat and meat products industry [1]. So in this reverence the leather industry is an environmentally friendly industry because it processes waste products from meat production industries [2]. However, the leather industry has frequently been allied with high pollution due to the bad smell, organic wastes and high water consumption caused during traditional manufacturing and processing [3]. Diverse types of waste in quality and quantity, which come out during the process of hides and skins into leathers in many leather factories, impacts negatively on the environment to all around the world [1].

Solid waste generated in leather industry is contribute mainly through processes like skin trimming keratin waste ,flesh waste chrome shaving waste and buffing waste. Its main component is protein. If these proteins and other chemical which are present in the chemical treated protein are not utilized properly it will pose hazardous pollution problem to the environment [4]. The microorganisms represent an exceptional source of protease owing to their extensive biochemical diversity and susceptibility to genetic manipulation. Among the various proteases, bacterial proteases are the generally most noteworthy than others such as animal and fungal proteases [5]. The proteolytic enzymes have many physiological functions, ranging from the generalized protein digestion to the more specific regulated processes [6,7]. Alkaline proteases, an important group of industrial enzymes (Figure 1), are produced by a wide range of microorganisms, including fungi and bacteria. *Bacillus spp.* have been reported to produce alkaline protease from various substrates, viz. green-gram husk [8], mustered-oil cake, wheat bran, rice bran, Impetrate cylindrical grass, banana leaves, potato peels and used tea leaves [9], wheat bran and lentil husk [10]. However, few reports are available on alkaline proteolytic enzymes from *Bacillus cereus* [11].

ANFL is one of the major solid wastes that are generated during pretanning process of leather industry operation [12] (Figure 1). It contains the highest protein content (50 to 60%) [13,14]. ANFL discharged from the tannery industries are processed by different types of chemical or thermal operation. This is an energy intensive and time consuming process [15].

In our study, alkaline proteases were produced from various microorganisms by using untanned (ANFL) as substrate. Microbial isolates species are *Bacillus licheniformis, Bacillus coagulans, Bacillus polymxa* and *Bacillus amyloliquefaciens.* The alkaline protease generally used as biocatalyst for industrial and laboratory purposes [16].

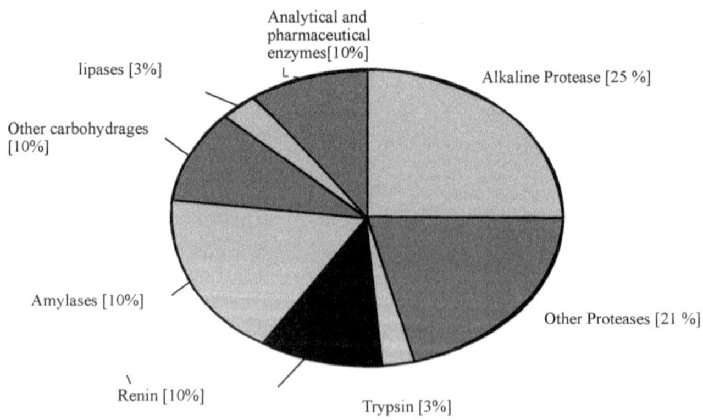

Figure 1: The contribution of different enzymes to the total sale of enzymes is indicated. The shaded portion indicates the total sale of proteases [12].

***Corresponding author:** Javed Ahmad, Department of Biochemical Engineering and Food Technology, Harcourt Butler Technological Institute, Nawabganj, Kanpur 208002, India, Tel: +91-09807121672

Proteases constitute at least 65% of the total industrial enzyme market [17]. They are used for various industrial applications, such as laundry detergents, protein recovery, leather preparation, solubilisation and organic synthesis [18]. Alkaline proteases have many physiological functions, ranging from generalized protein digestion to more specific regulated procedures [7]. *Bacillus licheniformis* and *Bacillus coagulans* can be used profitably for large scale production of alkaline protease to meet the present day demand of the industrial sector [19].

Chemical composition of tannery solid waste used in experiment

Chemicals		% in dry matter (mg/g)
Protein	-	79
Fat	-	7.57
P	-	1.476
K	-	0.47
Ca	-	0.59
Iron	-	2.24
Mg	-	1.84
Zn	-	0.116
Cu	-	0.002

Material and Method

Sample collection

Soil samples collection area was campus area of Chander Shekhar Azad Agriculture University, Kanpur (Near H.B.T.I., Kanpur Main Campus) Nawabganj, Kanpur District, and Uttar Pradesh in the month of July. The samples were collected at a depth of 8 cm to 10 cm in soil. The collected soil samples were brought to the laboratory in sterilized caped flask, air dried & air tied and stored at room temperature. 1(±0.04) gram of soil sample was taken in conical flask (250 ml) containing 100 ml sterile distilled water (DW).

The content was mixed well for homogenous mixture. This suspension was serially diluted up to 10^{-5} with sterile distilled water. The samples were transferred aseptically to Petri dishes containing sterile nutrient agar medium by pour plating technique. After inoculation, plates were incubated at 35-37°C, for 2 to 4 days. After that Bacterial colonies appeared. Then cultures were identified based on characteristics and they were identified by the help Gram staining methods and by biochemical tests as suggested by Bergey's manual [20] as well as Cappuccino and Sherman [21]. Colonies were examined by microscopic observations. The analysis of physiological and biochemical characteristics (Table 1) of colonies were done by using methods, as suggested by Joscy et al. [22] and Watanabe and Hayano [23] respectively. Here the biochemical tests have shown only for the Bacillus Species, although testing has been done on most of the colonies.

Inoculums preparation

B. *licheniformis, B. coagulans, B. Polymxa* and *B. amyloliquefaciens* were found predominantly in the soil sample. The pure culture of test organisms was maintained in nutrient agar slants for further studies for a period of 24 h.

Substrate preparation

The sufficient amount of ANFL, of Beef skins were collected from a commercial tannery, Jajmau Industrial Area, Kanpur, India. It was treated with ammonium chloride (1.25% w/v) for 3–4 h to remove the adsorbed calcium salts on the collagen fibres. The delimed ANFL fleshing was suspended in water for 1 to 2 h at neutral pH [23]. Very small pieces of ANFL were obtained by manual scissoring. The 24 h grown culture of *B. licheniformis, B. coagulans, B. Polymxa* and *B. amyloliquefaciens* were inoculated in sterilized minimal medium and poured into the ANFL containing flask initial moisture content maintained at 60-65%.

Optimization of enzyme activity

To optimize the alkaline protease production at various levels, as the effect of pH, temperature and harvest time on alkaline production protease production *by B.licheniformis, B. coagulans, B. polymxa* and *B. amyloliquefaciens* were monitored in the minimal salt medium containing ANFL. Protease activity was determined for 98 h with an interval of 14 h. Incubation periods were set as 14, 28, 42, 56, 84 and 98 h for a wide range study. Protease production efficiency obtained at wide range of temperatures (30, 40, 50, 60, 70 and 80°C) and pH (5.5, 6.5, 7.5, 8.5, 9.5, 10.5, and 11.5). Estimation of enzyme did as prescribed by Neheta et al. [24].

Results

Effect of pH levels on alkaline protease production

The effect of pH on enzyme production is shown in following Table 2 and Figure 2. All the microorganism were able to produce high levels of enzyme at pH 9.5.The *Bacillus licheniformis* was able to produce highest level of enzyme (0.170 ± 0.099 mg/l) at pH 9.5 with respect to others. All the steps done in triplicates.

Effect of temperatures on alkaline protease production

The effect of temperature on enzyme production is shown in fallowing Table 3 and Figure 3. All the microorganism were able to produce high levels of enzyme at 30°C, the optimal temperature for alkaline protease production. The *Bacillus licheniformis* was able to produce highest level of enzyme (0.214 ± 0.099 mg/l) at 30°C with respect to others. All the steps done in triplicates.

Effect of incubation periods on alkaline protease production

The effect of incubation periods on enzyme production is shown

Sr.No.	Biochemical Test	*B. licheniformis*	*B. coagulans*	*B. polymxa*	*B. amyloliquefaciens*
1	Gram's Staining	+	+	+	+
2	Nitrate reduction test	-	-	-	-
3	Catalase test	-		-	-
4	Starch hydrolysis test	+	+	+	+
5	Mac Conkey agar test	+	+	+	+
6	Citrate utilization test	+	+	+	+
7	Nitrate Reduction	-	-	-	-

Table 1: Biochemical Characteristic of isolated bacteria. (+positive result, -Negative result).

in fallowing Table 4 and Figure 4. All the microorganism were able to produce high levels of enzyme at incubation time period 98 h, the optimal time period for alkaline protease production. The *Bacillus licheniformis* was able to produce highest level of enzyme (0.186 mg/l) at 98 h with respect to others. All the steps done in triplicates.

Conclusion

The leather industries are widely known for its effluent problems and environment pollution. The biological approach discussed is one of the best ways to solve this problem as well as utilization. Our results

S.No	The level of pH	*B.licheniformis* (mg/l)	*B. Polymxa (mg/l)*	*B. amyloliquefaciens* (mg/l)	*B. coagulans* (mg/l)
1.	6.5	0.138 ± 0.099	0.132 ± 0.098	0.141 ± 0.099	0.135 ± 0.099
2.	7.5	0.146 ± 0.099	0.137 ± 0.098	0.148 ± 0.099	0.141 ± 0.099
3.	8.5	0.178 ± 0.099	0.145 ± 0.099	0.159 ± 0.099	0.151 ± 0.098
4.	9.5	0.191 ± 0.098	0.151 ± 0.099	0.171 ± 0.099	0.161 ± 0.099
5.	10.5	0.170 ± 0.099	0.145 ± 0.098	0.147 ± 0.098	0.143 ± 0.098
6.	11.5	0.168 ± 0.099	0.142 ± 0.099	0.145 ± 0.099	0.141 ± 0.099

Table 2: Effect of pH levels on alkaline protease production by *B. licheniformis, B. polymxa, B. amyloliquefaciens* and *B. coagulans.* (All the steps done in triplicates).

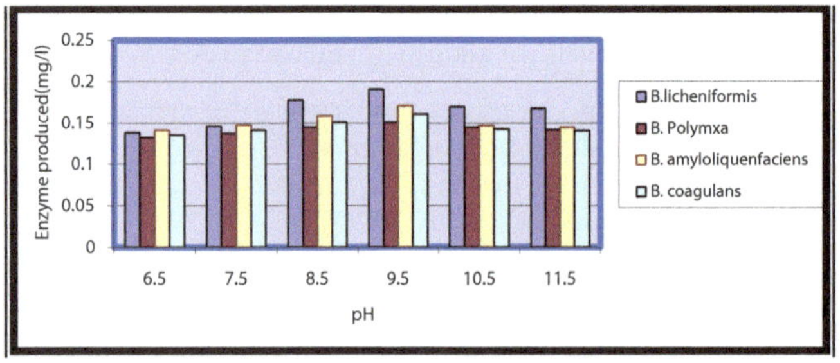

Figure 2: Effect of pH levels on alkaline protease production by *B. licheniformis, B. Polymxa, B. amyloliquefaciens* and *B. coagulans.* (All the steps done in triplicates).

S.No	Tempera ture (°c)	*B. licheniformis* (mg/l)	*B. Polymxa* (mg/l)	*B. amylolique faciens* (mg/l)	*B. coagulans* (mg/l)
1.	30	0.214 ± 0.099	0.198 ± 0.099	0.159 ± 0.099	0.172 ± 0.099
2.	40	0.199 ± 0.099	0.178 ± 0.098	0.156 ± 0.099	0.164 ± 0.099
3.	50	0.190 ± 0.099	0.175 ± 0.098	0.150 ± 0.099	0.160 ± 0.098
4.	60	0.185 ± 0.099	0.170 ± 0.098	0.145 ± 0.099	0.156 ± 0.098
5.	70	0.175 ± 0.099	0.165 ± 0.098	0.140 ± 0.099	0.149 ± 0.098
6.	80	0.164 ± 0.099	0.156 ± 0.098	0.131 ± 0.099	0.138 ± 0.098

Table 3: Effect of temperature levels on alkaline protease Production by *B. licheniformis, B. Polymxa, B. amyloliquefaciens and B. coagulans.* (All the steps done in triplicates).

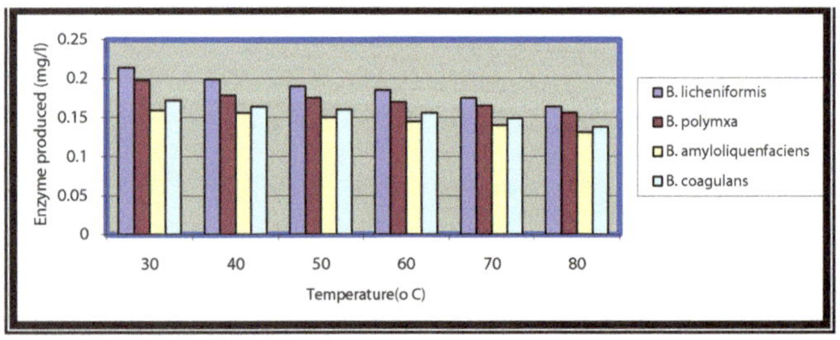

Figure 3: Effect of temperature levels on alkaline protease Production by *B. licheniformis, B. Polymxa, B. amyloliquefaciens* and *B. coagulans.* (All the steps done in triplicates).

S.No	Incubation time (h)	*B.licheniformis* (mg/l)	*B. polymxa* (mg/l)	*B. amylolique faciens* (mg/l)	*B.coagulans* (mg/l)
1	14	0.146 ± 0.103	0.136 ± 0.103	0.142 ± 0.103	0.132 ± 0.102
2	28	0.154 ± 0.102	0.146 ± 0.102	0.149 ± 0.102	0.138 ± 0.103
3	42	0.161 ± 0.103	0.151 ± 0.102	0.157 ± 0.103	0.147 ± 0.102
4	56	0.164 ± 0.102	0.156 ± 0.103	0.165 ± 0.102	0.154 ± 0.103
5	84	0.178 ± 0.102	0.169 ± 0.102	0.176 ± 0.103	0.168 ± 0.102
6	98	0.186 ± 0.102	0.179 ± 0.102	0.185 ± 0.102	0.172 ± 0.103

Table 4: Effect of incubation periods on alkaline protease production by *B. licheniformis, B. Polymxa, B. amyloliquefaciens* and *B. coagulans.* (All the steps done in triplicates).

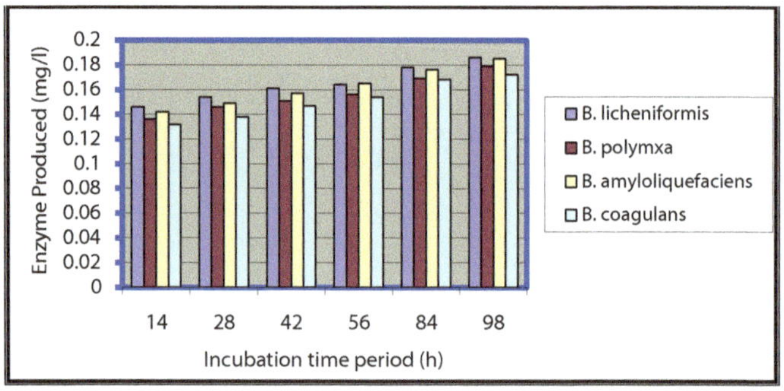

Figure 4: Effect of incubation periods on alkaline protease production by *B. licheniformis, B. Polymxa, B. amyloliquefaciens* and *B. coagulans.* (All the steps done in triplicates).

indicated that optimum condition for alkaline protease productivity on the basis of pH, temperature and pressure. Characterization as well as optimization for each feature of growth and production yield is essential requirements. *Bacillus licheniformis* gave best result for maximum efficiency. It can be use on industrial level. Further research is required on full purification of alkaline protease for its complete utilization and meeting the needs of the industry.

Acknowledgement

Authors thank Associate Professor S.J. Nagar and (H.B.T.I. KANPUR) for their support and encouragement.

Reference

1. Ozgunay H, Colak S, Mutlu MM, Akyuz F (2007) Characterization of Leather Industry Wastes. Polish J of Environ stud 16: 867-873.
2. Lang MF, Kolozmik K, Sukop S, Mla DM (1999) Products of enzymatic decomposition of chromium tanned leather waste. JSLTC 83: 187.
3. Taylor MM, Cabeza LF, Dimaio GL, Brown EM, Marmer WN, et al. (1998) Processing of leather waste: pilot scale studies on chrome shavings. 1. isolation and characterization of protein products and separation of chrome cake. JAL CA 93: 61-82.
4. Kanagaraj J, Velappan KC, Chanra BNK, Sadulla S (2006) Solid waste generation in leather industry and its utilization for cleaner environment- A Review. Journal of Scientific and Industrial Research 65: 541-548.
5. Ward OP (1985) Proteolytic enzymes. Comprehensive Biotechnology 3: 789–818.
6. Kalisz HM (1988) Microbial proteinases. Adv Biochem Eng Biotechnol 36: 1–65.
7. Kaminishi H, Hamatake H, Cho T, Tamaki T, Suenaga N, et al. (1994) Activation of blood clotting factors by microbial proteinases. FEMS Microbiol Lett 121: 327–332.
8. Prakasham RS, Rao CHS, Sarma PN (2006) Green gram husk, an inexpensive substrate for alkaline protease production by *Bacillus sp.* in solid state fermentation. Bioresour Technol 97: 1449–1454.
9. Mukherjee AK, Adhikari H, Rai SK (2008) Production of alkaline protease by

a thermophilic *Bacillus subtilis* under solid-state fermentation (SSF) condition using Imperata cylindrical grass and potato peel as low-cost medium: Characterization and application of enzyme in detergent formulation. Biochem Eng J 39: 353–361.

10. Uyar F, Baysal Z (2004) Production and optimization of process parameters for alkaline protease production by a newly isolated *Bacillus sp.* under solid state fermentation. Process Biochem 39: 1893–1898.

11. Ghorbel B, Sellami-Kamoun A, Nasri M (2003) Stability studies of protease from *Bacillus cereus BG1*. Enzyme Microb Technol 32: 513–518.

12. Rao MB, Tanksale AM, Mohini SG, Deshpande VV (1998) Molecular and biotechnological aspects of microbial proteases. Microbiol Mol Biol Rev 62: 597-635.

13. Ravindran B, Dinesh SL, John Kennedy L, Sekaran G (2008) Vermicomposting of solid waste generated from leather industries using epigeic earthworm *Eisenia foetida*. Appl Biochem Biotechnol 151: 480-488.

14. Ganesh Kumar A, Swarnalatha S, Sairam B, Sekaran G (2008) Production of alkaline protease by *Pseudomonas aeruginosa* using proteinaceous solid waste generated from leather manufacturing industries. Bioresour Technol 99: 1939-1944.

15. Bajza Z, Markovic I (1999) Influence of enzyme concentration on leather waste hydrolysis kinetics. J Soc Leath Tech Ch 83: 172-176.

16. Bajza Z, Vrcek IV (2001). Water Quality Analysis of Mixtures Obtained from Tannery Waste Effluents. Ecotox Environ Safe 50: 15-18.

17. Rao NM, Sastry TP, Sivaparvathy M, Srinivasan TS (1994) Evaluation of tannery by product meal for broiler. Ind J Poultry Sci 29: 37-41.

18. Anwar A, Saleemuddin M (1998) Alkaline proteases: A review. Bioresour Technolology 64: 175-183.

19. Asokan S, Jayanthi C (2010) Alkaline protease production by Bacillus licheniformis and Bacillus coagulans. Journal of cell and tissue research 10: 2119-2123.

20. Bergey D (1957) Bergey's Manual of Determinative Bacteriology. American Society for Microbiology: VII Edition. Williams & Willkins Co Publishers, Baltimore, USA.

21. Cappuccino JG, Sherman N (1999) Foloia Microbial. 6: 231-245.

22. Josey P, Beyho JL, Johnson AWB, Beringer JE (1979) J Appl Bacteriol 46: 343-350.

23. Witanabe K, Hayano L (1994) J Biol Fert Soil 18: 341-346.

24. Nehete PN, Shaj VD, Kothari RM (1986) Enzyme Microbial Technol 8: 370-374.

Asphaltene Stability in Crude Oil during Production Process

Hadi Belhaj[1,2*], Hadil Abu Khalifeh[1,2] and Nasser Al-Huraibi[2]

[1]*The Petroleum Institute, Abu Dhabi, UAE*
[2]*Regent Group, Calgary, Canada*

Abstract

Asphaltene instability may take place in the reservoir leading to permeability damage and contributing to flow restriction issues. It may also occur in production strings and surface facilities causing conduit clogging. Any change in oil composition or pressure and temperature at any stage of production will destabilize crude oil producing asphaltene precipitation. In this study, the stability of target crude oil under the influence of a direct current (DC) is investigated. The amount of the asphaltene deposit and its electrical charge at various operating conditions are investigated. The method consists of applying a DC voltage between two metal electrodes immersed in the crude oil sample. The amounts of deposits accumulated on the surfaces of the electrodes are recorded during experiment time utilizing a miniature two load cells connected to the electrodes.

Electric field strength up to 500V is applied. This study confirms that asphaltene colloids are electrically charged. The fact that deposits form on the anode surface proves that asphaltene particles possess a negative charge. The experiment shows that time and potential differences are sensitive parameters. At the first 24 hours, all the asphaltene in the crude oil were extracted. This phenomenon reflects either limited amount of asphaltene exists in the sample, or that the resin content plays a major role in asphaltene colloids charge switch.

Introduction

The procedural definition of asphaltene is the heavy organic components in crude oil that is soluble in toluene but insoluble in heptane/pentane. Asphaltenes are attached to similarly structured molecules, resins, which interact with asphaltenes to improve their solubility in aliphatic media to become stable in the crude oil [1]. A number of studies have indicated that the asphaltenes and the dispersion medium nature are important factors that decide the stability of crude oils [2-4].Changing composition of crude oil, pressure or temperature destabilizes the crude oil resulting flocculation and deposition of asphaltene molecules.The deposition of asphaltene is the utmost challenge faced during the production of oil reservoirs. When asphaltene deposits an array of problems emerges such as permeability reduction and wettability alteration in the formation, pipeline plugging, and pumps failure at the surface, catalyst poisoning and heat exchangers fouling at the refinery. Asphaltene deposition can be significant in many EOR processes including gas injection [5], CO_2 injection [6], nitrogen injection [7] and Vapex [8].

As an example, several major onshore fields in Abu Dhabi are challenged with asphaltene deposition in their producing wells. One asphaltene clean-up operation using organic solvents and coiled tubing unit costs US $200,000 [9]. In addition, during the existence of CO_2 pilot project three asphaltene removal jobs were conducted [10]. Asphaltene remediation techniques are very expensive with the potential impact on health and environment. Petroleum industry is seeking other alternatives to remediate or prevent asphaltene deposition.

One of the interesting approaches is using electricity. Eelectrical means have been suggested to enhance oil recovery by applying an electrical potential difference between a producing well and electrode well [11]. Also, it has been involved in technologies utilized to solve a specific problem [12] or as process-side effects (e.g., streaming potential induced by the oil flow).

It is believed that electrical effects play an important role on the asphaltene deposition. Application of electrical potential causes the precipitation of asphaltene particles on one of the electrodes (elecrtro-deposition) [13,14]. Thus controlling the electro-deposition of the asphaltene particles can lead to the development of preventive techniques [15,16].

There are limited publications devoted to the investigation of the asphaltenes electrical charge, how important it is to their stability and the behavior of asphaltenes when crude oil is under the influence of an electric field [17-21].

The effect of various resins on asphaltene precipitation from different petroleum fluids was examined by Goual et al. [4]. Various resins were added to three different petroleum fluids to measure precipitation with n-pentane. Results showed that the polarity of resins had a strong effect on precipitation. Resins with higher polarity were more effective than resins with lower one. The stability of asphaltene suspended in oil by resins is related to the electrical characteristics between asphaltene, resins and the remainder of the crude.

Alkafeef [3] found that the electrical charge of asphaltene in Marrat oil is very low and had a negligible impact on flocculation and deposition.

Khvostichenko et al. [22] investigated the effect of different parameters that may affect the net charge of asphaltene particles. They investigated the charge switching under various conditions in oil/diluent media. They concluded that application of a DC electric field to stable oil or a solution of asphaltenes in toluene did not causes aggregation or precipitation of asphaltenes.

***Corresponding author:** Hadi Belhaj, The Petroleum Institute, Abu Dhabi, UAE

Hashmi et al. [23] indicated that asphaltene colloids have a heterogeneous surface charging. Some particles carry a small net positive charge that can deposit on cathode while others carry a small net negative charge that can deposit on anode. Metallic components, Pi electron clouds, and acid-base surface groups could contribute to the surface charge.

As discussed above, the topic is still an active debate. More research is needed to look into the mechanisms that generate the asphaltene surface charge and the parameters that could affect it.

In this study, an intensive experimental program will be conducted to investigate the stability and electrical charges of asphaltene in crude oil at various operating conditions. Target crude oil samples from selected Abu Dhabi reservoirs will be collected. This project can be an inspiration for a potential pilot project to tackle asphaletene deposition problem from Abu Dhabi oil reservoirs. An optimum criterion or scheme can be developed to prevent or minimize asphaltene deposition in the reservoir or during production.

Methodology

The experimental setup is composed of a cylindrical container of 500 ml capacity holding two electrodes extending into the oil sample in the vessel. The length of the electrodes is in the range of 10-15 cm. The electrodes are 1 to 2 cm wide. The two electrodes are connected to a Direct Current (DC) electric field. The electric field strength can be up to 500 V. Each electrode is hanged from a load cell (capacity is up to 200 gram) to monitor the change in the mass of the electrode with time. This method is capable of measuring asphaltene mass with real time all through the experiment.

The temperature of the oil sample is measured by a thermocouple and maintained at the set temperature by a temperature controller connected to a heating tape surrounding the oil container. The experimental variables such as temperature, mass of electrodes, electrical field strength are connected to a data acquisition system to record the values as a function of time. An 8.0 Megapixels camera is used to take shots at specific time intervals.

As shown in Figure 1, the two electrodes (Cathode and Anode) connected to the power source are immersed in the oil container. The electrodes are 1.1 cm away from each other to prevent any short-circuiting.

To begin with, light crude oil was used at a voltage of 200 V for 24 hours. Asphaltene precipitation was observed. The load cell connected to the electrodes recorded the change in the mass of the electrodes with time. The electrodes removed out of the oil and left to dry to allow oil droplets to drain and light components to evaporate. Asphaltene deposition has been observed on the anode as can be seen in Figure 2a. But, no deposition was seen on the cathode as shown in Figure 2b. The data recorded by the load cells confirms the results. A continuous increase in the mass of the anode with time is observed by the load cell while the one connected to the cathode doesn't detect any change in the mass. Then, the same potential difference of 200V was applied to the left crude oil from the first run for additional 24 hours. In the end, 200 V and 400V were applied to heavy crude oil for 24 hours.

Results

Experiments involving electrical effects were conducted in order to relate these effects to the deposition problem as well as to define the type of charge that the asphaltene particle may possess.

The load cell sensed the increase in the mass of the electrode with time as deposits adhered to it. This mass increase was recorded by the data acquisition system and cumulative mass percentage was calculated from the load cell data recorded.

The temperature of the oil in the container throughout an experiment was controlled by the temperature control system. The

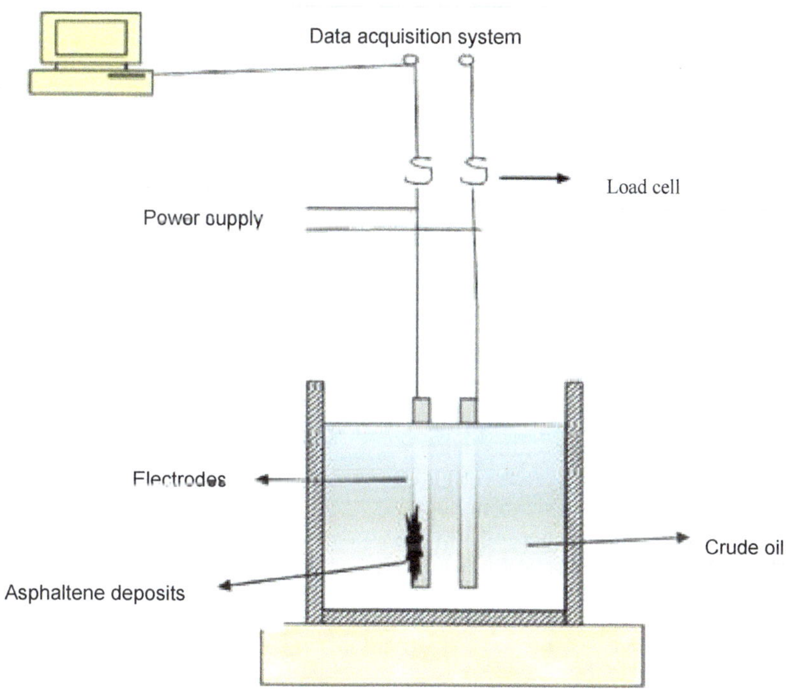

Figure 1: Schematic of the static experiment.

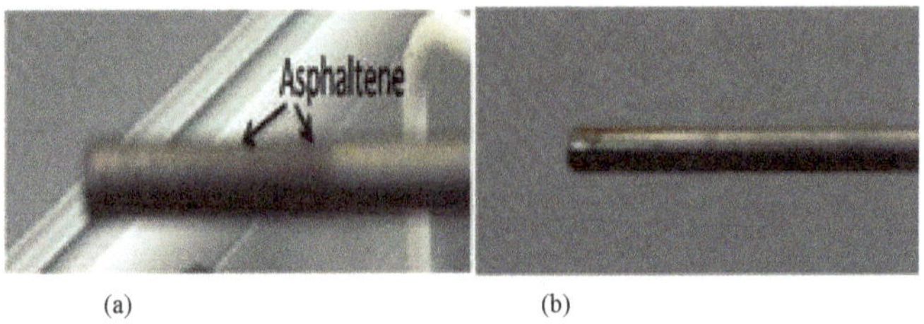

Figure 2: a)Asphaltene deposition on the anode; b) No asphaltene deposition.

relative deviations of the temperature lie between ± 2%, which indicates a relatively good temperature control.

As shown in Figure 2a, a significant amount of asphaltene deposition on the positive electrode (anode) was detected after applying an electric current on the crude oil sample. Moreover, since aphaltene deposited on the anode the electric charge on the asphaltene particles was predicted to be negative. Toluene was added to deposits on the anode to verify if the deposits were asphaltene. The deposits dissolved in toluene and then precipitated in n-heptane which confirmed that deposits were asphaltene particles.

Figure 2b showed that no deposits were found on the electrodes when the left crude oil from the first run was stimulated by a potential difference of 200 V for another 24 hours. This indicated that all the asphaltene in the crude oil were taken out in the first run producing asphaltene-free crude oil. Theoretically, if all the asphaltene can be extracted from the crude oil in the reservoir then no more asphaltene would be deposited onto the production well.

The two electrodes were immersed in heavier crude oil and a 200 V potential difference was applied to the heavier crude oil for 24 hours. No asphaltene deposition was observed on either of the electrodes. This can be due to the following reasons: 1- the heavier crude oil was very stable 2-the voltage was not high enough to destabilize the crude oil 3- the distance between the two electrodes was too large to allow the current to pass, 4- the time was not sufficient to induce deposition. Therefore, the voltage was increased to 400 V and left for 24 hours. After the 24 hours, asphaltene deposition was observed on the anode as shown in Figure 3. On the anode, there was a small swelling, which was a sign of the asphaltene deposits. On the cathode, there was no asphaltene deposition. Toluene was added to the deposits on the anode to verify if the deposits were asphaltene. The deposits dissolved in toluene and then precipitated in n-heptane which confirmed that deposits were asphaltene particles.

Discussion

The cumulative mass, as a percentage of the total asphaltene mass deposited on the anode, versus time for light crude oil experiment at 200 V is shown in Figure 4. These kinetic data suggests that deposition occurred over 120 hours, and no deposition is observed after this time. Level of deposition is initially rapid. Cumulative mass reached 60% of the total mass deposited over the first hour of the experiment but gone down as the time increases. A 40% cumulative mass was obtained through the remainder time. No accumulation was observed on the cathode.

Figure 5 shows the cumulative mass deposited on the anode as a function of time for heavy crude oil experiment at 400V.The time of deposition extended to 140 hrs. After 1 hour, the cumulative mass reached 41%.At 200 V the rate of movement of deposits to the anode is too slow to build up a significant amount. The electrical potential is insufficient to trigger the particles to stay on the electrode. However, above 200 V an increase in the field strength stimulates deposition. Higher potential is required to destabilize heavy crude oil. On the other hand, the application of high electric fields may trigger field-induced dissociation, charge reversal [19] and other related effects.

Comparing the two runs and in spite of the higher potential, less deposition is obtained over the same period of time.

These experimental results prove that the amount of asphaltene in crude oil is limited evident by the no increase of asphaltene deposit after 120 hours in the light crude case and no deposit after 140 hours in the heavy crude case. The fact that the largest amount of asphaltene deposited in the first one 1 hour in both cases indicate that all the nearby asphaltene colloids are affected very early and it took some time for the rest to be dragged to the anode. It will be very interesting to study the dynamic mode and compare to the static results.

In line with our experimentation, we believe that aggregation of asphaltene particles starts somewhere in the reservoir; therefore, in order to prevent asphaltene deposition at the surface facilities from taking place, aggregation of these molecules have to be stopped downhole.

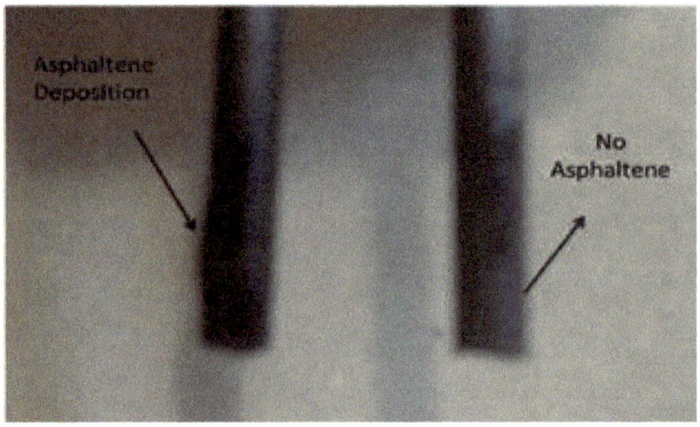

Figure 3: Asphaltene deposition on the anode for a heavy Abu Dhabi crude upon applying 400 V for 24 hours.

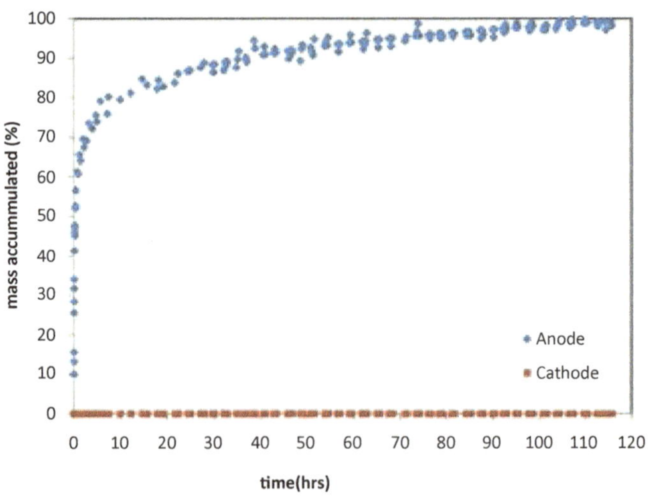

Figure 4: Mass accumulated with time for light crude oil at 200V.

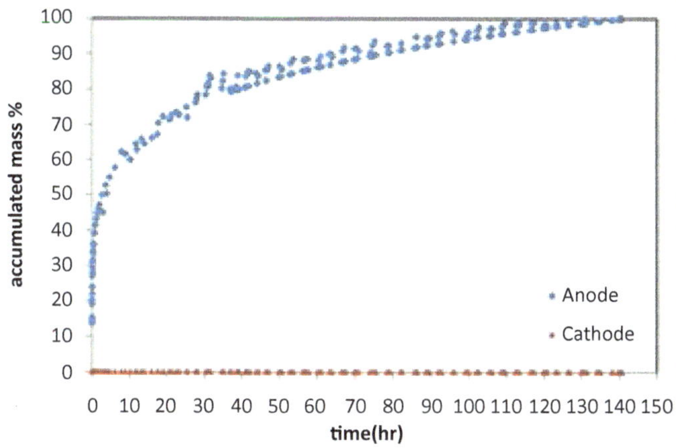

Figure 5: Mass accumulated with time for heavy crude oil at 400V.

One of the interesting approaches is using electricity to stop the aggregation or withdraw them into a dumpster well. The usage of electricity in lab scale has been done since 1945; however, there are many discrepancies among researchers, when dealing with asphaltene electric charge. Some researchers claimed that asphaltene is positively charged; conversely considersasphaltene particles to be negatively charged. However, in this study asphaltene molecules have proven to be negatively charged as demonstrated above, which confirms that results found by Taylor, Preckshot et al. and Khvostichenko et al. [13,21,22]. There is a possibility though that may be, some types of asphaltenes are naturally negatively charged and others are positively charged. Notably, most of the work on asphaltene has been on modeling and analyzing the chemical composition of the fractions, which have not contributed much to the prevention of asphaltene precipitation and deposition. This study provides a new approach to resolve the problem of asphaltene deposition by means of applying a constant voltage to remove asphaltene particles from the crude oil.

To implement this method in field scale, an observer well should be used as the anode and the production well as the cathode. In this way, asphaltene particles will be extracted from the oil by moving away from the production well and going toward the observer. The implementation of this experiment would be facilitated if a conductive chemical was used to accelerate the deposition rate without inducing corrosion into the tubing. This is possible by virtue of utilizing the observer well as a dumpster for asphaltene, so to speak. Consequently, if no more asphaltene particles existed in the crude that would intuitively imply no asphaltene deposition would occur. In addition to that, removing asphaltene from the oil will improve the quality of the crude and reduce the cost of processing. Having the ability to control the deposition rate or ultimately prevent it from occurring is the prime target for this study and the ones thereafter. The downside to using electricity is power consumption, which can appreciably increase the cost of oil production. However, the ability to understand how precipitation and deposition occur would set a new ground to research in this field. Before implementing this experiment on a full-scale, corrosion and consequently economic feasibility studies have to be conducted. Safety measures should be taken into account due to the potential danger involved with passing electric currents.

It has been repeatedly reported in the literature, that one of the main causes of asphaltene deposition is electrical streaming potential; however, not much has been done to, effectively, to tackle this issue. Due to the nature of the experiment, the electrokinetic effect caused by the flow of oil cannot be studied. This experiment is merely an attempt to study the effect of electricity on asphaltene deposition in static mode. The results found in the experiment however are the cornerstone for future work in this area. The next step is testing this idea on dynamic mode to try to resolve the electrokinetic negative effects caused by the flow of oil through any conduit.

Conclusions

Experimental runs showed that significant amounts of asphaltene deposits were obtained when light crude oil was destabilized by the application of a DC current. Destabilization of the heavier crude oil was achieved at higher voltages. Asphaltene particles possess negative charge in heavy and light Abu Dhabi crude oils. Application of a DC current on crude oil for a specified time may extract all asphaltene particles dissolved in crude oil. Therefore, time is a very critical factor. This study provides a new approach to resolve the problem of asphaltene deposition by means of applying a constant voltage to remove asphaltene particles from the crude oil.

Acknowledgment

The authors are thankful to the Petroleum Institute, ADNOC (Abu Dhabi National Oil Company), and OPCO's, Abu Dhabi, UAE for their support and enthusiasm towards this research study.

References

1. Andersen SI, Speight JG (2001) Petroleum Resins: Separation, Character, and Role in Petroleum. Pet Sci Technol 19: 1-34.
2. Hammami A, Ferworn KA, Nighswander JA, Overa S, Stange E (1998) Asphaltenic Crude-Oil Characterization - An Experimental Investigation of the Effect of Resins on the Stability of Asphaltenes. Pet Sci Technol 16: 227-249.
3. Alkafeef SF (2001) An Investigation of The Stability of Colloidal Asphaltene in Petroleum Reservoirs. SPE International Symposium on Oilfield Chemistry, 13-16 February 2001, Houston, Texas USA.
4. Goual L, Firoozabadi A (2004) Effect of resins and DBSA on asphaltene precipitation from petroleum fluids. AIChE J 50: 470-479.
5. Hossein R, Nabiollah M, Ali Reza N, Muhammad S (1997) Field-Scale Asphalt Precipitation during Gas Injection into a Fractured Carbonate Reservoir. SPE Western Regional Meeting, 25-27 June 1997, Long Beach, California USA.
6. Srivastava RK, Huang SS, Dong M (1999) Asphaltene Deposition During CO2 Flooding. SPE Prod Facil 14: 235-245.
7. Jamaluddin AKM, Joshi N, Iwere F, Gurpinar O (2002) An Investigation of Asphaltene Instability Under Nitrogen Injection. SPE International Petroleum Conference and Exhibition in Mexico, 10 12 February 2002, Villahermosa, Mexico.
8. Nghiem LX, Kohse BF, Sammon PH (2001) Compositional Simulation of the VAPEX Process. J Can Petrol Technol 40.
9. Misra S, Abdallah D, Nuimi S (2013) Successful Asphaltene Cleanout Field Trial in on-shore Abu Dhabi Oil Fields. 18th Middle East Oil & Gas Show and Conference (MEOS), Mar 10 - 13, 2013 2013, Bahrain International Exhibition Centre, Manama, Bahrain.
10. Abdallah D (2012) Impact of Asphaltenes Deposition on Completion Design for CO2 Pilot in an Onshore Abu Dhabi Field. Abu Dhabi International Petroleum Conference and Exhibition, 11-14 November 2012, Abu Dhabi, UAE.
11. Eow JS, Ghadiri M, Sharif AO, Williams TJ (2001) Electrostatic enhancement of coalescence of water droplets in oil: a review of the current understanding. Chem Eng J 84: 173-192.
12. Aggour MA, Muhammadain AM (1992) Investigation of waterflooding under the effect of electrical potential gradient. J Petrol Sci Eng 7: 319-327.
13. Preckshot GW, DeLisle NG, Cottrelland CE, Katz DL (1943) Asphaltic Substances in Crude Oils. Transactions of the AIME 151: 188-205.
14. Moore EW, Crowe CW, Hendrickson AR (1965) Formation, Effect and

Prevention of Asphaltene Sludges During Stimulation Treatments. J Petrol Technol 17: 1023-1028.

15. Sheu EY, Acevedo S (2006) A dielectric relaxation study of precipitation and curing of Furrial crude oil. Fuel 85: 1953-1959.
16. Goual L (2009) Impedance Spectroscopy of Petroleum Fluids at Low Frequency. Energ Fuel 23: 2090-2094.
17. Katz DL, Beu KE (1945) Nature of Asphaltic Substances. Ind Eng Chem 37: 195-200.
18. Dykstra H, Beu K, Katz D (1944) Precipitation of asphalt from crude oil by flow through silica. Oil Gas J 43: 79-102.
19. Eldib IA (1962) The solvation, ionic and electrophoretic properties of colloidal asphaltenes in petroleum. 7: 31-42.
20. Pierre LM, Lenin H (1975) Electrical and Other Effects Related to the Formation and Prevention of Asphaltene Deposition Problem in Venezuelan Crudes. SPE Oilfield Chemistry Symposium, 13-14 January 1975, Dallas, Texas USA.
21. Taylor SE (1998) The electrodeposition of asphaltenes and implications for asphaltene structure and stability in crude and residual oils. Fuel 77: 821-828.
22. Khvostichenko DS, Andersen SI (2009) Electrodeposition of Asphaltenes. 1. Preliminary Studies on Electrodeposition from Oil−Heptane Mixtures. Energ Fuel 23: 811-819.
23. Hashmi SM, Firoozabadi A (2012) Controlling Nonpolar Colloidal Asphaltene Aggregation by Electrostatic Repulsion. Energ Fuel 26: 4438-4444.

Assessment of Bioremediation of Oil and Phenol Contents in Refinery Waste Water via Bacterial Consortium

Abha Singh, Vinay Kumar and Srivastava JN*

Microbiology Lab, Botany Department, Dayalbagh Educational Institute, Agra, Uttar Pradesh, India

Abstract

The present study concerns with the biotreatment of waste water of petroleum refinery. The experiments were performed to assess the degradation potential of microbial species present in the waste water and using microbial consortium to treat the water in a suspension form. The four bacterial isolates were isolated and assessed for the biodegradation of crude oil and phenol in the waste water gravimetrically and by spectrophotometry respectively. The isolated bacterial strains were identified as *Alcaligenes odorans, Bacillus subtilis, Corynebacterium propinquum* and *Pseudomonas aeruginosa* which were inoculated into waste water in the form of bacterial consortium. After inoculation of consortium, it was found that oil and phenol contents in the waste water reduced up to 70% and 85% respectively via bioremediation. The search for cheaper and environment friendly options for enhancing degradation of petroleum contamination was the main research interest and bioremediation is showing a greater effective way of remediating natural ecosystem contaminated with crude oil from a number of decades.

Keywords: Bioremediation; Bacterial consortium; Bacterial isolation; Crude oil; Phenol

Introduction

Environmental pollution implies any alterations in the surroundings but it is restricted in use especially to mean any deterioration in the physical, chemical and biological quality of the environment [1]. Contamination of the environment with crude oil results in pollution. Wide scale production, transport use and disposal of petroleum globally have made it a major contamination in both prevalence and quantity in the environment. The oil gets mixed with the river or marine water by many ways as accidental spills or discharge of refineries in river or other water bodies. Multiple initiatives have been developed to resolve the problem of petroleum pollution. An array of procedures has been developed including physical, chemical and biological techniques. Among these procedures, bioremediation is currently associated to physicochemical procedures.

The demand of petroleum as a source of energy and as a primary raw material for chemical industries in recent years has resulted in an increase in world production. This dramatic increase in production, refining and distribution of crude oil has brought with it an ever increasing problem of environmental pollution [2]. The persistence of petroleum pollution depends on the quantity and characteristic of hydrocarbon mixture and on the properties of the affected ecosystem. The ability to isolate high numbers of certain oil-degrading microorganisms from oil polluted environment is commonly taken as evidence that these microorganisms are active degraders of that environment [3]. Virtually every refinery, from primary distillation to final treatment, contains various fractions of oils and other hydrocarbon compounds in their waste waters. The oil and grease in this waste water may appear as free oil, dispersed oil, emulsified oil, soluble oil or as a coating or suspended matter.

Other than oil and grease contamination, phenol and its derivatives are also among the most important contaminants present in the environment. Phenol is one of the major organic pollutants encountered in waste water produced by industrial and refinery activities. Phenols have been reported as highly toxic and hazardous to living organisms [1]. Due to the potential toxicity and persistence of this kind of contaminant in the environment, rapid removal and detoxification is the need of the hour.

Many treatment methods are used in the treatment of oil wastewater. Most of the physicochemical and thermal methods are expensive, as they require expensive equipment and machineries and expend good amount of energy. Bioremediation is a promising method, where wastewater adapted consortium of microbial species is used for the degradation of pollutants from water. The present investigation was carried out with primary objective to study oil and phenol degradation using the bacteria isolated from waste water itself in the form of a consortium.

Material and Methods

Collection of sample and isolation of bacteria

The water sample was collected from Mathura Refinery Unit of Indian Oil Corporation Limited in a pre sterilized glass bottle (5L capacity). The collection site inside refinery was DAF (Dissolved Air Floatation) outlet after physical separation in the refinery itself. The sample was brought to the laboratory at temperature maintained from 2-10°C in an insulated container. 6 samples were collected at the interval of 30 days each from January to June, 2011.

Serial dilution technique was used for the isolation of bacteria. In this technique, sample suspension was prepared by serially diluted upto 10^6. From this solution (10^6 ml^{-1} sample) 1 ml was pipetted on to the plates with nutrient medium-3 (composed of 2 g yeast, 1 g beef

***Corresponding author:** Srivastava JN, Microbiology Lab, Botany Department, Dayalbagh Educational Institute, Agra, Uttar Pradesh, India, Tel: +91-9456433062

extract, 5 g NaCl, 5 g peptone and 15 g agar in 1000 ml distilled water) spreaded by glass spreader and incubated at 30°C temperature for (24 hr) observation. Each single colony that appeared onto the plate was considered as one colony forming unit (c.f.u.). The colonies which grew vigorously and produced clear zone around them, were marked and were re-isolated onto a fresh nutrient agar plate and slants for further experiments.

Characterization and identification of isolated bacteria

Twelve strains were isolated from the water sample and were characterized by growing on nutrient medium enriched with oil and phenol, grams reaction, physiological properties and microscopic observations [4]. The ability to grow in different media was tested at 37°C and neutral pH. Results were recorded after 24 hr at OD_{600} by spectrophotometer.

The isolated strains were allowed to grow on the nutrient medium containing 5% crude oil. Out of 12, 4 strains showed clear zone around their colonies showing depletion of oil. These selected 4 strains were also cultured over the MS medium containing phenol as the only carbon and energy source. The bacterial growth was measured spectrophotometrically at 600nm. The selected bacterial isolates which were able to grow in the phenol as well as oil enriched medium. These isolates were identified by BD-BBL™ Crystal Mind Autoreader (an instrument for bacterial identification).

The pure selected strains were transferred to slants, preserved at 4°C in refrigerator and subcultured at an interval of 30 days.

Composition of media (phenol as carbon and energy source)

Cells were grown at 37°C in Mineral medium for all phenol degrading bacteria. The compositions were 12.8 g of $Na_2HPO_4.7H_2O$, 3 g of KH_2PO_4, 0.5 g of NaCl, 1g of $(NH_4)_2SO_4$ and 2 g of phenol per litre. The pH was adjusted to 7 with 6N NaOH [5].

Preparation of inoculum

The bacterial cultures were inoculated in nutrient broth medium and kept in the incubator shaker at 200rpm at 37°C for a period of five days. The growth was recorded depending on the extent of turbidity by nephlometer. Equal volumes of the culture (1 NTU: Nephlometric Turbidity Unit) broths from the above 4 isolates were mixed to prepare mixed bacterial consortium. The inoculum was mixed with the water sample (1:9) and incubated for 10 days at 37°C for Bioremediation studies.

Biodegradation studies

Laboratory bioremediation studies were carried out at optimized pH (7) and temperature (37°C) in aerobic condition for assessing the oil and phenol degradation potential of isolated strains. For extraction of crude oil, 50 ml of water sample was mixed with 50 ml petroleum ether: acetone (1:1) in a separating funnels and was shaken vigorously to get a single emulsified layer. Acetone was then added to it and shaken gently to break the emulsification, which resulted in three layers. Top layer was a mixture of petroleum ether, crude oil and acetone; clumping cells make the middle layer and the bottom aqueous layer contains acetone, water and biosurfactant in soluble form. The lower two layers were discarded while top layer was collected in a sterilized beaker. The extracted oil was passed through anhydrous sodium sulphate to remove moisture. The petroleum ether and acetone was evaporated on a water bath. The gravimetric estimation of residual oil left after biodegradation was made by weighing the quantity of oil in a tarred vial.

For estimation of phenol in water samples, chloroform extraction method was employed. The sample was taken in distillation flask. The pH was adjusted to 3 with phosphoric acid and distilled water was added to make up the volume upto 300 ml. then sample was distilled to collect 250 ml of distillate and 5 ml of ammonium chloride solution was added to it to adjust the pH to 10. 1.5 ml of potassium ferricyanide and 4-aminoantipyrene solution each were added to the contents and allowed to react for 3 minutes. Thereafter, the colour was extracted in 10ml chloroform and measured on spectrophotometer at 510nm wavelength. The quantity of phenol was calculated with the calibration curve prepared.

$$Phenol\left(mgL^{-1}\right) = \frac{mg\ of\ phenol\ in\ standard\ curve}{ml\ of\ sample} \times 1000$$

Results and Discussion

Characterization of oil and phenol degrading bacteria

The 4 isolates that were able to utilize oil and phenol as carbon and energy source could potentially degrade oil as well as phenol were studied morphologically and by biochemical characteristics for identification (Table 1). By comparing with those mentioned in Bergey's Manual of Systematic Bacteriology, the bacteria were identified as *Alcaligene sodorans, Bacillus subtilis, Corynebacterium propinquum* and *Pseudomonas aeruginosa*. Further, these were confirmed by BD-BBL™ Crystal Mind Autoreader. The isolated strains were allowed to grow on six different medium i.e., Nutrient Agar Medium (NAM), Trypticase Dextrose Agar (TDA), Trypticase Soy Agar (TSA), Sucrose medium (SM), Mineral Salt medium (MSM) and Yeast Extract medium (YEM). The isolates showed maximum growth in Nutrient Agar medium and Mineral Salt medium, while other media were not found suitable for further studies.

Determination of bacterial growth

Bacterial counts of four selected isolates were estimated by plating quadruplicate samples onto Mineral Salt Medium containing oil and phenol for estimation of bacterial strain capable of degrading phenol and oil. A large increase in biomass was observed in first 7 days of the study. After 7 days, count increased very slowly. The optical density was used to monitor the biomass growth. The plot of optical density in relation to time is shown in (Figure 1).

Characteristics	*Alcaligenes odorans*	*Bacillus subtilis*	*Corynebacterium propinquum*	*Pseudomonas aeruginosa*
Gram staining	Negative	Positive	Positive	Negative
Cell shape	Bacilli	Bacilli	Bacilli	Bacilli
Motility	Motile	Motile	Non motile	Motile
Colony	Thin, irregular	Circular	Slimy	Round, raised
Catalase reaction	Positive	Positive	Positive	Positive
Nitrate reduction	Negative	Positive	Positive	Positive
H2S production	Positive	Negative	Negative	Negative
Indole production	Negative	Negative	Negative	Negative
Fermentation Tests	Negative	Negative	Negative	Negative
Dextrose	Negative	Negative	Negative	Negative
Lactose	Negative	Negative	Negative	Negative
Sucrose	Negative	Positive	Negative	Negative
Starch Hydrolysis	Positive	Positive	Negative	Positive
Oxidase Citrate Utilization	Negative	Positive	Negative	Positive

Table 1: Morphological and Biochemical Characteristics of isolates.

Biodegradation study of oil and phenol

The oil and phenol contents in the water samples were determined monthly for six months i.e., 6 samples were evaluated at the time interval of 30 days. In the study, the efficiency of crude oil and phenol degradation of the designed bacterial consortium was tested and results clearly showed that mixed bacterial consortium can carry out 70% degradation of oil and 85% of phenol as well. Rahman et al. [4] also showed degradation of oil upto 78% after incubation of samples for 20 days inoculated with the bacterial consortium. The minimum degradation of oil and phenol was observed to be 35% and 75% respectively. Percentage reduction in the oil and phenol contents is shown in (Figure 2).

The oil contents were found to be 480-140 mgL^{-1} in the samples before treatment while after treatment, they ranged from 90-267 mgL^{-1}. The phenol concentration in the water samples was present in the range of 3.71-1.2 ppm. The consortium remediated the phenol and remained phenol in the sample was found to be 0.3-0.78 ppm (Table 2), that came under permissible limit according to the Environmental Assessment (EA). Since, all the microbes in the present study exhibited higher biodegradation of oil contents and phenols from the water. Survival of microorganisms in a medium containing petroleum after their inoculation is a key deciding factor in the rate of bioremediation of oil and phenol. They survived and adopted the oil contaminated liquid environment very easily as also reported by many scientists [6,7]. In agreement with the results obtained, some researchers earlier reported better degradation of phenol by the bacterial consortium upto 70% mainly consisting *Pseudomonas* sp., *Bacillus* sp., *Alcaligenes* sp. and *Corynebacterium* sp. [8-10].

Rahman et al. [4] also showed similar results in the case of oil degradation as the efficiency of bacterial consortium could carry out maximum of 78% degradation. This finding correlates with the present study as here is the observation of upto 70% degradation of oil from the waste water after 10 days of incubation. Several other workers [11,12] have described the ability of mixed bacterial consortia to degrade upto 60% crude oil. Apart from this, many other genera have also been reported for phenol degradation: *Acinetobacter, Agrabacterium, Burkholderia, Klebsiella, Ralstonia*and *Rhodococcus* [13]. Also the introduction of mineral medium helped in the process of biodegradation which is earlier supported by other workers [14].

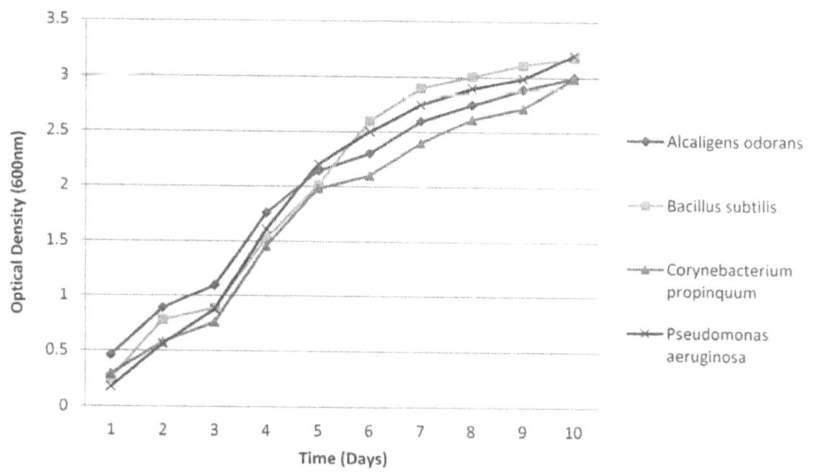

Growth profile of bacterial isolates.

Figure 1: Growth profile of bacterial isolates observed during 10 days at 600nm wavelength.

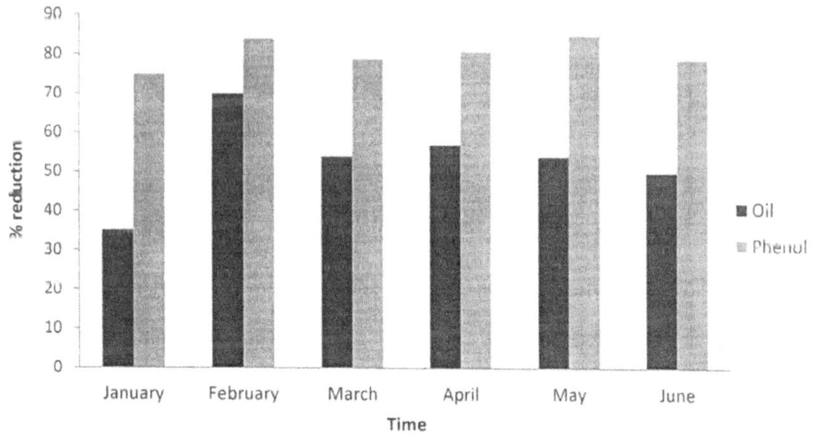

Percentage (%) reduction of oil and phenol in waste water.

Figure 2: Reduction percentage in the oil and phenol contents of waste water after treatment with bacterial consortium.

Samples	Oil contents (mgL^{-1})		Phenol (ppm)	
	Untreated	Treated	Untreated	Treated
January	140	90	1.20	0.30
February	480	148	2.67	0.42
March	580	267	3.71	0.78
April	260	114	1.73	0.33
May	344	194	2.15	0.32
June	378	186	2.39	0.49

Table 2: Determination of oil and phenol in waste water before and after treatment of refinery waste water via bacterial consortium.

Conclusion

Bioremediation has an edge over other treatment methods because it can efficiently degrade oil and phenol from the waste water and doesn't allow the contaminant to accumulate. The present study clearly showed that by using the mixed bacterial consortium which can efficiently degrade the crude oil components and phenols as well, temperature of 37°C and pH 7, maximum percentage of degradation can be achieved. The bacterial cells were able to utilize phenol and oil as the carbon and energy source. Furthermore, these strains show promise for possible utilization for waste water treatment. Hence we suggest use of the above optimized conditions and the mixed bacterial consortium for bioremediation of crude oil contamination.

Acknowledgements

Authors wishes to thank Prof. LD Khemani, Dean of Science Faculty and Prof DS Rao, Head of Department, Dayalbagh Educational Institute, for providing lab facilities and technical support. Authors are grateful to University Grants Commission for financial assistance as Senior Research Fellowship and are highly obliged to Mr. VS Dhakate, Senior Technical Manager, Mathura Refinery Unit, for permitting sample collection.

References

1. Moschella PS, Laane RPWM, Back S, Behrendt H, Bendoricchio G, et al. (2005) Group report: Methodologies to support implementation of the water framework directive. Managing European Coasts Environmental Science Springer Berlin Heidelberg 137-152.
2. Ojo OA (2006) Petroleum-hydrocarbon Utilization by native Bacterial Population from a Wastewater Canal Southwest Nigeria. African Journal of Biotechnology 5: 333-337.
3. Okerentugba PO, Ezeronye OU (2003) Petroleum degrading potentials of single and mixed microbial cultures isolated from rivers and refinery effluent in Nigeria. African Journal of Biotechnology 2: 288-292.
4. Rahman KS, Thahira-Rahman J, Lakshmanaperumalsamy P, Banat IM (2002) Towards efficient crude oil degradation by a mixed bacterial consortium. Bioresour Technol 85: 257-261.
5. Khleifat KM (2007) Biodegradation of Phenol by Actinobacillus Sp.: Mathematical Interpretation and Effect of Some Growth Conditions. Bioremediation Journal 11: 103-112.
6. Das K, Mukherjee AK (2006) Crude petroleum-oil biodegradation efficiency of Bacillus subtilis and Pseudomonas aeruginosa strains isolated from a petroleum-oil contaminated soil from North-East India. Bioresour Technol 98: 1339-1345.
7. Rahman KS, Rahman TJ, Kourkoutas Y, Petsas I, Marchant R, et al. (2003) Enhanced bioremediation of n-alkane in petroleum sludge using bacterial consortium amended with rhamnolipid and micronutrients. Bioresour Technol 90: 159-168.
8. Gayathri KV, Vasudevan N (2010) Enrichment of Phenol Degrading Moderately Halophilic Bacterial Consortium from Saline Environment. J Bioremed Biodegrad 1:104.
9. Dastager SG, Deepa CK, Pandey A (2009) Isolation and Characterization of High-Strength Phenol-Degrading Novel Bacterium of the Pantoea Genus. Bioremediation Journal 13: 171-179.
10. van Schie PM, Young LY (1998) Isolation and characterization of phenol-degrading denitrifying bacteria. Appl Environ Microbiol 64: 2432-2438.
11. Chhatre S, Purohit H, Shanker R, Khanna P (1996) Bacterial consortia for crude oil spill remediation. Water Sci Technol 34: 187-193.
12. Vasudevan N, Rajaram P (2001) Bioremediation of oil sludge-contaminated soil. Environ Int 26: 409-411.
13. Koutny M, Ruzicka J, Chlachula J (2003) Screening for phenol-degrading bacteria in the pristine soils of south Siberia. Appl Soil Ecol 23: 79-83.
14. Naik BS, Mishra IM, Bhattacharya SD (2011) Biodegradation of Total Petroleum Hydrocarbons from Oily Sludge. Bioremediation Journal 15: 140-147.

Bioconversion of Heavy Hydrocarbon Cuts Containing High Amounts of Resins by Microbial Consortia

Gholllami M[1]*, Roayaei M[1], Ghavipanjeh F[2] and Rasekh B[3]

[1]*Biology Department, Shahid Chamran University, Ahwaz, Iran*
[2]*Environment Group, Energy Department, Materials and Energy Research Center , Alborz, Iran*
[3]*Microbiology and Biotechnology Group, Research Institute of Petroleum Industry, Tehran, Iran*

Abstract

Microbial conversion of petroleum resins existing in a heavy fraction of crude oil was investigated by native isolated consortia. The heavy oil fraction was taken from an end cut of distillation tower of Abadan refinery in Iran. Several Soil samples were taken from different oil contaminated sites of Abadan oil refinery in Khuzestan province. Seven bacterial consortia were then prepared by enrichment method and were examined for their ability to convert resins. Amongst them, consortium AB6 reduced 37.3% of resins while, aliphatic and aromatic compounds increased by 86.8% and 6.7%, respectively. Three pure cultures were isolated from consortium AB6 which were identified as *Serratia sp., Raoultella sp. and Ochrobactrum sp.*

Keywords: Biological conversion; Resins; *Serratia; Raoultella; Ochrobactrum*

Introduction

Quality Improvement of crude oils has been one of the important problems in the recent 60 years. There are some chemical and physical methods for crude oil upgrading. But, due to high cost of the current oil processes together with their safety and environmental problems, attention to affordable and environmentally acceptable methods has gained significant importance [1,2]. In this regard, biotechnology as well as special ability of microorganisms for production of a wide range of biochemical materials has been an important technique in overcoming the defects of the other methods [3,4].

Crude oil is a mixture of hydrocarbons containing hydrogen, carbon and lesser amounts of oxygen, sulphur and nitrogen plus some trace metals such as vanadium, nickel, copper and iron [5]. The hydrocarbons can be divided into aliphatic or saturates (including paraffins, isoparaffins and cycloparaffins), aromatics, resins and asphaltenes. Short chain paraffins are more degradable substrates for microorganisms compared with long chain paraffins, isoparaffins, cycloparaffins, aromatics, resins and asphaltenes. Resins and asphaltenes contain several hetroatoms per molecules and typically account for 30-60% of the total sulphur, 70-90% of total nitrogen and 80-90% of the total vanadium and nickel present in a crude oil [6,7]. Resins are large molecules containing polar groups with molecular weights substantially lower than those of asphaltenes. Their percentage in heavy oil fractions is higher than that of asphaltenes which is more in bitumen.

Resins and asphaltenes are less biodegradable and insufficient literatures are presented in this field [8]. Venkateswaran et al. [9] found a consortium consisted of *pseudomonas sp* and several un-named strains, which could degrade 35% of resins fractionated from crude oil, presenting the first results on biological degradation of resins as the authors claimed. In 2006, Ghavipanjeh et al. [10] investigated the biological conversion of a heavy hydrocarbon cut named as Isomax feed (a heavy end of vacuum distillation column) by mixed cultures. Their results showed a reduction of more than 10% of the resin fraction of the feed [10]. Heavy fractions of crude oil and distillation residues contain more amounts of resins and asphaltenes [11]. However, there is a trend towards the use of heavier feedstock, which requires additional processes based on the destruction of heavy petroleum compounds and lead to an increase in the portion of light fractions in feedstock [12].

Middle East region makes up only 1% of Earth's surface. However, it contains 65.7% of the world's oil content. Iran is the first oil-rich country in this region with production capacity of 4 million barrels per day of crude oil and diesel fuel [13]. Khuzestan province contains massive oil and gas reservoirs and so is the most important oil region in Iran. However, most of the crude oils of the region contain a high level of resins. This makes process difficulties, clogging of transportation pipelines and hard environmental deterioration problems when spilled out.

Different reports during recent decades have shown that microbial consortia are more effective on degradation and conversion of hydrocarbons, compared with individual strains [9,14,15]. In this research, different soil samples were taken in order to investigate the effect of native consortia on bioconversion of petroleum resins. This could lead to biological upgrading of heavy petroleum fractions and help in removal of environmental oil spills. Resin Compounds are adhesive hard solids that precipitate in the bottom of distillation tower or in oil wells leading to oil recovery decreases. This paper investigates the biological conversion of resin using native mixed cultures and identifying the types of the bacteria involved in the process.

Materials and Methods

Samples and feed

Seven soil samples were collected from oil contaminated sites of Abadan oil refinery in Khuzestan province and named as AB1, AB2,

***Corresponding author:** Mahboobeh Ghollami, Biology Department, Shahid Chamran University, Ahwaz, Iran, Tel: 021- 44350761; Fax: 021- 44216661

AB3, AB4, AB5, AB6 and AB7. A Heavy oil fraction containing a high percentage of resin compounds was taken from an end cut of distillation tower of Abadan oil refinery and used as the hydrocarbon feed for enrichment of microorganisms. The characteristics of the heavy end feed are shown in Table 1.

Enrichment and bacterial growth

Batch cultures were conducted in 250 ml flasks containing 5% (w/v) hydrocarbon feed, 10 ml of supernatant prepared from each soil sample (10% (w/v) soil in distilled water) and a mineral salt medium (MSM) consisting of K_2HPO_4 2.75 g/l, KH_2PO_4 2.25 g/l, $(NH_4)_2SO_4$ 1 g/l, $MgSO_4.7H_2O$ 0.2 g/l, NaCl 0.1 g/l and $FeCl_3.6H_2O$ 0.02 g/l. The pH of the medium was adjusted at 6.8 ± 2. The flasks were kept in a shaker incubator at 30°C and 150 rpm for 10 days. The inoculation procedure from the previous flasks was repeated three times for preparation of the enrichment cultures. The enrichment cultures were named as AB1, AB2, AB3, AB4, Ab5, AB6 and AB7.

Duplicate batch cultures were conducted from the last enrichment cultures in contact with the heavy oil feed as the same conditions as above, in order to investigate the biological conversion of the feed. Also, a flask was prepared without microorganisms as the control sample. All the flasks were incubated at 30°C and 150 rpm for 10 days. The optical densities of the aqueous phases of the cultures were measured along time in order to investigate the growth of the microorganisms. At the end of the incubation, the samples were used for further investigations as follows.

Sample selection and bacteria isolation

Best sample selection was made upon bacterial growth and visual observations of the organic phases compared to that of the control sample. Strains of the consortia were isolated considering their bioconversion capacity. Selected isolates were purified by pour plate method and kept in refrigerator for further experiments.

Hydrocarbon analysis

After bio-treatment of the feed, the oil phases were separated from the aqueous phases in order to analyze the bioconversion of the organics especially the resins by the enriched cultures [16]. In this method, normal pentene was used as the solvent of the organic phase because of its ability to solve the organic phase and different molecular weight with water and its evaporation point is also low. So, it could be easily separated from the organic phase. 15 ml of pentane was added to sample in a decanter funnel and shake vigorously and then let to settle for 45 minute. The oil and aqueous phases were then separated. The procedure was repeated twice. The extracted organic phases were then dehydrated by passing through dehydrated $MgSO_4$ and evaporated in a water bath for the separation of the solvent. The separated bio-treated organic phases plus the control sample were divided into four defined fractions: Saturates (aliphatics), Aromatics, Resins and Asphaltenes (SARA fractions). The asphaltenes fraction was determined by precipitation with n-pentane according to ASTM method D-893. The remained organic phases after asphaltene separation were conveyed on adsorption chromatography to separate aliphatic, aromatic and resin fractions of the oils. The procedure was made according to ASTM D2549 with some modifications [10,17,18]. This was achieved by using n-pentane solvent (for elution of aliphatics), diethyl ether (for elution of aromatics) and, chloroform and ethanol (for elution of resins). An optical test of TLC paper under UV light was done at the same time for identification of hydrocarbon fractions during elution. A spot of the eluted on TLC paper would turn to purple under UV light by the aromatics while making no color for the aliphatics. The fractions were then evaporated to remove the solvents and weighed. The aliphatic, aromatic and resin fractions of the bio-treated oils were compared to that of the control sample.

Parameter	Value
Initial boiling point (°C)	384
Final Boiling Point (°C)	575
Sulfur (ppm)	18863
Carbon (Wt %)	65
Iron (ppm)	3.5
Nickel (ppm)	5.2
Resin (%)	48.95
Asphaltene (%)	7.37
Viscosity (g/cm³)	0. 597

Table 1: The characteristics of the heavy end feed.

Resin analysis

In order to further inspection of the effect of native cultures on the resin contents of the heavy end feed, the separated resin fractions were divided into soft resins and hard resins. The resin fractions were dissolved in 15 ml of a mixture of isobutyl alcohol and cyclohexane (80:20) in which the hard resins were precipitated. The soft resins were then precipitated in 12 ml isobutyl alcohol [6].

GC-MASS analysis

The separated bio-treated organic phases of the selected consortia as well as the control sample were analyzed for the compositional changes using gas chromatography mass spectroscopy (GC-MASS). Here, Agilent instrument model 7890A equipped with a mass selective detector model 5975C and column, BP-5 (50m×0.25mm) was used with initial temperature 50°C, final temperature 280°C, ramp rate 5°C per min and helium as the carrier gas at 20 ml/min.

Identification of the isolates

The pure cultures isolated from the selected consortium were identified by several morphological, biochemical and phylogenic methods [13,19-21]. The biochemical tests were include Gram stain, Methyl Red test, Voges-Proskauer test, Citrate test, Acid production (TSI), Urease, Indole production, Motility, producing hydrogen sulfide, and lysine decarboxylase. The molecular identification of bacteria was done by amplification of gene 16s rRNA by universal primers. The sequences (5'-3') of the forward and reverse primers were: AAGAGTTTGATCATGGCTCAG and AGGAGGTGATCCAACCGCA, respectively [22]. The PCR program for amplification of gene was 94°C the temperature of denaturation for 1 min, 62°C the temperature of bonding primers for 40 second and 150 second at 72°C. The number of amplification cycles was 30 [23]. After that the DNA sequence data were analyzed using BLAST-N (Local alignment search tool) [24].

Results and Discussion

Enrichment and bacterial growth

After 3 times sub-culturing of the samples at 30°C and 150 rpm, the visual observations were made on the bio-treated samples compared to the control sample. Amongst the seven samples, four samples had imposed obvious changes on the organic phases, while causing turbidity in the aqueous phases. These samples included AB4, AB5,

AB6, and AB7 Figure 1 shows the growth curves of the two selected consortia (AB4 and AB6) in the presence of the heavy end cuts.

Bacterial isolation

The isolated bacteria of the selected consortia were: 4 strains from AB4, 2 strains from AB5, 3 strains from AB6 and 3 strains from AB7.

Hydrocarbons analysis

The results of the gravimetric measurement of SARA fractions of the selected bio-treated samples are shown in Table 2. According to the table, the aliphatic fractions of the all bio-treated samples are increased compared with the control sample. Also, the aromatic fractions of AB4 and AB6 are increased compared with the control sample. The resin fractions of all bio-treated samples are decreased compared with the control sample but are more significant in AB4 and AB6. This emphasizes the ability of consortia AB4 and AB6 in declining the resin contents of the heavy end cuts as we expected. Also, the increase of aliphatics and aromatics are also a good sign. Considering the resin compounds reduction, we can say the enhancement in aliphatics and aromatics fractions amount that are lighter compounds than resins is probably due to conversion and changes of resin compounds.

Resin analysis

The resin compounds fractionated from the control and the superior bio-treated samples (ABR4 and ABR6) were separated into soft and hard resins. The results are shown in Table 3, which can be seen that the hard resins are entirely removed after bio-treatment of the feed. Though, 14.4% of total resins of the control sample are the hard resins. This indicates that larger molecules are converted to lighter ones via biological process.

GC-MASS analysis

The GC-MASS analyses of the mixtures of aliphatic and aromatic fractions of AB6 and the control sample are shown in Figure 1. As reveals from the figures, the differences between the chromatograms are not clearly obvious due to high complexity of the compounds. Some of the compounds are shown in the figures, as could be seen the peaks heights are changes.

The analysis of the GC-MASS digital library data revealed some differences as indicated in Table 4. The table shows that some compounds are disappeared from the control sample, while some others are appeared in the bio-treated samples. For example, toluene

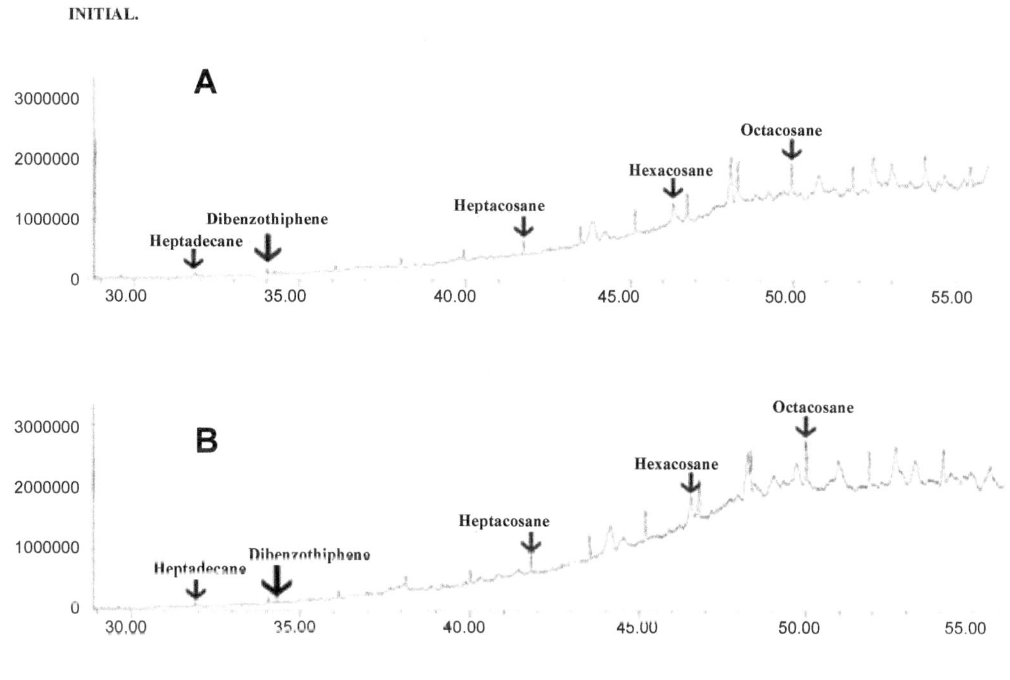

Figure 1: A: GC-MASS chromatogram of the mixture of aliphatic and aromatic fractions of the control sample. **B:** GC-MASS chromatogram of the mixture of aliphatic and aromatic fractions of AB6 sample.

	Hydrocarbon fractions						
	Saturates		Aromatics		Resins		Asphaltenes
Sample name	% in sample	% of changes	% in sample	% of changes	% in sample	% of changes	% in sample
AB4	26.85	49.75	30.09	18.09	34.58	-27.93	8.48
AB5	26.6	48.35	19.59	-23.12	45.21	-5.77	8.60
AB6	33.51	86.89	27.20	6.75	30.07	-37.3	8.51
AB7	23.99	33.8	23.15	-9.14	44.41	-7.44	8.45
Control	17.93	-	25.48	-	47.98	-	8.61

Table 2: The amounts of SARA fractions in the biotreated oil samples and the percent of changes compare to the control.

Sample name	Hard resin (%)	Soft resin (%)
Control	14.4	85.6
ABR4	0	100
ABR6	0	100

Table 3: The amounts of hard resins and soft resin of the resin fractions.

Time	Compounds	% in control sample	% in treated sample
4.014	1-propanol, 2- methyl	–	0.22
4.025	Propene, 2-methyl-3-Propoxy	9.59	–
9.856	Pentadecane	0.95	–
9.879	Toluene	–	0.54
28.630	Tetradecane	–	0.20
28.933	Dibromotetradecane-1-01acetate	0.26	–
36.028	Heptacosane, 1-chloro-	–	0.35
36.097	Dibenzothiphene, 4-methyl-	0.54	–
36.795	Imidazo(4,5-f)quinolone,2-amino-3-methyl-	2.68	–
	Anthracene, 9,butyl tetradecahydro		0.34
36.961	Octadecane	–	0.37
39.856	Cyclotriacontane	–	
39.953	Phenanthrene, 3,6-dimethyl	1.25	–
41.756	Heptacosane, 1-chloro-	1.35	–
41.807	Triacontane	–	3.04
45.149	Hexacosane	2.86	–
45.206	Octacosane	–	5.63
52.439	9.19-cyclolanost-23-ene-3, 25-diol,	15.00	–
52.679	3-acetate	–	12.32

Table 4: The compositional changes of the mixture of aliphatic and aromatic fractions of the sample treated by AB6 compared with that of control, according to GC-MASS Digital library.

	Strains		
Characteristics	AB6S1	AB6S2	AB6S3
Morphology	Bacilli	Cocobacill	Cocobacill
Gram stain	-	-	-
Urea hydrolysis	-	-	-
Citrate	+	+	+
MR	+	+	-
VP	-	-	+
TSI	Alk-Alk	Alk-Alk	Acid-Acid
Lysine decarboxylase	+	+	+
Indole	-	+	-
Hydrogen sulfide	-	-	-
Motility	+	-	-

Table 5: Biochemical characteristics of the isolated strains of AB6 consortium.

and tetradecane did not exist in the control but appeared in the bio-treated sample by AB6. Also, the amounts of short chain alkanes were decreased compared to the control. However, the amounts of alcoholic compounds, alkanes and polycyclic compounds were increased compared to the control.

Identification of the isolates

Three pure cultures were isolated from consortium AB6 as the best native consortium on conversion of resins. The biochemical characteristics of the isolated organisms were shown in Table 5. The strain AB6S1 produced red pigments on nutrient agar plates with a weak elevation after 24 hours of incubation. They were rod shaped, aerobic, gram negative and motile. They were also citrate positive, decarboxylases lysine, indole negative and hydrogen sulfide negative. Morphologically, the colonies of AB6S2 were isolated were pigmented in shades of green and cells were also round and small. The bacteria were citrate positive, decarboxylases lysine, indole positive, hydrogen sulfide negative, non- motile, aerobic, and gram negative. Another isolate, AB6S3 produced colonies, which were cream, circular, flat and rough edges. Microscopically, they were seen as gram negative and non-motile. The bacteria were citrate positive, indole negative, hydrogen sulfide negative and decarboxylases lysine. The results revealed the genus of the bacteria.

Conclusions

This research investigated the biological conversion of a heavy end cut containing high amounts of resins by means of native bacterial consortia. Seven consortia were adapted from which consortium AB6 had the highest activity on the feed. The analytical results on the bio-treated oil showed that the amounts of aliphatics and aromatics were increased by 86.9% and 6.7%, respectively. Also, the amounts of resins were decreased by 37.3%. This could be due to elimination of heteroatoms in molecules of resins by the bacteria that beaks the large molecule. Also, further analysis of the resins content of the feed showed that all the hard resins were eliminated after bio-treatment and the remained resins were all in the class of soft resins. This reinforces the idea of cleavage of hard resins into smaller molecules by bacteria, which again could be due to elimination of heteroatoms from the molecules.

The isolates of consortium AB6 were identified as *Serratia sp.*, strain PBCC12 (Gen bank Accession No, JN566121.1), *Raoultella sp.*, strain PBCC13 (Gen bank Accession No, JN566122.1) and strain PBCC14 (Gen bank Accession No, JN566123.1) *Ochrobactrum sp.*

Acknowledgements

This work was supported by Materials and Energy Research Center under a M.Sc. Thesis. The authors are so thankful to Mr. Shahab Aldin Zarekhafri for him numerous helps during the whole states of experiments.

References

1. Simanzhenkov V, Idem R (2003) Crude Oil Chemistry. CRC Press Newyork USA.
2. Mashreghi M, Marialigeti K (2005) Characterization of Bacteria Degrading Petroleum Derivatives Isolated from Contaminated Soil and Water. Journal of Sciences, Islamic Republic of Iran 16: 317-320.
3. Das N, Chandran P (2011) Microbial degradation of petroleum hydrocarbon contaminants: an overview. Biotechnol Res Int.
4. Obire O, Nwaubeta O (2001) BioDegradation of Refined Petroleum Hydrocarbons in Soil. J Appl Sci Environ Mgt 5: 43-46.
5. Grishchenkov VG, Townsend RT, McDonald TJ, Autenrieth RL, Bonner JS, et al. (2000) Degradation of petroleum hydrocarbons by facultative anaerobic bacteria under aerobic and anaerobic conditions. Process Biochem 35: 889-896.
6. Speight JG (2002) Handbook of Petroleum Product Analysis. John Wiley & Sons, Inc. USA.
7. Gailiute I, Grigiskis S, Zekaite G, Cipinyte V (2011) Selection of Microbes and Conditions That Induced Bio-Cracking of Branched Hydrocarbon Squalane. Environment Technology Resources 1: 46-52.
8. Van Hamme JD, Singh A, Ward OP (2003) Recent Advances in Petroleum Microbiology. Microbiol Mol Biol Rev 67: 503-549.
9. Venkateswaran K, Hoaki T, Kato M, Maruyama T (1995) Microbial degradation of resins fractionated from Arabian light crude oil. Can J Microbiol 41: 418-424.
10. Ghavipanjeh F, Pazouki M, Shayegan J, Hosseinnia A (2006) Evaluation of microorganism utilization in cracking of ISOMAX feed (project phase 1, final report). National Iranian Oil Refining and Distribution Company (NIORDC).
11. Bernard O, Michel M (2005) Petroleum Microbiology. American Society for Microbiology.
12. Grin'ko AA, Golovko AK (2011) Fractionation of resins and asphaltenes and investigation of their composition and structure using heavy oil from the Usa field as an example. Petrol Chem 51: 192-202.

13. Amiriyan A, Mazaheri Assadi M, Saggadian VA, Noohi A (2004) Bioemulsan Production by Iranian Oil Reservoirs Microorganisms. Iranian J Env Health Sci Eng 1: 28-35.

14. Diaz-Ramirez IJ, Escalante-Espinosa E, Favela-Torres E, Gutierrez-Rojas M, Ramirez-Saad H (2008) Design of bacterial defined mixed cultures for biodegradation of specific crude oil fractions, using population dynamics analysis by DGGE. Int Biodeter Biodegr 62: 21-30.

15. Ghazali FM, Abdul Rahman RNZ, Salleh AB, Basri M (2004) Biodegradation of hydrocarbons in soil by microbial consortium. Int Biodeter Biodegr 54: 61-67.

16. Korda A, Santas P, Tenente A, Santas R (1997) Petroleum hydrocarbon bioremediation: sampling and analytical techniques, in situ treatments and commercial microorganisms currently used. Appl Microbiol Biotechnol 48: 677-686.

17. Premuzic ET, Lin M Bohenek MS, Zhou WM (1999) Bioconversion Reactions in Asphaltenes and Heavy Crude Oils. Energy Fuels 13: 297-304.

18. ASTM committee(1982) Annual Book of ASTM Standards. Part 46.

19. Okoh AI (2006) Biodegradation alternative in the cleanup of petroleum hydrocarbon pollutants. Biotechnology and Molecular Biology Review 1: 38-50.

20. Mittal A, Singh P (2009) Isolation of hydrocarbon degrading bacteria from soils contaminated with crude oil spills. Indian J Exp Biol 47: 760-765.

21. Malatova K (2005) Isolation and Characterization of Hydrocarbon Degrading Bacteria from Environmental Habitats in Western New York State. Rochester Institute of Technology Newyork USA.

22. Weisburg WG, Barns SM, Pelletier DA, Lane DJ (1991) 16S ribosomal DNA amplification for phylogenetic study. J Bacteriol 173: 697-703.

23. Benson HJ, Brown AE (2006) Benson's Microbiological Applications: Laboratory Manual in General Microbiology. McGraw-Hill Higher Education USA.

24. Altschul SF, Gish W, Miller W, Myers EW, Lipman DJ (1990) Basic local alignment search tool. J Mol Biol 215: 403-410.

Biosurfactants from Fungi: A Review

Garima Bhardwaj[1], Swaranjit Singh Cameotra[2] and Harish Kumar Chopra[1]*

[1]Department of Chemistry, Sant Longowal Institute of Engineering and Technology, Longowal - 148106, Sangrur, Pubjab, India
[2]Institute of Microbial Technology, Sector-39-A, Chandigarh-160036, India

Abstract

Biosurfactants are the surfactants of microbial origin. They offer so many advantages over their synthetic counterparts due to their biodegradable and environmental friendly nature, that's why gaining much more attention in creating the era of green technology. Their applications range from cosmetic, pharmaceutical and food processes as emulsifiers, humectants, preservatives, detergents etc. The present review deals with the production, purification, characterization and applications of various fungal biosurfactants.

Keywords: Biosurfactants; Sophorolipids; Fungi; Purification; Characterization

Introduction

Biosurfactants are structurally and functionally diverse amphiphilic, surface active compounds which lower the surface and interfacial tension between individual molecules at respective surfaces and interfaces. Thus, these are very important in the living systems and can be regarded as the backbone of the biological membranes which promise the transport and exchange of the various important materials [1,2]. Biosurfactants are ecologically safe and can be applied in bioremediation processes. The microorganisms which produce biosurfactants can also be used in the various bioremediation technologies like solubilisation and removal of oil from contaminated soil, sludge in oil storage tank etc. [3]. The most important example of this is in Microbial Enhanced Oil Recovery (MEOR) which is an eco-friendly petroleum recovery process [4].

Variety of bacteria and few fungi are reported to produce biosurfactants using renewable sources [5,6]. Higher yields of 120 and 40 g/L of fungal surfactants has been reported using carbon sources like tallow fatty acid residues, animal fat, glycerol and oleic acid [7-9]. Furthermore fungi yield a good amount of biosurfactant when compared to bacteria. The reason may be presence of rigid cell wall in them [10]. The higher yield seems to make their use possible at industrial level and replacement of surfactants by biosurfactants.

Fungi Producing Biosurfactants

Where the field of production of biosurfactants by bacterial species is well explored, relatively fewer fungi are known to produce biosurfactants. Among fungi, *Candida bombicola* [11-14,7,8], *Candida lipolytica* [15,16], *Candida ishiwadae* [17], *Candida batistae* [18], *Aspergillus ustus* [19], *Ustilago maydis* [20] and *Trichosporon ashii* [21] are the explored ones. Many of these are known to produce biosurfactant on low cost raw materials. The major type of biosurfactants produced by these strains is sophorolipids (glycolipids). The structure of sophorolipids produced by *Candida batistae* [18], *Candida bombicola* [22] and *Candida sp.* SY16 [10] are given in Figure 1.

Patents on fungal biosurfactants

Due to wide industrial applications many authors have claimed for patents on biosurfactants. Some of the important patents in the last few years are listed in Table 1.

Physicochemical Parameters Affecting the Biosurfactant Production

The production of biosurfactants by the use of various culture conditions is an important aspect because a small alteration in the composition of important nutrients leads to the modification of the resulting biosurfactant. The various physicochemical factors are discussed as follows:

Carbon sources

Carbon source plays an important role in the growth as well as production of biosurfactants by the various microorganisms and it varies from species to species. When only one from glucose and vegetable oil was used for the production of biosurfactant by *Torulopsis bombicola,* a very low yield of biosurfactant was obtained but when both carbon sources were supplemented together yield increased to 70 g/L [23]. While at the concentration of 80 and 40 g/L of glucose and soybean oil, the maximum yield of sophorose lipids was obtained by *Torulopsis bombicola* [24], even higher yields of 120 g/L sophorolipids was produced with *Candida bombicola* in 8 days, when sugar and oil were used as carbon sources [13]. When canola oil and glucose were used as the carbon sources by *Candida lipolytica,* in the concentrations of 10% each, maximum yield of sophorolipids (8 g/L) was obtained [15]. Also, when the industrial residue were used for the production of biosurfactant by *Candida lipolytica,* yielded 4.5 g/L of protein-lipid-carbohydrate complex with the reduction in surface tension of distilled water from 71 to 32 mN/m [16]. Although with *Candida lipolytica* higher production of bioemulsifier was obtained, when supplemented with 1.5% glucose (w/v) [25]. *Candida antarctica* and *Candida apicola* yielded 13.4 and 7.3 g/L of sophorolipids respectively, when soap stock was supplemented in 5% v/v concentration [26]. The resting cells of

***Corresponding author:** Harish Kumar Chopra, Professor, Department of Chemistry, Sant Longowal Institute of Engineering and Technology, Longowal - 148106, Sangrur, Pubjab, India, Tel: 91-1672-305204; Fax: 91-1672-280072

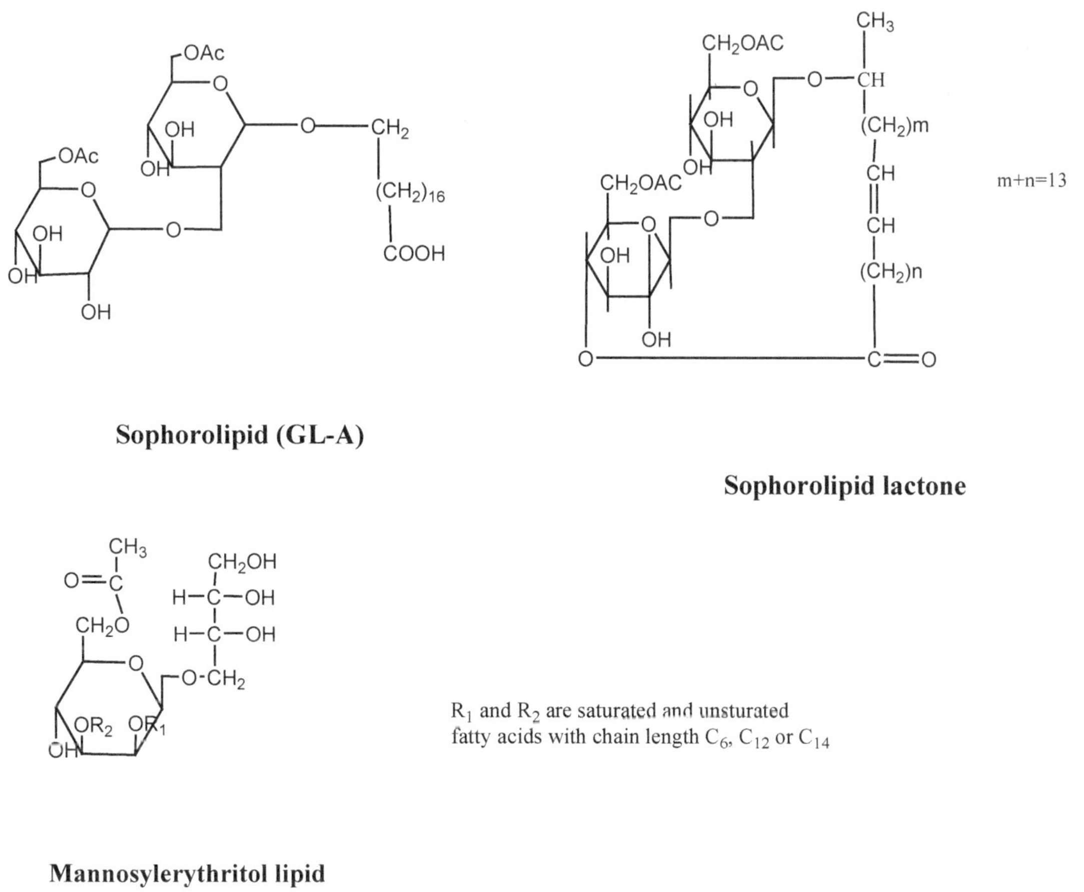

Figure 1: Structure of sophorolipids.

S. No.	Fungi/ type of biosurfactant	Inventor	Title of the patent	Publication No.	Publication Date/ Year
1.	Sophorolipid producer	Borzeix F	Sophorolipids as stimulating agent of dermal fibroblast metabolism	US 6057302 A	2 May 2000
2.	Sophorolipid producer	Borzeix F, Concaix	Use of sophorolipids comprising diacetyl lactones as agent for stimulating skin fibroblast metabolism	US 6596265 B1	22 July 2003
3.	*C. albicans,* *C. rugosa,* *C. tropicalis,* *C. lipolytica,* *C. torulopsis*	Awada S, spendlove R, Awada M	Microbial biosurfactants as agents for controlling pests	US 20050266036 A1	1 Dec 2005
4.	Sophorolipid producer	Gross RA, Shah V, Doncel GF	Spermicidal and virucidal properties of various forms of sophorolipids	WO 2005089522 A2	29 Sept. 2005
5.	Sophorolipid producer	Cox TF, Crawford RJ, Gregory LG, Hosking SL, Kotsakis P	Mild to the skin, foaming detergent composition	WO2011120776 A1	6 Oct 2011
6.	*Candida bombicola* ATCC 22214.	Soetaert W, De MS, Saerens K, Roelants S, Van Bl	Yeast strains modified in their sophorolipid production and uses thereof	EP 2580321 A1	17 Apr 2013

Table 1: Patents on biosurfactants produced by fungi.

Pseudozyma (Candida antarctica) was able to covert C_{12} to C_{18} n-alkanes into Mannosylerythritol Lipids (MEL), the yield of MEL was 140 g/L after 4 weeks and the produced biosurfactant was able to emulsify soybean oil [27]. The modification in the fatty acid constitution of final biosurfactant was observed when the fatty acid composition was changed in the fermentation media of *Candida ingens* [28].

Nitrogen sources

This is the second most important supplement for the production of biosurfactants by microorganisms. Various organic and inorganic nitrogen sources were used in the production of biosurfactants. The higher yields of sophorose lipids, biosurfactant by the fungi *Torulopsis bombicola* and *Candida bombicola* were observed using yeast extract and urea as the nitrogen source [13,8,24]. Although, the higher yields of mannosylerythritol lipid by *Candida sp.* SY16, *Candida lipolytica* and *Candida glabrata* UCP 1002 were observed with ammonium nitrate and yeast extract [10,29,15,16].

pH

pH plays an important role in the production of biosurfactants. *Candida* species produced maximum yields of biosurfactants in a wide pH range, as seen in *Candida glabrata* UCP 1002 which produced maximum biosurfactant at pH 5.7, *Candida sp.* SY16 at pH 7.8, *Candida lipolytica* at pH 5.0, *Candida batistae* at pH 6.0 [29,10,30,18]. While *Aspergillus ustus* and *Pichia anamola* produced maximum yield of biosurfactant at the pH 7.0 and 5.5 respectively [19,31].

Temperature

Various microbial processes are temperature dependent and get affected by a little change. The most favorable temperature for the production of biosurfactants by various fungi is 30°C as observed in various *Candida species* viz. *Candida sp.* SY16, *Candida bombicola*, *Candida batistae* and *Torulopsis bombicola* [10,8,18,23]. While, in case of *Candida lipolytica*, 27 °C was found to be the best temperature.

Incubation time

Incubation time also have a significant effect on the production of biosurfactant. Different microorganisms are able to produce biosurfactants at different time intervals. The maximum biosurfactant production by *Aspergillus ustus* MSF3 was observed after 5 days of incubation while in case of *Candida bombicola*, the incubation periods were 7, 8 and 11 days [7,13,12]. Also, the maximum biosurfactant production by *Candida bombicola* using animal fat was observed after the 68 h of incubation [8].

Purification and Analytical Methods for the Characterization of Biosurfactants

After the production of biosurfactants the most important step is their purification from the fermentation media so as to make them available for various industrial applications. This section deals with the purification and characterization methods employed for the extraction of biosurfactants from various fungal species. The purification and

S. No.	Fungi	Type of biosurfactants	Solvents used in the purification processes	Determination of various parameters by analytical methods	References
1.	*Torulopsis bombicola*	Sophorose lipid	Hexane, Chloroform, Ethyl acetate	**Glucose**- Dinitrosalicylic acid [32] **Soybean oil**- Wjis method [33] **Sophorose lipid**- Phenol sulphuric acid [34] **Cell growth**- Plotting absorbance at 650 nm against dry cell weight	[24]
2.	*Candida bombicola*	Sophorolipids (SL)	Not mentioned	**Biomass**- Measured as optical density at 650 nm **Nitrogen**- With an Orion 720A free ammonia electrode. **Phosphorus and glucose**- HPLC	[11]
3.	*Candida bombicola*	Sophorolipid	Pentane and Ethyl acetate	**Hydroxyl fatty acid –** Methanolysis [35]	[12]
4.	*Candida ingens*	Not mentioned	Ethyl acetate extraction [36]	**Biomass**- [37] **Surface tension**- Fisher tensiometer	[28]
5.	*Candida bombicola*	Sophorolipid	n-Hexane and Ethyl acetate [38]	**Biomass**- O.D. at 650nm in a spectrophotometer **Oil**- Dry weight after n-hexane removal by evaporation. **Sophorolipid**- Conc. is checked by dry weight after evaporation of ethyl acetate and structure was analysed by HPLC **Glucose**- HPLC	[13]
6.	*Candida bombicola*	Sophorolipid	Not mentioned	Not mentioned	[14]
7.	*Candida bombicola*	Sophorolipid	Ethyl acetate and Hexane [39]	**Ni^{2+} estimation**- Inductively coupled plasma mass spectrometry [40]	[7]
8.	*Candida bombicola*	Sophorolipid	Hexane or Ethyl acetate	**Protein** -Lowry method [41] **Total carbohydrates**- anthrone assay [42] **Total intracellular lipids**- [43] **Amino acid**- HPLC	[8]
9.	*Candida lipolytica*	Protein-lipid-polysaccharide complex	Ethyl acetate extraction	**Protein**- Total protein test kit from Lab test Diagnostica, S.A., Brazil. **Sugars**- Phenol–sulfuric acid method [34] **Lipid**- Gravimetric estimation [44]	[15]
10.	*Candida lipolytica*	Protein-lipid-carbohydrate complex	Chloroform [30]	**Protein**- Total protein test kit from Lab test Diagnostica, S.A., Brazil. **Sugars**- Phenol–sulfuric acid method [34] **Lipid**- Gravimetric estimation [44]	[16]
11.	*Candida ishiwadae*	Glycolipid	Ethyl acetate and Hexane [45]	**Fatty acid analysis**- GC-MS	[17]
12.	*Candida batistae*	Sophorolipid	Ethyl acetate extraction	**Acid form of SLs/ Lactone form**- HPLC	[18]
13.	*Aspergillus ustus*	Glycolipoprotein	Ethyl acetate, diethyl ether and dichloromethane	**Protein**- [46] **Carbohydrate**- Phenol–sulphuric acid method [47] **Lipid**- For free fatty acid using the method of [48]	[19]
14.	*Ustilago maydis*	Glycolipid	Chloroform:Methanol extraction [49]	**Biomass**- [49] **Free fatty acids**- [50] **Lipase activity**- [51]	[20]
15.	*Trichosporon ashii*	Sophorolipid	Chloroform and methanol extraction	**Carbohydrate**- Anthrone reagent method [52] **Lipid**- [53] **Protein**- [46]	[21]

Table 2: Purification and analytical methods for the characterization of biosurfactants.

analytical methods for the characterization of fungal biosurfactants are given in Table 2.

Characterization of Biosurfactants by Various Chromatographic and Spectroscopic Techniques

For the complete structure elucidation of biosurfactants, various chromatographic and spectroscopic techniques were used. A combination of these techniques is very helpful in the characterization of the compound. These techniques are discussed as follows:

Thin layer chromatography (TLC)

This is the most important and preliminary technique for the characterization of various types of biosurfactants. The various solvent systems and developer employed in thin layer chromatography are given in Table 3.

Gas chromatography- mass spectroscopy (GC-MS)

In the structure elucidation of sophorolipids by *Candida batistae,* for the confirmation of fatty acids, the sample was investigated for GC-MS analysis. The major peak at Retention Time (RT) 51.5 min. was supposed due to be 18-hydroxyoctadecenic acid and this was confirmed by comparing the sample with the authenticated sample produced by *Starmerella bombicola*. The Fragmentation pattern of the sample were observed at m/z 312 $[M]^+$ (relative intensity, 1.3); m/z 294 $[M\text{-}H_2O]^+$ (relative intensity, 2.4), m/z 31 $[CH_2CH{=}CH_2]^+$ (relative intensity, 52). In addition to the major peaks additional peaks were also observed (e.g. m/z 31, m/z 55, m/z 67, m/z 81, m/z 95, m/z 110, m/z 123, m/z 137, m/z 151, m/z 165, m/z 213, m/z 237, m/z 262 and m/z 280). The major peak for the standard sophorolipids was observed at 36.9 min. and this reveals the presence of 17-hydroxyoctadecenic acid. The Fragmentation pattern of the standard peak were observed at m/z 312 $[M]^+$ (relative intensity 0.6), m/z 294 $[M\text{-}H_2O]^+$ (relative intensity, 18), m/z 45 $[CH_3\text{-}CH{=}OH]^+$ (relative intensity, 57). Here, also the additional peaks were observed (e.g. m/z 29, m/z 55, m/z 67, m/z 81, m/z 95, m/z 109, m/z 123, m/z 137, m/z 151, m/z 165, m/z 213, m/z 265 and m/z 279) [18]. For the structure elucidation of monoacylglycerols produced by *Candida ishiwadae*, the fatty acid moieties methyl oleate and linoleyl of the compounds (a & b) having Retention Factor (Rf) values 0.23 and 0.17 was determined by the methanolysis. From the combined data of ^{1}H-NMR and mass spectral data, the compounds (a & b) were determined to be 1-oleylglycerol and 1-linoleylglycerol. Further confirmation was done from their molecular weights, 356 and 354 [17]. In the structure elucidation of sophorolipids produced by the *Candida bombicola*, hydroxyl-acid methyl esters were liberated by the methanolysis and were confirmed by GC-MS. The 16-hydroxydecanoic acid was confirmed by comparing the fragmentation pattern with the standard 16-hydroxyhexadecanoic acid purchased from Sigma Aldrich. Also, an isomer 15-hydroxyhexadecanoic acid was confirmed because of the availability of the same fragmentation pattern in the library [12].

High performance liquid chromatography (HPLC) and Liquid chromatography-mass spectroscopy (LC-MS)

In the structure elucidation of mannosylerythritol lipid from *Candida sp.* SY16, the acid hydrolysate of glycoside gave two spots on TLC at Rf values 0.38 and 0.46 respectively and corresponded to D-mannose and meso-erythritol. Also, in the HPLC chromatogram the peaks at RT 10.5 and 7.8 min. corresponded to D-mannose and meso-erythritol respectively. The molar ratio of D-mannose and meso-erythritol was calculated by integration of HPLC peak areas against the known concentrations of the authenticated standards and hence calculated to be 1:1 [10]. The comparative studies of sophorolipids from two carbon sources soybean and glucose by *Pichia anamola* were done. The HPLC of the biosurfactant from glucose and soybean medium showed the peaks at retention time of 9.269 (m/z 675.687, 691.880 and 708.062) and 9.779 min. (m/z 659.499, 675.627 and 691.830). These were compared with the standard sophorolipids sample having a peak at RT of 9.646 (m/z 648.760 and 650.816). The comparative study is given in Table 4 [31] .

Proton nuclear magnetic resonance spectroscopy (^{1}H-NMR)

In the ^{1}H-NMR spectrum of sophorolipids produced by *Starmellela bombicola,* the duplet at 1.20 ppm was assigned as $-CH_3$ at

Fungi	Biosurfactant type	Solvent system	Identification of functional group	Developer	References
Candida bombicola	Sophorolipid	Chloroform:Methanol:Water (65:15:2) [54]	Diacylated lactone	Not mentioned	[12]
Candida lipolytica	Protein-lipid-polysaccharide complex	(i) Hexane:Isopropyl ether: Acetic acid (15:10:1) (ii) Chloroform:Methanol:water (65:24:4)	**Protein**-40.2%, **Carbohydrate**-24.0% **Lipid**- 19.5%	**Ninhydrin**-Free amino groups **Iodine vapors**-Lipids **α-napthol-H_2SO_4** Sugar [55]	[16]
Torulopsis petrophilum	Glycolipid	Chloroform:Methanol:Water (65:15:2)	Not mentioned	α-napthol	[53]
Candida batistae	Sophorolipid	Chloroform:Methanol (8:2)	Not mentioned	Anthrone/sulfuric acid: [34]	[18]
Pichia anamola	Sophorolipid	Chloroform:Methanol:Acetic acid (65:25:4)	Not mentioned	Iodine vapour and Molisch reagent	[31]

Table 3: Various solvent systems and developer employed in TLC method.

Sample	RT from HPLC	$[M+H]^+$	Molecular Weights (Da)	Hydroxy carboxylic acid of sophorolipid
Sophorolipid (standard)	9.646	648.760 650.816	647 650	C18+Na C20:1
Biosurfactant from glucose medium	9.269	675.687 691.880 708.062	675 691 707	C20+Na C20+Na ^{Ox} C18:1+Ac+Ac
Biosurfactant from glucose medium	9.779	659.499 675.627 691.830	658 675 691	$[C20]_{Lactone}$ C20+Na C20+Na ^{Ox}

Ox is the oxidized form of sophorolipids

Table 4: Retention times and $[M+H]^+$ of mass spectroscopy for peaks from standard sophorolipid, glucose medium and soybean oil medium samples.

ω-end of fatty acid (C-18) and the sextet at 3.74 ppm was assigned as hydroxylated methine group (-CHOH-) at ω-1 position (C-17). While, the sophorolipids produced by *Candida batistae* showed two multiplets at 3.58 and 3.80 ppm and were assigned as the hydroxylated methylene group (-OCH_2-) at ω-end of fatty acid (C-18) [18]. For the structure elucidation of monoacylglycerols produced by *Candida ishiwadae*, the compounds with Rf values 0.23 and 0.17 were subjected for ^{1}H-NMR. For Rf 0.17, the five protons at δ 3.5-4.2 corresponded to the carbinol protons which indicated the presence of glycerol moiety. The glycerol moiety contains a substituent at -OH group attached to Cl-atom as indicated by high values at δ 4.10 and 4.18. Also, the additional protons at δ 0.9-2.5 and two olefinic protons at δ 5.38 suggested the presence of monounsaturated fatty acid moiety. Also, for the Rf value 0.23 the compound elucidated was monoacylglycerol with an additional double bond [17]. For the structure elucidation of mannosylerythritol lipid by *Canida sp.* SY16, the protons resonating at 5.45, 4.93, 4.26 and 4.44 ppm confirms the acylation positions of sugar moiety. They had lower chemical shifts in comparison to the lipid-free sugar moiety protons which resonate at 4.00, 3.62, 3.72 and 3.88 ppm. The triplet at 0.90 ppm and singlet at 2.08 was assigned due to methyl (-CH_3), broad peak at 1.28-1.40 ppm was assigned due to -CH_2- groups in fatty acids, multiplet at 5.30-5.38 ppm was assigned =CH_2- in unsaturated fatty acids [10].

Carbon nuclear magnetic resonance spectroscopy (^{13}C-NMR)

In ^{13}C-NMR spectrum of sophorolipids produced by *Starmellela bombicola*, peak at 20.7 were assigned as the ω-end carbon (C-18) and the one at 77.4 ppm were assigned as ω-1 carbon (C-17) of fatty acid respectively. While the peak at 69.7 ppm in the ^{13}C-NMR spectrum of sophorolipids produced by *Candida batistae* was assigned as the ω-end carbon (C-18) [18]. In the structure elucidation of mannosylerythritol lipid from *Candida sp.* SY16, peaks at 23.7-33.1 ppm were assigned as -CH_2- groups in fatty acids and peaks at 128.8, 129.7, 130.2 and 131.1 ppm were assigned as four =CH- groups in unsaturated fatty acids [10].

Applications

Biosurfactants have numerous applications in the bioremediation processes, food industries, cosmetic industries and biomedical fields. The various reported applications of the fungal biosurfactants are as follows:

Microbial enhanced oil recovery and cleaning of oil tanks

The Sophorolipids from *Candida lipolytica* and *Candida bombicola* are very promising in the cleaning of oil tanks, decontamination of polluted areas, microbial enhanced oil recovery, industrial cleaning, low-end consumer products and house-hold applications [16,7]. The biosurfactants from *Torulopsis bombicola* and *Aspergillus ustus* MSF3 were used for the release of bitumen from the contaminated soil and for the degradation of hydrocarbons [23,19]. Mannosylerythritol lipids from *Candida antarctica* have potential applications in the removal and biodegradation of hydrocarbons in oil-contaminated soil and were also used to rinse oil and grease from the contaminated soil [27,55].

Food and oil industry

Biosurfactants are able to stabilise various types of emulsions, so are valuable for the food industry. The biosurfactants from the *Candida lipolytica* and *Saccharomyces cerevisiae* are good choices for the food and oil industries [15] further, the biosurfactant from *Saccharomyces cerevisiae* used as a single cell protein [56]. The bioemulsifier, liposan from *Candida lipolytica* was able to stabilize the emulsions of vegetative oils and water. It was also able to stabilize the cottonseed oil, corn oil, soybean oil and peanut oil emulsions [57-60].

Biomedical field

The biosurfactants are extensively useful in the biomedical fields. They possess significant anti-biological activities. The biosurfactant from the *Aspergillus ustus* MSF3 have significant antimicrobial activity against the *Candida albicans* and gram-negative bacterium [19].

Cosmetic industry

The biosurfactants are capable of their usage in the cosmetic industry due to their skin friendly properties. Sophorolipids from the mutant strain *Candida bomobicola* ATCC 22214 have great uses in the cosmetic industries due to their anti-radical properties, stimulation of fibroblast metabolism and hygroscopic properties to support healthy skin physiology [14].

References

1. Lang S (2002) Biological amphiphiles (microbial biosurfactants). Curr Opin Colloid Interphase Sci 7: 12-20.
2. Satpute SK, Banpurkar AG, Dhakephalkar PK, Banat IM, Chopade BA (2010) Methods for investigating biosurfactants and bioemulsifiers: a review. Crit Rev Biotechnol 30: 127-144.
3. Banat IM (1995) Biosurfactants production and possible uses in microbial enhanced oil recovery and oil pollution remediation: A review. Bioresource Technol 51: 1-12.
4. Sen R (2008) Biotechnology in petroleum recovery: The microbial EOR. Prog Energ Combust 34: 714-724.
5. Satpute SK, Banat IM, Dhakephalkar PK, Banpurkar AG, Chopade BA (2010) Biosurfactants, bioemulsifiers and exopolysaccharides from marine microorganisms. Biotechnol Adv 28: 436-450.
6. Patil JR, Chopade BA (2001) Studies on bioemulsifier production by Acinetobacter strains isolated from healthy human skin. J Appl Microbiol 91: 290-298.
7. Felse PA, Shah V, Chan J, Rao KJ, Gross RA (2007) Sophorolipid biosynthesis by Candida bombicola from industrial fatty acid residues. Enzyme Microb Tech 40: 316-323.
8. Deshpande M, Daniels L (1995) Evaluation of sophorolipid biosurfactant production by Candida bombicola using animal fat. Bioresource Technol 54: 143-150.
9. Kim HS, Jeon JW, Lee HW, Park YI, Seo WT, et al. (2002) Extracellular production of a glycolipid biosurfactant, mannosylerythritol lipid, from Candida antarctica. Biotechnol Lett 24: 225-229.
10. Kim HS, Yoon BD, Choung DH, Oh HM, Katsuragi T, et al. (1999) Characterization of a biosurfactant mannosylerythritol lipid produced from Candida sp. SY 16. Appl Microbiol Biotechnol 52: 713-721.
11. Casas JA, Garcia de Lara S, Garcia-Ochoa F (1997) Optimization of a synthetic medium for Candida bombicola growth using factorial design of experiments. Enzyme Microb Tech 21: 221-229.
12. Cavalero DA, Cooper DG (2003) The effect of medium composition on the structure and physical state of sophorolipids produced by Candida bombicola ATCC 22214. J Biotechnol 103: 31-41.
13. Casas JA, García-Ochoa F (1999) Sophorolipid production by Candida bombicola: medium composition and culture methods. J Biosci Bioeng 88: 488-494.
14. Williams K (2009) Biosurfactants for cosmetic applications: Overcoming production challenges. MMG 445 Basic Biotechnology 5: 78-83.
15. Sarubbo LA, Farias CB, Campos-Takaki GM (2007) Co-utilization of canola oil and glucose on the production of a surfactant by Candida lipolytica. Curr Microbiol 54: 68-73.
16. Rufino RD, Sarubbo LA, Campos-Takaki GM (2007) Enhancement of stability of biosurfactant produced by Candida lipolytica using industrial residue as substrate. World J Microb Biot 23: 729-734.

17. Thanomsub B, Watcharachaipong T, Chotelersak K, Arunrattiyakorn P, Nitoda T, et al. (2004) Monoacylglycerols: glycolipid biosurfactants produced by a thermotolerant yeast, Candida ishiwadae. J Appl Microbiol 96: 588-592.

18. Konishi M, Fukuoka T, Morita T, Imura T, Kitamoto D (2008) Production of new types of sophorolipids by Candida batistae. J Oleo Sci 57: 359-369.

19. Kiran GS, Hema TA, Gandhimathi R, Selvin J, Thomas TA, et al. (2009) Optimization and production of a biosurfactant from the sponge-associated marine fungus Aspergillus ustus MSF3. Colloids Surf B Biointerfaces 73: 250-256.

20. Alejandro CS, Humberto HS, María JF (2011) Production of glycolipids with antimicrobial activity by Ustilago maydis FBD12 in submerged culture. African Journal of Microbiology Research 5: 2512-2523.

21. Chandran P, Das N (2010) Biosurfactant production and diesel oil degradation by yeast species Trichosporon asahii isolated from petroleum hydrocarbon contaminated soil. International Journal of Engineering Science and Technology 2: 6942-6953.

22. Zhou QH, Klekner V, Kosaric N (1992) Production of sophorose lipids by Torulopsis bombicola from safflower oil and glucose. J Am Oil Chem Soc 69: 89-91.

23. Cooper DG, Paddock DA (1984) Production of a Biosurfactant from Torulopsis bombicola. Appl Environ Microbiol 47: 173-176.

24. Kim SY, Oh DK, Lee KH, Kim JH (1997) Effect of soybean oil and glucose on sophorose lipid fermentation by Torulopsis bombicola in continuous culture. Appl Microbiol Biotechnol 48: 23-26.

25. Sarubbo LA, Marçal MC, Neves ML, Silva MP, Porto AL, et al. (2001) Bioemulsifier production in batch culture using glucose as carbon source by Candida lipolytica. Appl Biochem Biotechnol 95: 59-67.

26. Bednarski W, Adamczak M, Tomasik J, PÅ‚aszczyk M (2004) Application of oil refinery waste in the biosynthesis of glycolipids by yeast. Bioresour Technol 95: 15-18.

27. Kitamoto D, Ikegami T, Suzuki GT, Sasaki A, Takeyama YI, Idemoto Y, Koura N, Yanagishita H (2001) Microbial conversion of n-alkanes into glycolipid biosurfactants, mannosylerythritol lipids, by Pseudozyma (Candida antarctica). Biotechnol Lett 23: 1709-1714.

28. Amézcua-Vega C, Poggi-Varaldo HM, Esparza-García F, Ríos-Leal E, Rodríguez-Vázquez R (2007) Effect of culture conditions on fatty acids composition of a biosurfactant produced by Candida ingens and changes of surface tension of culture media. Bioresour Technol 98: 237-240.

29. Sarubbo LA, de Luna GM, de Campos-Takaki GM (2006) Production and stability studies of the bioemulsifier obtained from a new strain of Candida glabrata UCP 1002. Electronic Journal of Biotechnology 9: 400-406.

30. Cirigliano MC, Carman GM (1984) Isolation of a bioemulsifier from Candida lipolytica. Appl Environ Microbiol 48: 747-750.

31. Thaniyavarn J, Chianguthai T, Sangvanich P, Roongsawang N, Washio K, et al. (2008) Production of sophorolipid biosurfactant by Pichia anomala. Biosci Biotechnol Biochem 72: 2061-2068.

32. Miller GL (1959) Use of Dinitrosalicylic Acid Reagent for Determination of Reducing Sugar. Anal Chem 31: 426-428.

33. Helrich K (1990) Official methods of analysis of the Association of Official Analytical Chemists. 15th edn Association of Official Analytical Chemists.

34. DuBois M, Gilles KA, Hamilton JK, Rebers PA, Smith F (1956) Colorimetric Method for Determination of Sugars and Related Substances. Anal Chem 28: 350-356.

35. Davila AM, Marchal R, Monin N, Vandecasteele JP (1993) Identification and determination of individual sophorolipids in fermentation products by gradient elution high-performance liquid chromatography with evaporative light-scattering detection. J Chromatogr 640: 139-149.

36. Sim L, Ward OP, Li ZY (1997) Production and characterisation of a biosurfactant isolated from Pseudomonas aeruginosa UW-1. J Ind Microbiol Biotechnol 19: 232-238.

37. Rau U, Manzke C, Wagner F (1996) Influence of substrate supply on the production of sophorose lipids by Candida bombicola ATCC 22214. Biotechnol Lett 18: 149-154.

38. Garcia-Ochoa F, Casas JA (1999b) Unstructured kinetic model for sophorolipid production by Candida bombicola. Enzyme Microb Tech 25: 613-621.

39. Singh SK, Felse AP, Nunez A, Foglia TA, Gross RA (2003) Regioselective enzyme-catalyzed synthesis of sophorolipid esters, amides, and multifunctional monomers. J Org Chem 68: 5466-5477.

40. Huang CC, Yang MH, Shih TS (1997) Automated on-line sample pretreatment system for the determination of trace metals in biological samples by inductively coupled plasma mass spectrometry. Anal Chem 69: 3930-3939.

41. Herbert D, Phipps PJ, Strange RE (1971) Chapter III Chemical Analysis of Microbial Cells. Method Microbiol 5: 209-344.

42. Gerhardt P (1981) Manual of Methods for General Bacteriology. American Society for Microbiology, Washington DC USA.

43. Esders TW, Light RJ (1972) Glucosyl- and acetyltransferases involved in the biosynthesis of glycolipids from Candida bogoriensis. J Biol Chem 247: 1375-1386.

44. Manocha MS, San-Blas G, Centeno S (1980) Lipid composition of Paracoccidioides brasiliensis: possible correlation with virulence of different strains. J Gen Microbiol 117: 147-154.

45. Kitamoto D, Akiba S, Hioki C, Tabuchi T (1990) Extracellular Accumulation of Mannosylerythritol Lipids by a Strain of Candida antarctica. Agr Biol Chem Tokyo 54: 31-36.

46. LOWRY OH, ROSEBROUGH NJ, FARR AL, RANDALL RJ (1951) Protein measurement with the Folin phenol reagent. J Biol Chem 193: 265-275.

47. Chaplin MF, Kennedy JF (1994) Carbohydrate Analysis. A Practical Approach, 2 edn IRL Press, UK.

48. Sadasivam S, Manickam A (2004) Biochemical Methods. 2 edn New Age International India.

49. Monteiro SA, Sassaki GL, de Souza LM, Meira JA, de Araújo JM, et al. (2007) Molecular and structural characterization of the biosurfactant produced by Pseudomonas aeruginosa DAUPE 614. Chem Phys Lipids 147: 1-13.

50. AOCS (1998) Method Ca 5a-40: Free Fatty Acids. In: Firestone D (Editor), Official Methods and Recommended Practices of the American Oil Chemists Society. 5 edn, American Oil Chemists Society, Champaign, IL, USA.

51. Smeltzer MS, Hart ME, Iandolo JJ (1992) Quantitative spectrophotometric assay for staphylococcal lipase. Appl Environ Microbiol 58: 2815-2819.

52. Spiro RG (1966) Analysis of sugars found in glycoproteins. Method Enzymol 8: 3-26.

53. Folch J, Lees M, Stanly SGH (1957) A simple method for the isolation and quantification of total lipids from animal tissues. J Biol Chem 226: 497-509.

54. Asmer HJ, Lang S, Wagner F, Wray V (1988) Microbial production, structure elucidation and bioconversion of sophorose lipids. J Am Oil Chem Soc 65: 1460-1466.

55. Chaplin MF (1986) Carbohydrate Analysis: A Practical Approach. The practical approach series IRL Press UK.

56. Cameron DR, Cooper DG, Neufeld RJ (1988) The mannoprotein of Saccharomyces cerevisiae is an effective bioemulsifier. Appl Environ Microbiol 54: 1420-1425.

57. Cirigliano MC, Carman GM (1985) Purification and Characterization of Liposan, a Bioemulsifier from Candida lipolytica. Appl Environ Microbiol 50: 846-850.

58. Cooper DG, Paddock DA (1983) Torulopsis petrophilum and Surface Activity. Appl Environ Microbiol 46: 1426-1429.

59. Konishi M, Morita T, Fukuoka T, Imura T, Kakugawa K, et al. (2007) Production of different types of mannosylerythritol lipids as biosurfactants by the newly isolated yeast strains belonging to the genus Pseudozyma. Appl Microbiol Biotechnol 75: 521-531.

60. Adamczak M, Bednarski W (2000) Influence of medium composition and aeration on the synthesis of biosurfactants produced by Candida antarctica. Biotechnol lett 22: 313-316.

Diversity of Hydrocarbon-Related Catabolic Genes in Oil Samples from Potiguar Basin (RN, Brazil)

Leandro Costa Lima Verde[1], Tiago Rodrigues e Silva[1], Bruna Martins Dellagnezze[1], Eugênio Vaz dos Santos Neto[2] and Valéria Maia de Oliveira[1]*

[1]*Research Center for Chemistry, Biology and Agriculture (CPQBA), Campinas University - UNICAMP, CP 6171, CEP 13081-970, Campinas, SP, Brazil*
[2]*Petrobras R&D Center, Cidade Universitária, Quadra 7, Rio de Janeiro, RJ, CEP: 21949-900, Brazil*

Abstract

Biodegradation may result in physicochemical changes in crude oil and natural gas properties, being responsible for the decrease of saturated hydrocarbons and yielding heavy oil with low economic value. Studies on the diversity of microbial catabolic genes in oil reservoirs are scarce and could help to predict the potential of a petroleum sample to be biodegraded. The aim of this study was to evaluate the diversity of genes involved in hydrocarbon degradation in Brazilian petroleum samples (biodegraded and non-biodegraded) through the construction and analysis of gene libraries (alkane monooxygenase – *alk*, dioxygenase – ARHDs and 6-oxocyclohex-1-ene-1-carbonyl-CoA hydrolase - *bamA*). The results showed a differential distribution of catabolic genes between the sites, being the biodegraded oil more diverse for the *alk* and *bamA* genes. Sequences were similar to the *alkB* genes from *Geobacillus thermoleovorans* and several species of *Acinetobacter,* to ARHD genes from *Pseudomonas* spp. and two species of *Burkholderia*, and to *bamA* genes from deltaproteobacteria. Interestingly, most of the catabolic sequences recovered from both petroleum reservoirs grouped together forming distinct clusters in the phylogenetic tree reconstruction and may correspond to potentially new genes, possibly harbored by yet uncultivated microorganisms. This is the first report on the detection of *alk*, ARHD and *bamA* genes in petroleum reservoir environments, demonstrating the genetic potential of such microbial communities to biodegrade the oil.

Keywords: Petroleum microbiology, Microbial diversity; Oil degradation; Catabolic genes; Biodegradation, Hydrocarbons, Geobacillus, Oil

Introduction

Most of the world's oil is biodegraded and while the effects of biodegradation on the molecular composition and physical properties of crude oil and natural gas are relatively well known empirically, the actual processes that occur during the biodegradation of oil in deep reservoirs (below some hundred meters) remain unclear [1]. During biodegradation, the hydrocarbon content is transformed, with a consequent increase in oil density, sulfur content, acidity and viscosity. All these factors interfere with the extraction and refining operations, resulting in significant economic losses [2]. On the other hand, the mechanisms of oil degradation in petroleum reservoirs, as well as the microorganisms involved are still poorly understood.

Oil is a complex mixture of hydrocarbons such as saturated, unsaturated, linear, monoaromatic and polycyclic hydrocarbons [3,4]. Each of these compounds is biodegraded through different routes of several steps. Many studies have already demonstrated the existence of large and diverse populations of microbes with different metabolic activities in petroleum systems [5,6].

Nonetheless, knowledge of the diversity of catabolic genes involved in the processes of degradation of oil is still scarce. Information on community composition of bacteria and genes involved in the biodegradation process can bring light on the understanding of microbial metabolic pathway preferences, as well as a better application of specific microorganisms in biodegradation or bioremediation processes. Techniques able to identify the genes involved in the catabolic degradation of hydrocarbons are valuable tools for elucidating the structure of the microbial community that is truly functional in the environment.

This study examined the diversity of the functional gene involved in the degradation of hydrocarbons, the alkane hydroxylase (*alk*) gene, a membrane-bound enzyme that catalyzes the terminal or sub terminal oxidation of n-alkanes (C_5-C_{16}) for primary or secondary alcohols, which is the initial step in the aerobic degradation of n-alkanes [7]. Alkane monoxygenase proteins have had their genetics deciphered in a wide variety of bacteria, including Gram positive and negative such as *Rhodococcus erythropolis* NRRL B-16531, *Rhodococcus* sp. Q15, *Mycobacterium tuberculosis* H37Rv, *Pseudomonas putida* GPO1, *Acinetobacter* sp. ADP1 and *Acinetobacter* sp. M-1 [8-11]. Alkane monoxygenases were found in bacteria that inhabit diverse environments, including sediments in Alaska [12], contaminated soils [13,14], cold ecosystems [15,16], shallow aquifer [17] and soils of the Arctic and Antarctica [18-21].

The functional gene involved in the aerobic degradation of aromatic hydrocarbons, the dioxygenase protein coding gene, was also analyzed in this study. The first step in the degradation of aromatic hydrocarbons usually occurs through the incorporation of molecular oxygen in aromatic nucleus by a multicomponent enzyme system, forming cis-dihydrodiol. In this system, the terminal dioxygenase is composed of one large α subunit and one small β subunit [22]. The α-subunit contains two conserved regions: the Rieske center [Fe2-S2] and the

***Corresponding author:** Valéria Maia de Oliveira, Research Center for Chemistry, Biology and Agriculture (CPQBA), Campinas University - UNICAMP, CP 6171, CEP 13081-970, Campinas, SP, Brazil, Tel. : +55-19-2139 2874

catalytic domain containing mononuclear iron [23]. The dioxygenases belong to a large family known as aromatic-ring-hydroxylating dioxygenases (ARHDs) [24]. These genes are located in chromosomal or plasmid DNA and were identified in bacterial strains belonging to α-Proteobacteria (*Sphingomonas*) [25], β-Proteobacteria (*Alcaligenes, Burkholderia, Commamonas, Polaromonas, Ralstonia*) [26-28] and γ-Proteobacteria (*Pseudomonas*) [29].

The process of biodegradation in subsurface reservoirs is often anaerobic in nature [2,6]. Most degraded oils contain metabolic markers for the anaerobic degradation, such as specific reduced naftoic acids, indicating that the anaerobic metabolism of oil is the mechanism by which most biodegraded oil in the world were produced [30].

Therefore, research on the anaerobic degradation of hydrocarbons is also necessary for the understanding of the whole process of oil degradation in reservoirs. In this sense, this study also examined *bamA*, the functional gene involved in the degradation route of benzyl-CoA, by which most aromatic substrates (benzene, toluene, ethylbenzene, xylenes and also other aromatic compounds) are anaerobically degraded [31]. *bamA* gene encodes the protein 6-oxociclohex-1-ene-1-carbonyl-CoA hydrolase (6-OCH-CoA) and has been used for the detection of a wide range of anaerobic microorganisms able to degrade hydrocarbons [32]. This functional marker was recently used to detect anaerobic hydrocarbon-degrading micro-organisms in environmental samples [32], as well as to study the microbial community structure of sulfate reducing enrichment cultures growing on petroleum hydrocarbons, where *Desulfosarcina ovata* was showed to be dominant [33]. *bamA* was also used as a biomarker to investigate the diversity of hydrocarbon–degrading bacteria under iron-reducing conditions in a leachate contaminated aquifer, with *bamA* sequences found to be closely related to the ones from *Geobacter* species and *Georgfuchsia toluolica* [31].

This study aimed at investigating the presence and diversity of catabolic genes involved in the hydrocarbon degradation process in biodegraded and non-biodegraded oil samples from petroleum reservoirs through the construction and analyses of *alk*, ARHD and *bamA* gene libraries. The results allowed us to evaluate the effectiveness of the primer sets used as a tool to analyze the community structure of bacteria that have the genetic potential to degrade hydrocarbons by aerobic and anaerobic pathways.

Materials and Methods

Sampling

Oil samples (naturally mixed with formation water) were collected in July 2008 from two production wells, named GMR75 and PTS1, at the onshore Potiguar Basin (Northeast, Brazil), with logistic support from CENPES/Petrobras. Sampling details as well as oil characteristics and geological settings are given in Silva et al. [34].

Total DNA extraction and PCR amplification *alk*, ARHD and *bamA* genes

DNA extraction from oil/formation water samples was carried out using the PowerSoil™ DNA Isolation Kit (MoBio Laboratories, California), according to the manufacturer´s instructions. Five tubes from the kit were used to start with the DNA extraction. At the end of the procedure, DNA samples obtained from the five tubes were pooled, concentrated and used for subsequent PCR amplification. Crude oil/formation water samples were not filtered or centrifuged before DNA extraction. The PCR amplifications of the three functional genes were performed independently using the respective set of primers (Table 1) [35]. Twenty five microliter reaction mixtures contained 5 µl of total DNA, 1 U *Platinum Taq* DNA polymerase (Invitrogen), 0.2 mM dNTP mix (GE Healthcare), 1.2 µM of each primer, 1X *Taq* buffer and 1.5 mM $MgCl_2$. PCR amplifications were done using an Eppendorf Mastercycler Gradient (Eppendorf Scientific, New York, USA) with the programs shown in table 2. The amplification products were checked by electrophoresis in 1.2% (wt vol^{-1}) Agarose gels. Four replicate PCR reactions were done for each sample with each pair of primers. The corresponding PCR replicates were then pooled and concentrated in a speed vacuum concentrator 5301 Eppendorf, A-2-VC rotor.

Clone library construction and sequencing

The PCR products amplified with functional gene-targeted primer pairs were cloned using the pGEM-T Easy Vector (Promega), according to manufacturer's guidelines, and transformed into *Escherichia coli* JM 109 cells. The functional gene-targeted inserts were amplified from plasmid DNA of selected clones using the universal the M13 forward (5'-CGCCAG GGT TTT CCC AGT CAC GAC-3') and reverse (5'-TTT CAC ACA GGA AAC AGC TAT GAC-3') primers. PCR was performed in a 50 µL-reaction volume, containing 1 µL of an overnight clone culture, 0.4 µM each primer, 0.2 mM dNTP mix, 2 U *Taq* DNA polymerase (Invitrogen), 1X *Taq* buffer, and 1.5 mM $MgCl_2$. The amplification program consisted of an initial denaturation step at 94°C for 3 min, followed by 30 cycles of 94°C/20 s, 60°C/20 s, and 72°C/90 s. PCR products were purified as previously described for automated sequencing in the MegaBace DNA Analysis System 500 (GE Healthcare) and ABI Prism 377 DNA Sequencer™ (Applied Biosystems). The sequencing was carried out with the corresponding primers used for PCR amplification. The sequencing programs consisted of 30 cycles of

Primer	Gene target	Sequence (5´→ 3´)	Direction	Expected Amplicon size (bp)	Reference
alkF	*alk*	GCICAIGARITIRKICAYAA	forward	524	[19]
alkR	*alk*	GCITGITGITCISWRTGICGYTG	reverse		
ARHDf	α-ARHD	TTYRYITGYAIITAYCAYGGITGGG	forward	329	[35]
ARHDr	α-ARHD	AAITKYTCIGCIGSIRMYTTCCA	reverse		
SP9f	*bamA*	CAGTACAAYTTCTACACVACBG	forward	300	[32]
ASP1r	*bamA*	CMATGCCGATYTCCTGRC	reverse		

Table 1: Oligonucleotide primers used for PCR assays.

Primer set	Initial denaturation	Cycle number	Denaturation	Annealing	Extension (72°C)	Final extension (72°C)
alkF/alkR	97°C, 3 min	35	94°C, 1 min	48°C, 1 min	1 min	5 min
ARHDf/ARHDr	97°C, 3 min	35	94°C, 1 min	58°C, 1 min	1 min	5 min
SP9f/ASP1r	94°C, 10 min	35	94°C, 30 s	59°C, 45 s	1 min	10 min

Table 2: Polymerase chain reaction programs used in this study.

94°C/20 s, 43°C/20 s, and 72°C/90 s for *alk* gene, 30 cycles of 94°C/20 s, 53°C/20 s, and 72°C/90 s for ARHD gene and 30 cycles of 94°C/20 s, 54°C /20 s, and 72°C/90 s for *bamA* gene.

Phylogenetic analysis

The partial sequences from GMR75 and PTS01 clones were assembled in a *contig* (consensus sequence) using the phred/Phrap/CONSED program [36]. Nucleotide sequences were translated into protein sequences using the translate tool in the ExPASy (Expert Protein Analysis System) proteomics server of the Swiss Institute of Bioinformatics (http://us.expasy.org/tools/dna.html). Deduced protein sequences from each library were grouped into Operational Protein Families (OPFs) using MOTHUR version 1.14.0 [37]. The cutoff value was selected after initial examination of phylogenetic trees that included all protein sequences generated by neighbor-joining analysis [38] using MEGA5 software [39], with 1,000 bootstrap replicates. Deduced protein sequences were aligned using ClustalX, version 1.83 [40], with related reference sequences identified from BLASTp searches, in addition to other sequences of several representative *alkB*, ARHD and *bamA* sequences identified from the literature.

Diversity indices and statistics

Diversity index calculations (α-diversity measures) were performed individually for each catabolic gene library with the program MOTHUR [37]. Rarefaction curves were calculated using the OPF cutoff values established based on MOTHUR analyses in combination with phylogenetic clustering of the catabolic sequences. In addition, the Shannon (H') diversity index, Simpson index [41], and the nonparametric richness estimators ACE (abundance-based coverage estimator) [42] and Chao1 [43] were calculated for each community based on each target gene.

Nucleotide sequences accession numbers

The partial sequences of *alk*, ARHD and *bamA* genes determined in this study for the environmental clones were deposited at the Genbank database under the accession numbers: *alk* - JX171303-JX171314, ARHD - JX512729-JX512738 and *bamA* - JX512739-JX512747. It is worth to note that only one representative sequence of each OPF was deposited at the database.

Results and Discussion

The composition of microbial communities in samples from Brazilian petroleum fields was determined by the construction and analysis of three functional gene libraries: *alk*, ARHDs and *bamA*. In the case of *alk* gene libraries, 63 clones from GMR75 oil sample (biodegraded) and 87 clones from PTS01 oil sample (non-biodegraded) were obtained. For the ARHDs gene libraries, 88 clones from GMR75 sample and 80 clones from PTS01 sample were obtained, and about 90 clones from GMR75 library and 93 clones from PTS01 library were obtained for the *bamA* gene.

Operational Protein Family (OPF) was defined based on cutoff values at 93% dissimilarity level between sequences [44] in the case of *alk* gene, and 96% and 97% for the ARHDs and *bamA* genes, respectively. These cutoff values were selected based on the analysis of the phylogenetic trees of total clones in combination with MOTHUR analysis. Shannon, ACE and CHAO1 scores indicated that the GMR75 library is more diverse than the PTS01 library for the two functional genes analyzed (Table 3).

In both libraries of the *alk* gene, all 150 sequences showed the presence of the two internal conserved regions (motif B, EHXXGHH and motif C, NYXEHYG) [45,46] and matched with *alk* genes from GenBank using the BLASTX program. Seven OPFs in GMR75, 2 OPFs in PTS01 and 3 OPFs composed by sequences from both *alk* gene libraries were found (Figure 1).

One of these OPFs (OPF 11) matched with the *alkB* gene from *Geobacillus thermoleovorans*, which has the ability to grow in oil and its

Libary (n[a])	Richennes[b]	ACE[c]	Chao1[d]	Shannon[e]	1/Simpson[f]
***alk* (cut-off 93%)**					
GMR75 (63)	10	14	12	1.69	4.52
PTS01 (87)	5	9.21	6.5	0.80	1.60
ARHD (cut-off 96%)					
GMR75 (88)	4	4	4	1.15	2.90
PTS01 (80)	7	13	9	1.34	3.48
***bamA* (cut-off 97%)**					
GMR75 (90)	7	20	9	1.16	2.66
PTS01 (93)	6	7	6	0.97	1.99

[a] Number of gene sequences analyzed.

[b] Richness is based on observed unique OTUs.

[c] Nonparametric statistical prediction of total richness of different OTUs and OPFs based on distribution of abundant (10) and rare (10) OTUs.

[d] Nonparametric statistical predictions of total richness of OTUs and OPFs based on distribution of singletons and doubletons.

[e] Shannon diversity index. A higher number represents more diversity.

[f] Reciprocal of Simpson's diversity index. A higher number represents more diversity.

Table 3: Diversity indices and richness estimators of the functional gene clone libraries.

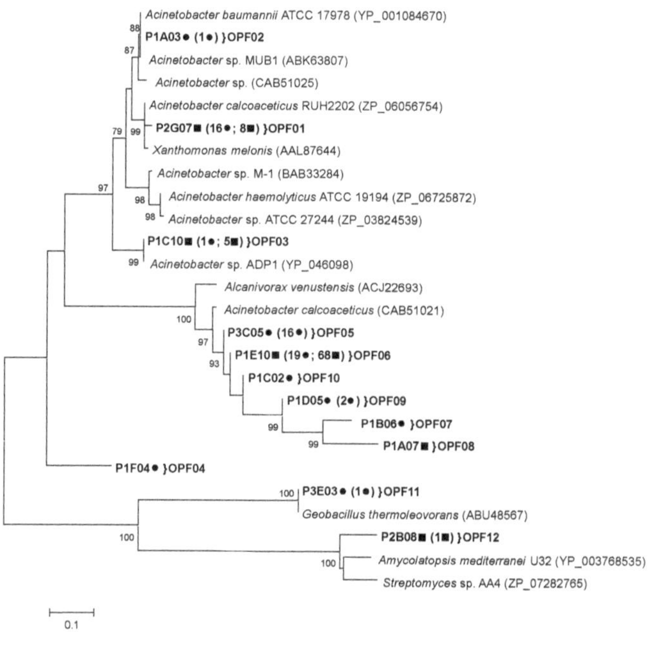

Figure 1: Phylogenetic tree based on the alignment of amino acid sequences (about 190 aa) of membrane-bound alkane monooxygenases. The tree was constructed by using the neighbor-joining (NJ) method in MEGA5 program with substitution model PAM Matrix (Dayhoff). Bootstrap values (1000 replicate runs, shown as %) >70% are listed. The scale bar denotes divergence percentage between sequences. Symbols: clones from GMR75 (●); clones from PTS01 (■). Access numbers of genes from GenBank database reference strains are in parentheses.

hydrocarbon degradation capacity was reported in an extensive study of the genus *Geobacillus* [47]. This bacterium has been commonly isolated from petroleum reservoirs, especially high temperature oil fields [48,49]. *Geobacillus* spp. has attracted industrial interest due to their potential applications in biotechnological processes as sources of various thermostable enzymes [50]. Shestakova and co-authors [51] also detected the presence of *Geobacillus*-like organisms in the *alk* gene RNA library and in the 16S rRNA gene library from hydrocarbon-oxidizing aerobic enrichments from a high-temperature petroleum reservoir in China, showing the predominance and functional activity of *geobacilli*.

Three of the *alk* OPFs (01, 02 and 03) matched with genes from *Acinetobacter baumanii*, *Acinetobacter calcoaceticus* and some strains of *Acinetobacter* spp. *Acinetobacter* spp. are widespread in nature and can be obtained from water, soil and living organisms. They can use various carbon sources for growth [52]. Sette et al. [53] characterized and compared the bacterial community structure of two distinct samples (non-biodegraded and highly biodegraded oils) from a petroleum field in Brazil and showed that the genus *Acinetobacter* was exclusive to and predominant in the biodegraded oil sample. Bacteria belonging to this genus are known to be involved in biodegradation and are considered one of the most efficient oil degraders [9,11,52]. The OPF04 was represented by only one sequence, which was recovered in a clearly separate cluster, not related to any previously identified *alk* sequences present in the public database (Figure 1). This OPF is a great candidate for a putative new *alk* gene from a bacterium indigenous to the oil reservoir. A great number of OPFs (05, 06, 07, 08, 09 and 10) were grouped together and were somewhat distantly related to sequences of organisms represented in the databases. These results indicate the presence of novel putative alkane monooxygenase genes in petroleum reservoirs not previously identified in any other environment or isolate. Kuhn and colleagues [19] found a differential distribution of *alk* genes between two sites of Antarctic marine environments, and the predominant presence of new *alk* genes, mainly in the pristine site. Recently, Miqueletto et al. [54] also observed a higher diversity of an alkane hydroxylase gene in petroliferous soil sample in comparison to a non-petroliferous soil. The authors showed that most of the alkane hydroxylase genes corresponded to potentially new genes.

Diversity analyses clearly showed profound differences between GMR75 and PTS01 libraries. ACE and Chao estimators yielded higher richness values of *alk* genes in the GMR75 library (Table 3). Both

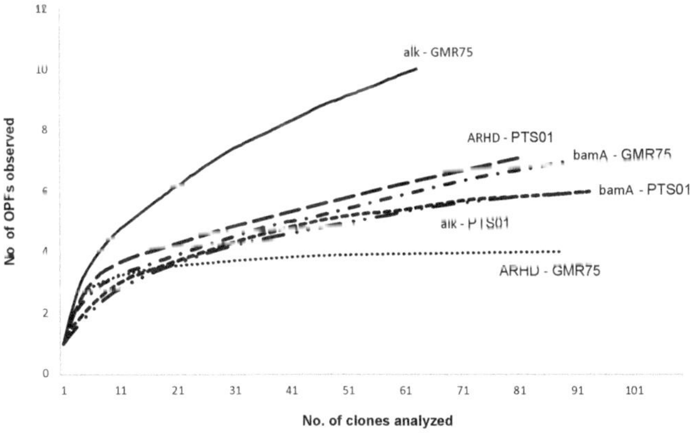

Figure 2: Rarefaction analysis of *alk*, ARHD and *bamA* deduced protein sequences from the GMR75 and PTS01.

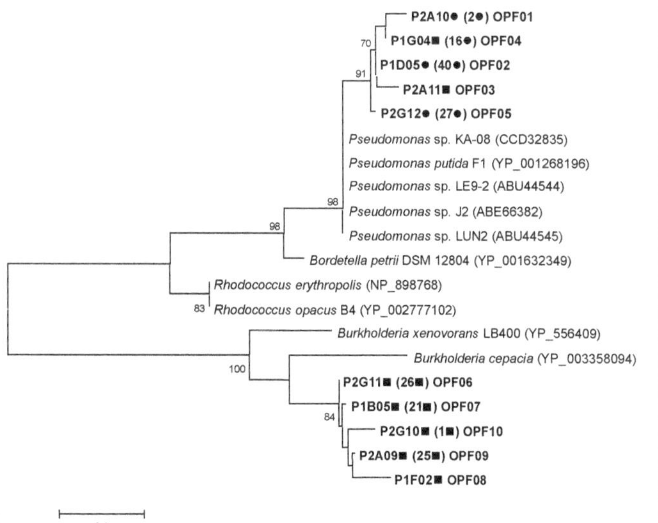

Figure 3: Phylogenetic tree based on the alignment of amino acid sequences (about 86 aa) of aromatic ring hydroxylating dioxygenase. The tree was constructed by using the neighbor-joining (NJ) method in MEGA5 program with substitution model PAM Matrix (Dayhoff). Bootstrap values (1000 replicate runs, shown as %) >70% are listed. The scale bar denotes divergence percentage between sequences. Symbols: clones from GMR75 (●); clones from PTS01 (■). Access numbers of genes from GenBank database reference strains are in parentheses.

Shannon and Simpson diversity indices were consistent with rarefaction analyses, showing higher diversity of *alk* sequences in GMR75library than in the PTS01 library (Table 3 and Figure 2).

In the ARHD gene libraries, the majority of the sequences from both libraries matched with ARHD genes from GenBank using the BLASTx program. Phylogenetic analysis revealed 3 OPFs exclusive to the GMR75 library, 6 OPFs exclusive to the PTS01 library and 1 OPF composed by sequences from both libraries (Figure 3). Five OPFs (OPFs 01, 02, 03, 04 and 05) were related with dioxygenase genes from several species of *Pseudomonas*. The enzymes for the degradation of aromatic hydrocarbons have been reported to be encoded by genes located on plasmids (e.g., TOL, NAH, SAL, OCT, and CAM) in *Pseudomonas* spp. [55]. A new genetic organization and co-regulation of a cluster of genes involved in the first steps of phenol and benzene catabolic pathways was described in *Pseudomonas* spp. M1 [56]. The authors showed that, differently from the established models for *Pseudomonas* upper pathway, the *Pseudomonas* spp. M1 exhibited exceptional biodegradation ability towards a wide range of toxic and/or recalcitrant compounds. On the other hand, 5 OPFs (OPFs 06, 07, 08, 09 and 10) were related with dioxygenase genes from *Burkholderia cepacia* and *Burkholderia xenovorans*. These two species have been extensively studied in terms of degradation of the low-molecular-weight PAHs naphthalene and phenanthrene [57] and detected in many crude oil samples [34,58,59].

Cunha et al. [60] isolated bacteria of the genus *Bacillus* from the rock of an oil reservoir in Brazil and detected the presence of catechol dioxygenase genes in 4 *Bacillus* isolates by PCR. Interestingly, in the present study none of the ARHD gene sequences analyzed matched with ARHD genes from *Bacillus* strains represented in the databases. In addition, rarefaction analysis of ARHD genes (Figure 2) yielded plateau-shaped curves, indicating that the ARHD gene diversity in the environment was close to saturation and the sampling effort was satisfactory. These results might suggest that ARHD genes belonging to the genus *Bacillus* are simply not present in Potiguar oil samples,

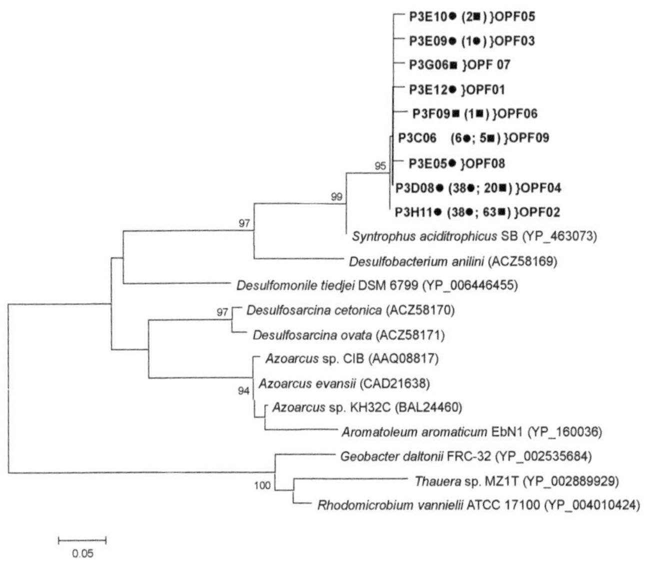

Figure 4: Phylogenetic tree based on the alignment of amino acid sequences (about 97 aa) of to 6-oxocyclohex-1-ene-1-carbonyl-CoA. The tree was constructed by using the neighbor-joining (NJ) method in MEGA5 program with substitution model PAM Matrix (Dayhoff). Bootstrap values (1000 replicate runs, shown as %) >70% are listed. The scale bar denotes divergence percentage between sequences. Symbols: clones from GMR75 (●); clones from PTS01 (■). Access numbers of genes from GenBank database reference strains are in parentheses.

or that they could not be detected due to extremely low abundance or to bias inherent to the DNA extraction (differential lysis) and/or PCR techniques (preferential amplification).

Phylogenetic analysis of *bamA* gene libraries obtained from samples GMR75 and PTS01 (Figure 4) showed that nine OPFs were grouped together and were somewhat distantly related to sequences of organisms represented in the databases, being *bamA* genes from *Synthrophus aciditrophicus* and *Desulfobacterium anilini* the closest described sequences. *Syntrophus aciditrophicus* is a bacterium that degrades a wide range of organic compounds, including alcohols, fatty acids, and hydrocarbons, including methane, which are degraded syntrophically under anaerobic conditions [61]. *Desulfobacterium anilini* is a sulfate-reducing bacterium known to be involved in phenol degradation [62]. Kuntze et al. [32] observed that the *bamA* gene was highly conserved among organisms studied by the group. In the present study, most of the clone sequences from the oil samples GMR75 and PTS01 were clustered into OPF02 and OPF04 (Figure 4), demonstrating that the gene under study has indeed a conserved region. In addition, all of the 183 sequences obtained from GMR75 and PTS01, showed none or low similarity level to the sequences from reference organisms. This suggests that the sequences corresponding to the *bamA* gene recovered from the petroleum reservoirs are potentially new sequences.

In spite of the higher diversity indices observed for the GMR75 library in the case of *alk* and *bamA* genes, the same was not observed for the ARHD gene. ACE and Chao indices indicated that the GMR75 library contained more distinct OPFs (more richness) than the PTS01 library for both catabolic genes (Table 3). Shannon and Simpson diversity indices also showed higher diversity of *alk* and *bamA* sequences in library GMR75 than in library PTS01. These results suggested that the bacterial communities of the biodegraded and non-biodegraded oil reservoirs are different also on the distribution of such catabolic genes.

A recent paper published by our research team [34] reported the microbial communities present in the same oil samples used in the present work, originated from the biodegraded (GMR75) and non-biodegraded (PTS1) terrestrial reservoirs from Potiguar Basin, by using 16S rRNA gene libraries. The authors demonstrated likewise the presence of the genera *Acinetobacter, Geobacillus, Pseudomonas* and *Syntrophus* in the microbial community. These results are in agreement with the ones found in the present work, confirming the presence of bacteria bearing catabolic genes involved in the process of biodegradation. However, Silva et al. [34] found the predominance of *Geobacillus* in the microbial community of PTS1 sample, but no representative of this genus was found in the GMR75 sample. These results are in contrast with those found in the present work, since *alk* genes similar to *Geobacillus* alkane monooxygenases were found only in the GMR75 oil. These contrasting results could be explained by the fact that the sampling effort was not sufficient to cover both phylogenetic [34] and functional diversity (Figure 2) present in these oil samples. On the other hand, Silva et al. [34] suggested that anaerobic metabolism is the dominant process for hydrocarbon degradation in GMR75 oil field, corroborating with the result that the biodegraded oil sample was more diverse for the *bamA* gene.

In conclusion, the combined results obtained from the catabolic gene libraries showed that both oil samples contained *alk*, ARHD and *bamA* genes related to genes present in bacterial species known to be hydrocarbon-degraders, such as *Acinetobacter*, *Pseudomonas, Geobacillus, Burkholderia* and deltaproteobacteria. Nonetheless, subsets of the catabolic sequences recovered from both petroleum reservoirs may correspond to new genes endemic to such environments, possibly harbored by yet uncultivated and unidentified microorganisms. Moreover, it is important to point out that this was the first work using degenerate primers to detect the presence of functional genes involved in the degradation of hydrocarbons in petroleum reservoirs, demonstrating the genetic potential of such microbial communities to biodegrade the oil. However, such biodegradation will only occur if the environmental conditions are appropriate for the metabolism of the degrading microorganisms. Curiously, the non-biodegraded oil sample showed to be more diverse in terms of the catabolic genes ARHD, suggesting that the environmental parameters that control biodegradation are playing a major role in differentiating both biodegraded and non-biodegraded oil reservoirs. As previously detailed in Silva et al. [34], the oil sample GMR75 has been classified as moderately biodegraded, produced from a reservoir with an *in situ* temperature of 42.2°C at a depth of 535.5-540.5 m. On the other hand, the PTS1 sample has been classified as non-biodegraded, collected from a deeper and hotter reservoir, with an *in situ* temperature of 48.3°C at a depth of 801-803 m [63]. Both oils were collected in the same basin, but they are related to different petroleum systems, so different source rocks and reservoirs. These discrepancies *in situ* temperature, depth and, possibly, in source rock composition and water salinity might explain the functional differences of the hydrocarbon degrading populations between both reservoirs. It is worth to mention that studies carried out with DNA sequences shows only the bacteria with potential to degrade oil, requiring further studies at the level of gene expression and GC–MS assays for the precise evaluation of the degradation process.

Finally, the results obtained allowed us to confirm the effectiveness of the primer sets used for the analysis of the community structure of bacteria that have the genetic potential to degrade hydrocarbons by aerobic and anaerobic pathways. The application of molecular methods for the rapid detection of specific microorganisms or genes in environmental samples is a valuable tool for studies focusing on diversity, abundance and distribution of hydrocarbon degradation populations, providing new insights into the diversity of catabolic genes in the oil reservoir environment and allowing a better understanding of the role of microorganisms on the biodegradation process.

Acknowledgements

We thank PETROBRAS for the financial support, donation of samples and authorization to publish this work. L. C. L. Verde was supported by grants from FAPESP (Fundação de Amparo à Pesquisa do Estado de São Paulo). The authors are grateful to Rubens T. D. Duarte for technical assistance.

References

1. Röling WF, Head IM, Larter SR (2003) The microbiology of hydrocarbon degradation in subsurface petroleum reservoirs: perspectives and prospects. Res Microbiol 154: 321-328.
2. Connan J (1984) Biodegradation of crude oils in reservoirs. Advances in Petroleum Geochemistry. Academic Press, London, UK.
3. Van Hamme JD, Singh A, Ward OP (2003) Recent advances in petroleum microbiology. Microbiol Mol Biol Rev 67: 503-549.
4. Higashioka Y, Kojima H, Sato S, Fukui M (2009) Microbial community analysis at crude oil-contaminated soils targeting the 16S ribosomal RNA, xylM, C23O, and bcr genes. J Appl Microbiol 107: 126-135.
5. Magot M, Ollivier B, Patel BK (2000) Microbiology of petroleum reservoirs. Antonie Van Leeuwenhoek 77: 103-116.
6. Aitken CM, Jones DM, Larter SR (2004) Anaerobic hydrocarbon biodegradation in deep subsurface oil reservoirs. Nature 431: 291-294.
7. van Beilen JB, Li Z, Duetz WA, Smits THM, Witholt B (2003) Diversity of Alkane Hydroxylase Systems in the Environment. Oil & Gas Sci Technol – Rev IFP 58: 427-440.
8. Watkinson RJ, Morgan P (1990) Physiology of aliphatic hydrocarbon-degrading microorganisms. Biodegradation 1: 79-92.
9. Smits TH, Balada SB, Witholt B, van Beilen JB (2002) Functional analysis of alkane hydroxylases from gram-negative and gram-positive bacteria. J Bacteriol 184: 1733-1742.
10. van Beilen JB, Smits TH, Whyte LG, Schorcht S, Röthlisberger M, et al. (2002) Alkane hydroxylase homologues in Gram-positive strains. Environ Microbiol 4: 676-682.
11. van Beilen JB, Funhoff EG (2007) Alkane hydroxylases involved in microbial alkane degradation. Appl Microbiol Biotechnol 74: 13-21.
12. Sotsky JB, Greer CW, Atlas RM (1994) Frequency of genes in aromatic and aliphatic hydrocarbon biodegradation pathways within bacterial populations from Alaskan sediments. Can J Microbiol 40: 981-985.
13. Knaebel DB, Crawford RL (1995) Extraction and purification of microbial DNA from petroleum-contaminated soils and detection of low numbers of toluene, octane and pesticide degraders by multiplex polymerase chain reaction and Southern analysis. Mol Ecol 4: 579-591.
14. Guo C, Sun W, Harsh JB, Ogram A (1997) Hybridization Analysis of Microbial DNA from Fuel Oil-Contaminated and Noncontaminated Soil Microb Ecol 34: 178-187.
15. Whyte LG, Greer CW, Inniss WE (1996) Assessment of the biodegradation potential of psychrotrophic microorganisms. Can J Microbiol 42: 99-106.
16. Yergeau E, Arbour M, Brousseau R, Juck D, Lawrence JR, et al. (2009) Microarray and real-time PCR analyses of the responses of high-arctic soil bacteria to hydrocarbon pollution and bioremediation treatments. Appl Environ Microbiol 75: 6258-6267.
17. Stapleton RD, Sayler GS (1998) Assessment of the Microbiological Potential for the Natural Attenuation of Petroleum Hydrocarbons in a Shallow Aquifer System. Microb Ecol 36: 349-361.
18. Whyte LG, Smits TH, Labbé D, Witholt B, Greer CW, et al. (2002) Gene cloning and characterization of multiple alkane hydroxylase systems in Rhodococcus strains Q15 and NRRL B-16531. Appl Environ Microbiol 68: 5933-5942.
19. Kuhn E, Bellicanta GS, Pellizari VH (2009) New alk genes detected in Antarctic marine sediments. Environ Microbiol 11: 669-673.
20. Panicker G, Mojib N, Aislabie J, Bej AK (2010) Detection, expression and quantitation of the biodegradative genes in Antarctic microorganisms using PCR. Antonie Van Leeuwenhoek 97: 275-287.
21. Bell TH, Yergeau E, Martineau C, Juck D, Whyte LG, et al. (2011) Identification of nitrogen-incorporating bacteria in petroleum-contaminated arctic soils by using [15N]DNA-based stable isotope probing and pyrosequencing. Appl Environ Microbiol 77: 4163-4171.
22. Kauppi B, Lee K, Carredano E, Parales RE, Gibson DT, et al. (1998) Structure of an aromatic-ring-hydroxylating dioxygenase-naphthalene 1,2-dioxygenase. Structure 6: 571-586.
23. Cébron A, Norini MP, Beguiristain T, Leyval C (2008) Real-Time PCR quantification of PAH-ring hydroxylating dioxygenase (PAH-RHDalpha) genes from Gram positive and Gram negative bacteria in soil and sediment samples. J Microbiol Methods 73: 148-159.
24. Gibson DT, Parales RE (2000) Aromatic hydrocarbon dioxygenases in environmental biotechnology. Curr Opin Biotechnol 11: 236-243.
25. Romine MF, Stillwell LC, Wong KK, Thurston SJ, Sisk EC, et al. (1999) Complete sequence of a 184-kilobase catabolic plasmid from Sphingomonas aromaticivorans F199. J Bacteriol 181: 1585-1602.
26. Goyal AK, Zylstra GJ (1996) Molecular cloning of novel genes for polycyclic aromatic hydrocarbon degradation from Comamonas testosteroni GZ39. Appl Environ Microbiol 62: 230-236.
27. Jeon CO, Park M, Ro HS, Park W, Madsen EL (2006) The naphthalene catabolic (nag) genes of Polaromonas naphthalenivorans CJ2: evolutionary implications for two gene clusters and novel regulatory control. Appl Environ Microbiol 72: 1086-1095.
28. Laurie AD, Lloyd-Jones G (1999) The phn genes of Burkholderia sp. strain RP007 constitute a divergent gene cluster for polycyclic aromatic hydrocarbon catabolism. J Bacteriol 181: 531-540.
29. Simon MJ, Osslund TD, Saunders R, Ensley BD, Suggs S, et al. (1993) Sequences of genes encoding naphthalene dioxygenase in Pseudomonas putida strains G7 and NCIB 9816-4. Gene 127: 31-37.
30. Head IM, Jones DM, Larter SR (2003) Biological activity in the deep subsurface and the origin of heavy oil. Nature 426: 344-352.
31. Staats M, Braster M, Röling WF (2011) Molecular diversity and distribution of aromatic hydrocarbon-degrading anaerobes across a landfill leachate plume. Environ Microbiol 13: 1216-1227.
32. Kuntze K, Shinoda Y, Moutakki H, McInerney MJ, Vogt C, et al. (2008) 6-Oxocyclohex-1-ene-1-carbonyl-coenzyme A hydrolases from obligately anaerobic bacteria: characterization and identification of its gene as a functional marker for aromatic compounds degrading anaerobes. Environ Microbiol 10: 1547-1556.
33. Higashioka Y, Kojima H, Fukui M (2011) Temperature-dependent differences in community structure of bacteria involved in degradation of petroleum hydrocarbons under sulfate-reducing conditions. J Appl Microbiol 110: 314-322.
34. Silva TR, Verde LCL, Santos Neto EV, Oliveira VM (2012) Diversity analyses of microbial communities in petroleum samples from Brazilian oil fields. Int Biodeter Biodegr.
35. Bellicanta GS (2005) Diversity of catabolic genes in sediment samples of Santos and São Vicente Estuarine Systems, SP. Biota Neotropica 5.
36. Ewing B, Hillier L, Wendl MC, Green P (1998) Base-calling of automated sequencer traces using phred. I. Accuracy assessment. Genome Res 8: 175-185
37. Schloss PD, Westcott SL, Ryabin T, Hall JR, Hartmann M, et al. (2009) Introducing mothur: open-source, platform-independent, community-supported software for describing and comparing microbial communities. Appl Environ Microbiol 75: 7537-7541.
38. Saitou N, Nei M (1987) The neighbor-joining method: a new method for reconstructing phylogenetic trees. Mol Biol Evol 4: 406-425.
39. Hall BG (2013) Building Phylogenetic Trees from Molecular Data with MEGA. Mol Biol Evol .
40. Thompson JD, Gibson TJ, Plewniak F, Jeanmougin F, Higgins DG (1997) The CLUSTAL_X windows interface: flexible strategies for multiple sequence alignment aided by quality analysis tools. Nucleic Acids Res 25: 4876-4882.
41. Magurran AE (2004) Measuring Biological Diversity. Blackwell Publishing, Oxford, UK.

42. Chao A, Ma MC, Yang MCK (1993) Stopping rules and estimation for recapture debugging with unequal failure rates. Biometrika 80: 193-201.

43. Chao A (1984) Non-parametric estimation of the number of classes in a population. Scand J Stat 11: 265-270.

44. Schloss PD, Handelsman J (2008) A statistical toolbox for metagenomics: assessing functional diversity in microbial communities. BMC Bioinformatics 9: 34.

45. Smits TH, Röthlisberger M, Witholt B, van Beilen JB (1999) Molecular screening for alkane hydroxylase genes in Gram-negative and Gram-positive strains. Environ Microbiol 1: 307-317.

46. van Beilen JB, Smits TH, Roos FF, Brunner T, Balada SB, et al. (2005) Identification of an amino acid position that determines the substrate range of integral membrane alkane hydroxylases. J Bacteriol 187: 85-91.

47. Turova TP, Nazina TN, MikhaÄlova EM, Rodionova TA, Ekimov AN, et al. (2008) AlkB homologues in thermophilic bacteria of the genus Geobacillus. Mol Biol (Mosk) 42: 247-257.

48. Onstott TC, Phelps TJ, Colwell FS, Ringelberg D, White DC, et al. (1998) Observations pertaining to the origin and ecology of microorganisms recovered from deep subsurface Taylorsville Basin, Virginia. Geomicrobiol J 15: 353–385.

49. Kato T, Miyanaga A, Kanaya S, Morikawa M (2009) Alkane inducible proteins in Geobacillus thermoleovorans B23. BMC Microbiol 9: 60.

50. McMullan G, Christie JM, Rahman TJ, Banat IM, Ternan NG, et al. (2004) Habitat, applications and genomics of the aerobic, thermophilic genus Geobacillus. Biochem Soc Trans 32: 214-217.

51. Shestakova NM, Korshunova AV, Mikhailova EM, Sokolova DSH, Tourova TP, et al. (2011) Characterization of the aerobic hydrocarbon-oxidizing enrichments from a high-temperature petroleum reservoir by comparative analysis of DNA and RNA derived clone libraries. Microbiology 80: 60-69.

52. Barbe V, Vallenet D, Fonknechten N, Kreimeyer A, Oztas S, et al. (2004) Unique features revealed by the genome sequence of Acinetobacter sp. ADP1, a versatile and naturally transformation competent bacterium. Nucleic Acids Res 32: 5766-5779.

53. Sette LD, Simioni KC, Vasconcellos SP, Dussan LJ, Neto EV, et al. (2007) Analysis of the composition of bacterial communities in oil reservoirs from a southern offshore Brazilian basin. Antonie Van Leeuwenhoek 91: 253-266.

54. Miqueletto PB, Andreote FD, Dias AC, Ferreira JC, Dos Santos Neto EV, et al. (2011) Cultivation-independent methods applied to the microbial prospection of oil and gas in soil from a sedimentary basin in Brazil. AMB Express 1: 35.

55. Irie S, Doi S, Yorifuji T, Takagi M, Yano K (1987) Nucleotide sequencing and characterization of the genes encoding benzene oxidation enzymes of Pseudomonas putida. J Bacteriol 169: 5174-5179.

56. Santos PM, Sá-Correia I (2007) Characterization of the unique organization and co-regulation of a gene cluster required for phenol and benzene catabolism in Pseudomonas sp. M1. J Biotechnol 131: 371-378.

57. Laurie AD, Lloyd-Jones G (2000) Quantification of phnAc and nahAc in contaminated New Zealand soils by competitive PCR. Appl Environ Microbiol 66: 1814-1817.

58. Yoshida N, Yagi K, Sato D, Watanabe N, Kuroishi T, et al. (2005) Bacterial communities in petroleum oil in stockpiles. J Biosci Bioeng 99: 143-149.

59. Korenblum E, Souza DB, Penna M, Seldin L (2012) Molecular analysis of the bacterial communities in crude oil samples from two brazilian offshore petroleum platforms. Int J Microbiol 2012: 156537.

60. da Cunha CD, Rosado AS, Sebastián GV, Seldin L, von der Weid I (2006) Oil biodegradation by Bacillus strains isolated from the rock of an oil reservoir located in a deep-water production basin in Brazil. Appl Microbiol Biotechnol 73: 949-959.

61. McInerney MJ, Rohlin L, Mouttaki H, Kim U, Krupp RS, et al. (2007) The genome of Syntrophus aciditrophicus: life at the thermodynamic limit of microbial growth. Proc Natl Acad Sci U S A 104: 7600-7605.

62. Ahn YB, Chae JC, Zylstra GJ, Häggblom MM (2009) Degradation of phenol via phenylphosphate and carboxylation to 4-hydroxybenzoate by a newly isolated strain of the sulfate-reducing bacterium Desulfobacterium anilini. Appl Environ Microbiol 75: 4248-4253.

63. dos Santos Neto EV, Hayes JM (1999) Use of hydrogen and carbon stable isotopes characterizing oils from the Potiguar Basin (Onshore), Northeastern Brazil. AAPG Bulletin 83: 496-518.

Effectiveness of Calcium Sulfate Scale Inhibitors in Spent Hydrochloric Acid/Seawater System

He J*, Arensman D and Nasr-El-Din HE

SPE, Texas A&M University, Texas, USA

Abstract

Previous results have reported the precipitation of calcium sulfate during matrix acid treatments when seawater was used to prepare HCl-based acid. Precipitations of calcium sulfate causes severe damage on the permeability of carbonate reservoirs and negatively impacts the performance of acid treatments. Typically, scale inhibitors are applied in the field to prevent the formation of calcium sulfate scale, and the object of this study is to evaluate the effectiveness of different types of scale inhibitors to inhibit the formation of calcium sulfate scale during the acid stimulation process. Scale inhibition efficiency was determined in batch tests under different temperatures (77 to 210°F) and different degrees of spent acid conditions. Sulfate ions were analyzed in the samples of the batch tests to detect the precipitations of calcium sulfate. Results showed that application of scale inhibitors can successfully mitigate calcium sulfate scale formation up to 210°F. At higher temperatures, the rate of calcium sulfate precipitation increased and the effectiveness of all types of scale inhibitors decreased greatly. Most of the scale inhibitors were not effective in acidic conditions and some of the scale inhibitors even precipitated out of the solution in the presence of high concentration of calcium ions. Methylene phoshponic acid scale inhibitor was the most effective one under various experimental conditions. The findings in this study provided information for better calcium sulfate scale control under acid stimulation conditions.

Keywords: Calcium sulfate; Scale inhibitor; Acid Treatment; Formation Damage; Seawater

Introduction

The potential of calcium sulfate precipitation is greatly increased when seawater is used for matrix acidizing treatments [1-3]. In those processes, calcium concentration is greatly increased, due to the reaction of the hydrochloric acid with calcite in the carbonate reservoir rocks. High concentrations of calcium ions will combine with high levels of sulfate ions in seawater, and significantly increase the scaling tendency of calcium sulfate. Moreover, the solubility of calcium sulfate drops greatly once the acid is spent [4-6]. Those two factors account for the precipitation of calcium sulfate once it exceeds the critical scaling tendency of calcium sulfate [7-9]. Calcium sulfate will deposit in the stimulated flow channels [10], possibly causing blockage of pore throats and severely impairing the formation permeability [11]. Hence, decrease well injectivity or productivity [12,13]. Even worse, the nucleation and precipitation process of calcium sulfate during acidizing treatments may affect the acid reaction with carbonate rocks, and cause a limited acid stimulation effect [3].

Arabian Gulf at Kuwait	Ion Concentration, mg/l
Chloride (Cl^-)	23,000
Sodium (Na^+)	15,850
Sulfate (SO_4^{2-})	4,000
Magnesium (Mg^{2+})	1,765
Calcium (Ca^{2+})	500
Bicarbonate (HCO_3^-)	142
Total dissolved solids (TDS), mg/l	45,157
pH	7.8
Density*, g/cm^3	1.028

*Measured at 77°F

Table 1: Composition of synthetic seawater.

Typically, the application of scale inhibitors has been an economical method used to mitigate calcium sulfate [14-16]. The use of scale inhibitors may act either as chelating agents to form a soluble complex [17-19], or as threshold inhibitors, which block the development of the supercritical nuclei [20,21], or as retarders of the growth of the calcium sulfate crystals [22-24]. By simply adding an acid-soluble and effective scale inhibitor into the acid system, mineral scale formation can be prevented during the process of acid stimulation [25-27]. In addition, combining scale inhibition and acid stimulation into a single package inherently reduces well intervention costs and well downtime, and therefore achieves significant economic benefits [28]. In this study, the effectiveness of various scale inhibitors to prevent the formation of calcium sulfate was evaluated in acid stimulation process, and the specific challenges of calcium sulfate scale control were discussed.

Experimental Studies

Fluids preparation

A synthetic seawater solution was prepared according to the composition of seawater in the Arabian Gulf at Kuwait [29], Table 1. The synthetic seawater with a density of 1.028 g/cm3 at 77°F was

***Corresponding author:** Jia He, PhD Student of Petroleum Engineering Department, Texas A&M University, Texas, USA, Tel: 9798452241; Fax: 9798451307

Abbreviated Name	Active Ingredient	Approxiamate Activity, %
A	Phosphinopolycarboxylic acid	29
B	Sulfonated polymer	31
C	Polyacrylic acid	19
D	P tagged sulfonated polymer	45
E	Diethylenetriamine penta(methylene phosphonic acid)	18
F	Bis-hexamethylene triamine-penta(methylene phosphonic) acid	12

Table 2: Scale inhibitors used in this study.

Figure 1: Visual cell.

Degree of Spent Acid	Case	Density, mg/l	pH	HCl Concentration, wt%	Sulfate Concentration, mg/l
15 wt% Live HCl	Acid	1.07	0	15.42	2300
	Mixtures	1.04	0	3.86	3510
10 wt% Partially Spent HCl	Acid	1.1	0	9.58	2820
	Mixtures	1.05	0.08	2.65	3600
5 wt% Partially Spent HCl	Acid	1.17	0	5.03	3350
	Mixtures	1.06	0.4	1.37	3810
1 wt% Partially Spent HCl	Acid	1.18	0.5	1.11	3600
	Mixtures	1.07	1.1	0.28	3850
0 wt% Completely Spent HCl	Acid	1.2	7.7	0	3920
	Mixtures	1.19	7.8	0	3940

Table 3; Properties of spent acids and mixtures of acid with seawater.

used to prepare HCl acid. All the salts used to prepare the seawater were obtained from Mallinckrodt Chemical Inc. Hydrochloric acid at 15 wt% was prepared using seawater through dilution of 36 wt% HCl, which was obtained from Mallinckrodt Chemical Inc., and 0.45 vol% corrosion inhibitor was added to the acid. The corrosion inhibitor was received from a local service company. To prepare spent-acid solutions, the required amount of calcium chloride was added to prepare spent-acid solutions of 10,5,1, and 0 wt% HCl from live 15 wt% HCl.

Types of scale inhibitors

Six different types of scale inhibitors were tested for performance comparison. All the scale inhibitors were received from a local service company and the chemistry of all the scale inhibitors were shown in Table 2.

Visual cell

A Hassler-type visual cell was used to observe the precipitation of calcium sulfate at different temperatures shown in Figure 1. The cell is manufactured by Temco, Inc., and can withstand up to 1000 psi and 300°F.

Experimental procedures

In general, the objective of this study is to investigate the effectiveness of scale inhibitors to prevent the formation of calcium sulfate scale during core flood experimental conditions. In our previous work, 10 cm^3 of 15 wt% HCl was injected into a limestone core with around 30 cm^3 of pore volume [3]. Therefore, the experimental procedures used to determine the efficiency of scale inhibitors are as follows:

- 15, 10, 5, 1, and 0 wt% HCl acids were first prepared using seawater.
- Then 10 cm^3 of acid and 30 cm^3 (approximate one pore volume of core sample) of seawater was mixed, and scale inhibitor was added as well, taking sample #1. This is the mixture of seawater and spent acid without adding $CaCl_2$.
- Then the amount of $CaCl_2$ produced by acid reaction with calcite was added into solution, taking sample #2. The properties of different degrees of acid and mixtures of acid and seawater are shown in Table 3.
- The solution was put into the visual cell to be heated up to 150, 210, and 250°F and samples #3, #4, and #5 were taken.
- A spectrophotometer (Orbeco SP600) was used to measure SO_4^{2-} concentration, using the turbidity method (precipitation with barium chloride).

Results and Discussion

Precipitation of calcium sulfate without scale inhibitors

In the first set of batch tests, 10 cm^3 of 0 wt% completely spent HCl was mixed with 30 cm^3 of seawater. The visual cell was heated up to 250°F with a confining pressure of 300 psi. The sulfate concentration profiles were shown in Figure 2. Sample #1 was the mixture of acid and seawater without adding $CaCl_2$. Calcium sulfate began to precipitate out of solution even at 77°F (sample #2) and sulfate concentration dropped sharply to 1,300 mg/l at 210°F. However, sulfate concentration changed only a little from 210 to 250°F, possibly to due to the less change in

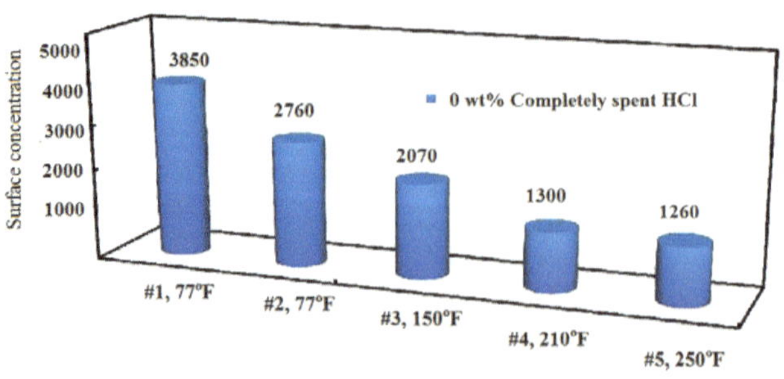

Figure 2: Sulfate concentrations in the acid and seawater mixtures at different temperatures without scale inhibitors.

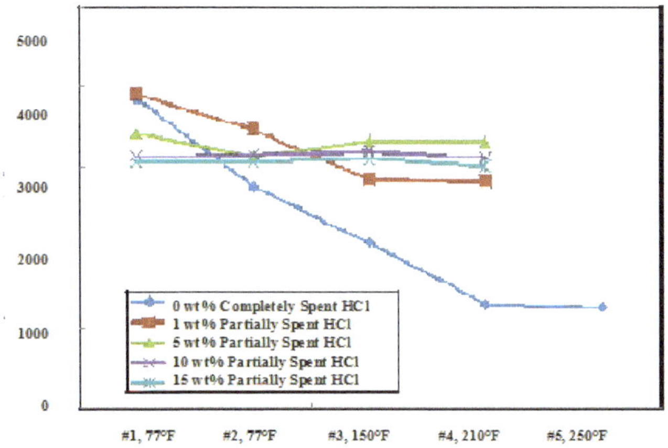

Figure 3: Sulfate concentrations in the acid and seawater mixtures at different temperatures without scale inhibitors.

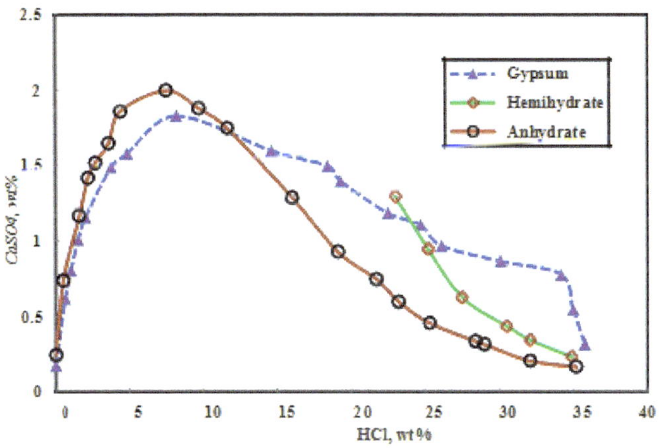

Figure 4: Solubility of calcium sulfate at 77°F (Kruchenko and Beremzhanov 1976).

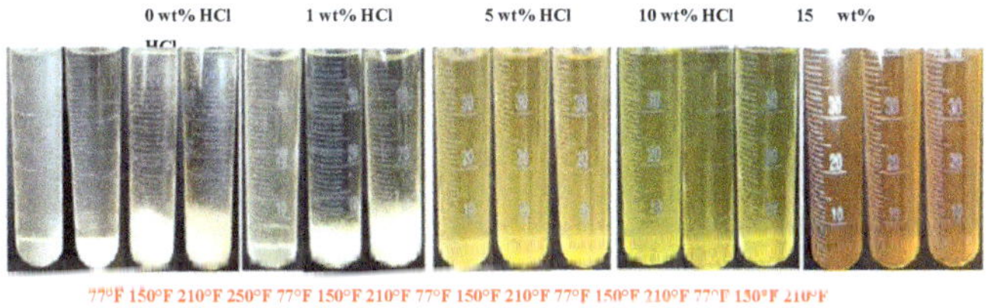

Figure 5: Precipitations were observed in the mixtures of 0 and 1 wt% spent HCl and seawater.

solubility of calcium sulfate from 210 to 250°F. Therefore, the following batch tests were done in the temperatures ranging from 77 to 210°F.

Figure 3 showed the sulfate concentration profiles under different degrees of acid spent conditions. The sulfate concentrations maintained almost constant in the mixtures of 5, 10, and 15 wt% acids with seawater. In those cases, the calcium ions produced by the reaction of HCl with calcite was not high enough to combine with sulfate ions and precipitate calcium sulfate out of solution. Besides, the solubility of calcium sulfate was more than that in completely spent acid, shown in Figure 4. The more reaction of acid with calcite was, the more calcium ions produced and the less solubility of calcium sulfate was in spent acid. Then calcium sulfate had much scaling tendency to precipitate.

Figure 5 clearly showed the precipitation of calcium sulfate at different temperatures in the mixtures of 0 and 1 wt% spent acids and seawater. The precipitations were collected at each temperature and were analyzed using Scanning Electron Microscope (SEM). Calcium sulfate was identified and the SEM images were shown in Figure 6. Further batch tests were conducted to test the effectiveness of different scale inhibitors in the mixtures of 0 and 1 wt% spent acids and seawater.

Performance of scale inhibitor in 0 wt% completely spent acid

The effectiveness of different scale inhibitors shown in Table 2 in completely spent acid was tested at different temperatures. Figure 7 showed the sulfate concentration profiles in the presence of 50 ppm different scale inhibitors. 50 ppm scale inhibitor was selected based on the previous results He et al. [3]. Contrary to the sharp decrease of sulfate concentration with the increment of temperature without adding scale inhibitors, the sulfate levels in the presence of different scale inhibitors kept almost constant at different temperatures, indicating no calcium sulfate precipitation in the solutions. This was also confirmed by the clear solutions at different temperatures shown in Figure 8. Those results showed that all the scale inhibitors performed very well in the mixtures of 0 wt% completely spent acids and seawater. The pH value of the mixtures were generally around 7.8 and the calcium ions concentration was the most among all the mixtures of different degree of spent acids and seawater. The main concern for this part is to investigate the effect of pH value on the effectiveness of scale inhibitors. That is, whether the scale inhibitors would function well in acidic conditions.

Performance of scale inhibitor in 1 wt% completely spent acid

Precipitation of calcium sulfate occurred in the mixtures of 1 wt%

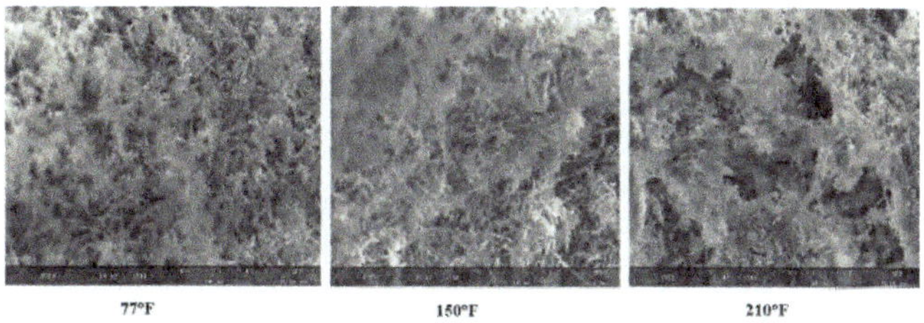

Figure 6: SEM images of calcium sulfate in the mixtures of completely spent acid and seawater without scale inhibitors.

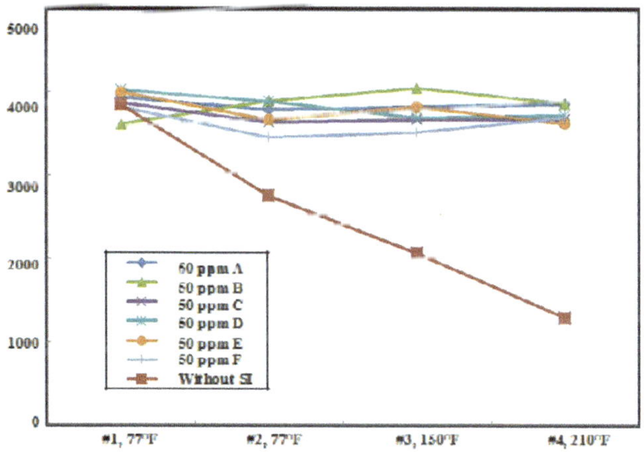

Figure 7: Sulfate concentrations in the mixtures of 0 wt% completely spent acid and seawater with 50 ppm different scale inhibitors.

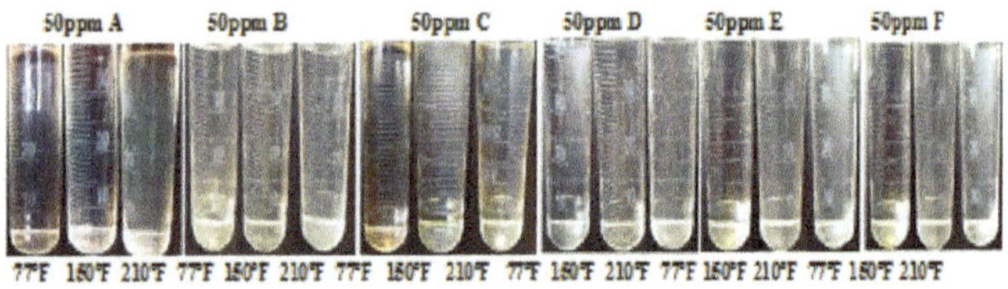

Figure 8: Clear solutions were observed in all the mixtures of 0 wt% completely spent acid and seawater.

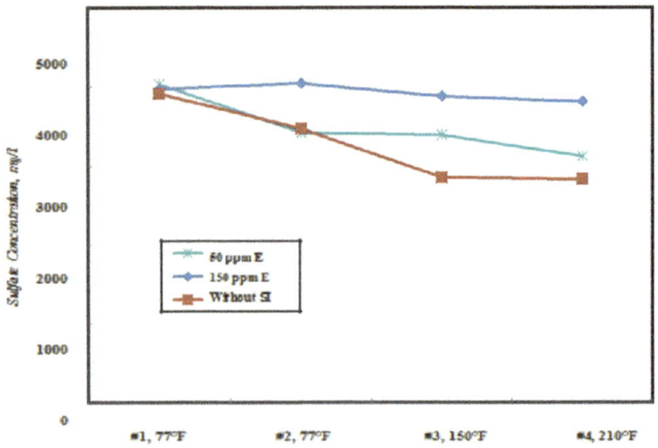

Figure 9: Sulfate concentrations in the mixtures of 1 wt% spent acid and seawater with E.

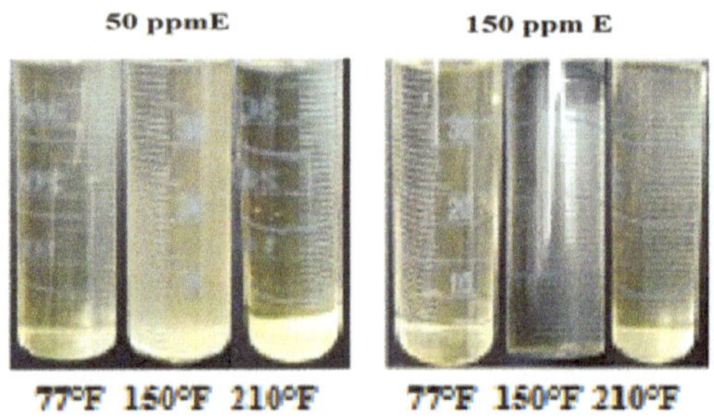

Figure 10: Performance of E as the calcium sulfate inhibitor.

spent acid and seawater, as shown in Figure 5. The pH value of the mixtures is around 1.1. The performance of all the 6 scale inhibitors can be divided into 3 categories in the very acidic conditions. With the addition of 50 ppm E, less reduction in the sulfate concentration level was observed at high temperatures in Figure 9, but there were still some precipitations of calcium sulfate out of solution (Figure 10). When the concentration of E was increased from 50 to 150 ppm, the sulfate level kept almost constant and clear solutions were shown in Figure 10. It was concluded that scale inhibitor E could function well in acidic conditions. However, higher concentrations of E were needed even the calcium ions concentration was less than that of mixtures of completely spent acid and seawater.

Figure 11 indicated that 50 ppm F and 50 ppm C were not effectively to prevent the precipitation of calcium sulfate out of the solutions at high temperatures (150 and 210°F). However, when the concentration of F was increased from 50 to 150 ppm, the sulfate level concentration almost kept constant while there were still some precipitations out of solutions at high temperatures (Figure 12). F is a methylene phosphonic-based scale inhibitor. It could be probably that F was not compatible with the solution in the presence of high calcium ions concentration and acidic conditions at high temperatures. The performance of C was different from that of F. When the concentration of C was increased from 50 to 150 ppm, the sulfate concentration dropped sharply, even more that without adding scale inhibitors. Cloudy solution was even observed with the addition of 150 ppm C at 77°F. Those findings concluded that some scale inhibitors may function well in neutral conditions, but not certainly function well in acidic conditions. Some scale inhibitors may even be not compatible with the solution at high temperatures while some scale inhibitors may cause more severe damage.

Figure 13 showed the performances of the other 3 scale inhibitors in the mixtures of 1 wt% spent acid and seawater. B behaved like the performance of 150 ppm F. The sulfate concentration levels kept almost constant but there were precipitation out of solutions shown in Figure 14. When the concentration of B was increased from 50 to 150 ppm, more precipitations and more reduction in the sulfate levels were observed at 77°F. In the presence of high calcium ion concentration and acidic conditions, B was not compatible with the solutions even at 77°F. For A and D, they did not function well in the acidic conditions, confirmed by the reduction in sulfate level and precipitations observed in the tubes. It was interesting to note that more reduction in sulfate level occurred even at 77°F when the concentrations of the 3 scale inhibitors were increased from 50 to 150 ppm.

Figure 13 and 14 showed that more damage occurred if the concentration of the scale inhibitor was increased even at 77°F. Further

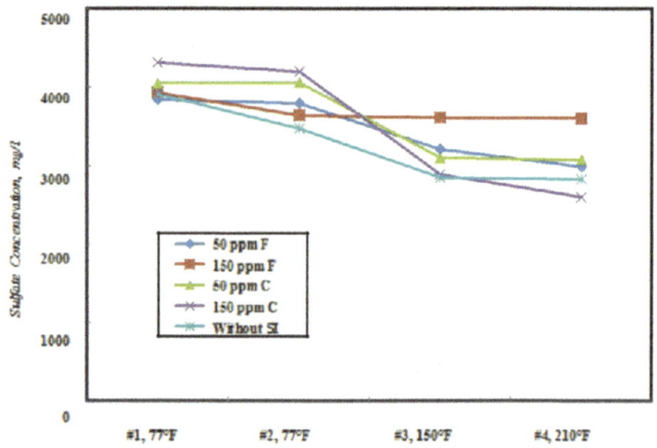

Figure 11: Sulfate concentrations in the mixtures of 1 wt% spent acid and seawater with F and C.

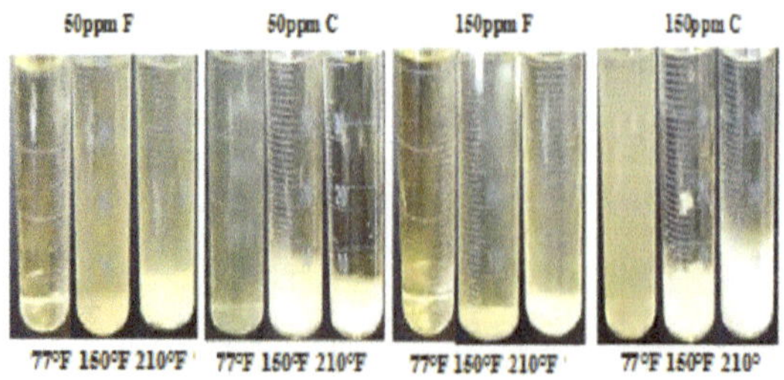

Figure 12: Performance of F and C as the calcium sulfate inhibitors.

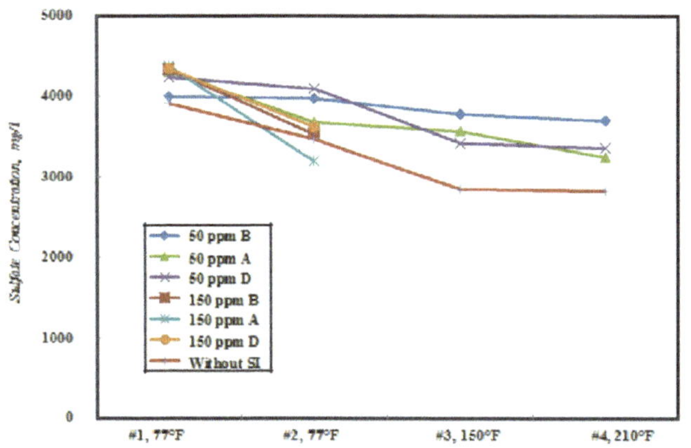

Figure 13: Sulfate concentrations in the mixtures of 1 wt% spent acid and seawater with A, B and D.

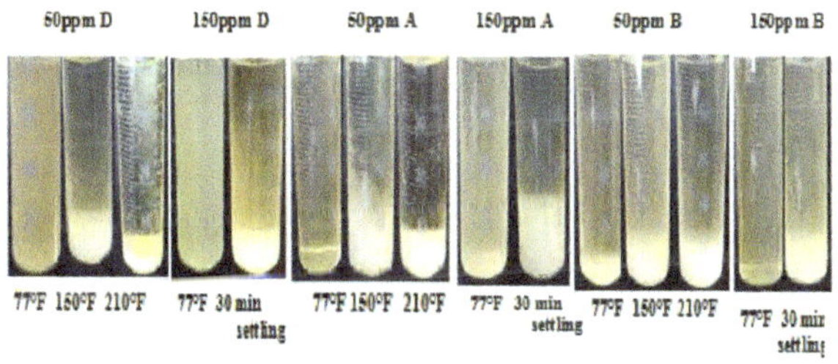

Figure 14: Performance of A, B and D as the calcium sulfate inhibitors.

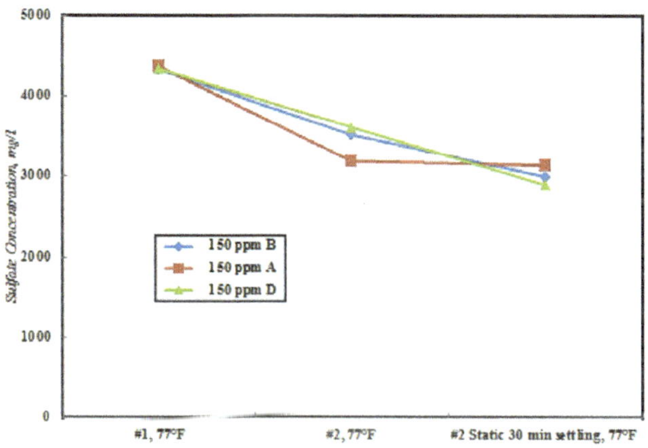

Figure 15: Sulfate concentrations in the mixtures of 1 wt% spent acid and seawater with A, B and D after 30 min static settling.

studies were conducted to investigate settling time effect instead of temperature effect. After statically settling for 30 min, more reduction in sulfate level and more precipitations were observed shown in Figure 15. Those findings concluded that there is one optimum concentration of scale inhibitors used for scale control in acidic conditions. Less scale inhibitor concentration is not enough to prevent formation of scale, but more scale inhibitor usage can cause more damage. All the results of batch tests were summarized in Table 4 and E was the most effective scale inhibitor under different conditions.

Conclusions

Based on the batch tests on the effectiveness of 6 different scale inhibitors, the following conclusions can be drawn:

1. In the acid stimulation process, calcium sulfate began to precipitate out of solution when acid was almost completely spent. 50 ppm of different scale inhibitors can successfully prevent the formation of calcium sulfate in the mixtures of completely spent acid and seawater under various temperature conditions.

2. However, most scale inhibitors did not perform well in acidic conditions. Methylene phoshponic acid scale inhibitor (scale inhibitor E) was the only effective one that can prevent the formation of calcium sulfate under various temperature conditions. The other scale inhibitors either were not sufficiently to prevent the formation of calcium sulfate or were not compatible with the solutions in the presence of high calcium ion concentrations and acidic conditions.

3. The findings in this study confirmed that the selections of appropriate scale inhibitors for field use depends on lots of factors, including temperature, pH value of the solution, and compatibility with solution.

Acknowledgement

The authors would like to thank members of the Middle East Carbonate Stimulation project for funding this research.

Acid Mixing with Seawater	Type of Scale Inhibitor	Concentration, ppm	#1, 77°F	#2, 77°F	#3, 150°F	#4, 210°F	#5, 250°F
0 wt% Completely Spent HCl	A	50	No	No	No	No	
	B	50	No	No	No	No	
	C	50	No	No	No	No	
	D	50	No	No	No	No	
	E	50	No	No	No	No	
	F	50	No	No	No	No	
	No SI	0	No	Yes	Yes	Yes	Yes
1 wt% Partially Spent HCl	A	50	No	No	Yes	Yes	
		150	No	Yes	Yes*		
	B	50	No	No	Yes	Yes	
		150	No	Yes	Yes*		
	C	50	No	No	Yes	Yes	
		150	No	No	Yes	Yes	
	D	50	No	No	Yes	Yes	
		150	No	Yes	Yes*		
	F	50	No	No	Yes	Yes	
		150	No	No	No	No	
	E	50	No	No	Yes	Yes	
		150	No	No	Yes	Yes	
	No SI	0	No	No	Yes	Yes	
5 wt% Partially Spent HCl	No SI	0	No	No	No	No	
10 wt% Partially Spent HCl	No SI	0	No	No	No	No	
15 wt% Live HCl	No SI	0	No	No	No	No	

Table 4: Summary of batch tests.

References

1. Oddo JE, Smith JP, Tomson MB (1991) Analysis of and Solutions to the $CaCO_3$ and $CaSO_4$ Scaling Problems Encountered in Wells Offshore Indonesia. SPE Annual Technical Conference and Exhibition, Dallas, Texas, USA.
2. Nasr-El-Din HA, Rosser HR, Hopkins JA (1996) Stimulation of Injection Water Supply Wells in Central Arabia. SPE International Petroleum Exhibition and Conference, Abu Dhabi, United Arab Emirates.
3. He J, Mohamed IM, Nasr-Ei-Din HA (2011) Mixing Hydrochloric Acid and Seawater for Matrix Acidizing: Is it a Good Practice? SPE European Formation Damage Conference, 7-10 June Noodrdwijk, Netherlands.
4. Flint O (1968) Increased Solubility of Calcium Sulphate in Seawater Containing Hydrochloric Acid. Desalination 4: 328-335.
5. Kruchenko VP, Beremzhanov VA (1976) The Solubilities of the Salts in the Calcium Sulphate – Hydrochloric Acid. Water System at 25°C. Russian Journal of Inorganic Chemistry 21: 152-153.
6. Li Z, Demopoulos GP (2002) $CaSO_4$ Solubilities in Concentrated Aqueous Chloride Solutions. Chloride Metallurgy 2: 561-574.
7. Yeboah YD, Samuah SK, Saeed MR (1993) Prediction of Carbonate and Sulfate Scales in Oilfields. Middle East Oil Show, Bahrain.
8. Yeboah YD, Somuah SK, Saeed MR (1993) A New and Reliable Model for Predicting Oilfield Scale Formation. SPE International Symposium on Oilfield Chemistry, New Orleans, Louisiana.
9. Raju KUG, Nasr-El-Din HA (2004) Calcium Sulfate Scale: Field Tests and Model Predictions. NACE Corrosion Conference, New Orleans, USA.
10. Delorey JR, Allen S, McMaster L (1996) Precipitation of Calcium Sulphate During Carbonate Acidizing: Minimizing the Risk. Society of Petroleum Engineers.
11. Tahmasebi HA, Kharrat R (2007) Prediction of Permeability Reduction Rate Due to Calcium Sulfate Scale Formation in Porous Media. SPE Middle East Oil and Gas Show and Conference, Kingdom of Bahrain.
12. Smith CF, Nolan III TJ, Crenshaw PL (1968) Removal and Inhibition of Calcium Sulfate Scale in Waterflood Projects. Journal of Petroleum Technology 20: 1249-1256.
13. Shen J, Crosby CC (1983) Insight into Strontium and Calcium Sulfate Scaling Mechanisms in a Wet Producer. J Petrol Technol 35: 1249-1255.
14. Yuan M, Smith JK, Cooley C, Williamson DA (2004) Effective Mineral Scale Control in Canyon Express, Gulf of Mexico. SPE International Symposium on Oilfield Scale, Aberdeen, United Kingdom.
15. Hoang TA, Ang HM, Rohl AL (2009) Effects of Organic Additives on Calcium Sulfate Scaling in Pipes. Aust J Chem 62: 927-933.
16. Fan C, Kan A, Fu G, Tomson MB, Dong S (2010) Quantitative Evaluation of Calcium Sulfate Precipitation Kinetics in the Presence and Absence of Scale Inhibitors. SPE Journal 15: 977-988.
17. Moore RE, Bischof AE, Robins JD, Brennema DR (1972) One-Step Anhydrite Scale Removal. Materials Protection and Performance 11: 41-41.
18. Jamialahmadi M, Mueller-Steinhagen H (1991) Reduction of Calcium Sulfate Scale Formation during Nucleate Boiling by Addition of EDTA. Heat Transfer Eng 12: 19-26.
19. Al-Khaldi MH, Al-Juhani A, Al-Mutairi SH, Gurmen MN (2011) New Insights into the Removal of Calcium Sulfate Scale. SPE European Formation Damage Conference, Noordwijk, Netherlands.
20. He S, Oddo JE, Tomson MB (1994) The Inhibition of Gypsum and Barite Nucleation in NaCl Brines at Temperatures from 25 to 90°C. Appl Geochem 9: 561-567.
21. Tomson MB, Fu G, Watson MA, Kan AT (2003) Mechanisms of Mineral Scale Inhibition. SPE Production & Operations 18: 192-199.
22. Liu ST, Nancollas GH (1973) The Crystal Growth of Calcium Sulfate Dihydrate in the Presence of Additives. J Colloid Interf Sci 44: 422-429.
23. Amjad Z, Hooley J (1986) Influence of Polyelectrolytes on the Crystal Growth of Calcium Sulfate Dehydrate. J Colloid Interf Sci 111: 496-503.
24. Amjad Z (1988) Calcium Sulfate Dihydrate (Gypsum) Scale Formation on Heat Exchanger Surfaces: The Influence of Scale Inhibitors. J Colloid Interf Sci 123: 523-536.
25. Smith PS, Jr. CCC, Rojas AM (2000) Combined Scale Removal and Scale Inhibition Treatments. International Symposium on Oilfield Scale, Aberdeen, United Kingdom.
26. Samy M (2003) Scalemat Restores Production in a Gas Well in GOM, Sugar Land, Texas, Schlumberger Case Study. Sugar Land, Texas: Schlumberger Technology Corporation.
27. Nasr-El-Din HA, Al-Saiari HA, Al-Hajji HH, Samy M, Garcia M, et.al. (2004) A Single-Stage Acid Treatment to Remove and Mitigate Calcium Carbonate Scale in Sandstone and Carbonate Reservoirs. SPE International Symposium on Oilfield Scale, Aberdeen, United Kingdom.
28. Smith PS, Cowie LG, Bourne HM, Grainger M, Heath SM (2001) Field Experiences with a Combined Acid Stimulation and Scale Inhibition Treatment. International Symposium on Oilfield Scale, Aberdeen, United Kingdom.
29. Cotruvo JA (2005) Water Desalination Processes and Associated Health and Environmental Issues. Water Conditioning and Purification 47: 13-17.

Energy-Return-On-Energy-Invested (EROEI) For Crude Oil and Other Sources of Energy

Jones JC*

School of Engineering, University of Aberdeen, UK

Abstract

The term Energy-Return-On-Energy-Invested (EROEI) is self-explanatory and this topic has been the author's primary research interest since about 2007. Many publications in journals including 'Fuel' (an Elsevier journal) have resulted. This paper comprises a synthesis and analysis of this work together with some new ideas.

Keywords: Energy-return-on-energy-invested; Crude oil; Sources of energy

Introduction and Literature Review

The matter of EROEI can be expressed very directly by means of the following rhetorical question. What is the point of getting a barrel of oil out of the ground if the energy required to get it out is equivalent to that released on the burning of two barrels of oil? The author on searching the literature for work on EROEI other than that with which he has himself been involved went to ScienceDirect and put in EROEI as a keyword. ScienceDirect can scan back over many decades but all of the publications on EROEI were from about the last five years. A third to a half of these has been selected for consideration herein. Discussion is in tabular form and comments follow the Table 1.

The information in Figure 1 is fairly well known. What is less widely known is that it does not even remotely signify oil depletion [1]. Over the period covered by the graph wells have increased in depth from tens or at most a hundred or so metres to thousands of metres and the energy required to raise oil from such depths obviously exceeds that required to raise it from shallower depths, other things being equal. Also, at the time when Figure 1 commences there was no offshore oil production anywhere. So the first major point being made in the present paper is that facile identification of increasing EROEI with depletion should be avoided.

The statement from reference [2] which has been quoted in the table raises another point of possibly questionable validity. The energy requirement for drilling in exploration is quite negligible in comparison with that in subsequent production [3], so initial inaccessibility of the oil is probably not a factor in the EROEI.

Publication [4] touches on some important points but in a way which is incomprehensible if not defective. In dismissing energy from 'natural sources' the authors are saying in effect that EROEI is not an energy balance, therefore energy having no potential use other than that to which it is being applied need not be considered. This is true, and the most obvious example is the wind turbine. In this the kinetic energy of the wind is converted to mechanical at the turbine and in analysis by the steady flow equation the kinetic energy of the wind must feature. It is however inconceivable that this energy would have found another engineering application.

The information in Figure 1 is fairly well known. What is less widely known is that it does not even remotely signify oil depletion. Over the period covered by the graph wells have increased in depth from tens or at most a hundred or so metres to thousands of metres and the energy required to raise oil from such depths obviously exceeds that required to raise it from shallower depths, other things being equal. Also, at the time when Figure 1 commences there was no offshore oil production anywhere. So the first major point being made in the present paper is that facile identification of increasing EROEI with depletion should be avoided.

The statement from reference [2] which has been quoted in the

Publication	Points of interest.
[1]	Features a figure for EROEI of US oil over the period 1930-2000 from a web source, which is reproduced as Figure 1 below.
[2]	'The reason [for a declining EROEI of oil] is that effort is redirected to less easily accessible reserves . . . .'
[4]	' EROEI = Produced energy/Consumed energy where produced energy is every useful energy form, while in the calculation of the consumed energy from natural sources is not considered. The result is that energy sources with an EROEI less than 1 are not sustainable.'
[6]	'The EROEI of a particular source is the ratio of the total amount of energy produced by this source to the amount of energy required as an input to produce it.'
[7]	'Carbon intensities of fuels are related to their EROEI being the ratio of MJ energy output to MJ energy input for generating the output . . . . ' 'For enhanced recovery additional processes are required, decreasing the EROEI and increasing upstream carbon emissions.'
[9]	'EROEI is defined as the ratio of gross energy produced by an energy supply process to the total (direct plus indirect) energy cost of its production . . . . ' 'Thus the EROEI approach provides the net energy analysis of an energy-based production process and can lead to completely different results from a pure financial assessment'

Table 1: Selected literature on EROEI.

***Corresponding author:** Jones JC, School of Engineering, University of Aberdeen, UK, Tel: +44 (0)1224 272793

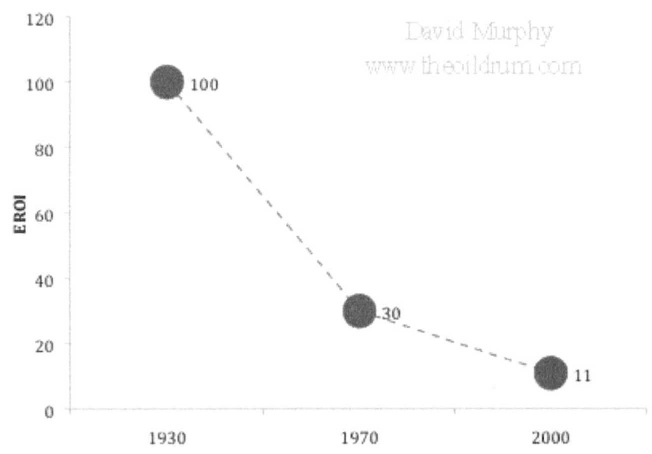

Figure 1: EROEI for US oil over the period 1930-2000. Reproduced from [1].

table raises another point of possibly questionable validity. The energy requirement for drilling in exploration is quite negligible in comparison with that in subsequent production [3], so initial inaccessibility of the oil is probably not a factor in the EROEI.

Publication [4] touches on some important points but in a way which is incomprehensible if not defective. In dismissing energy from 'natural sources' the authors are saying in effect that EROEI is not an energy balance, therefore energy having no potential use other than that to which it is being applied need not be considered. This is true, and the most obvious example is the wind turbine. In this the kinetic energy of the wind is converted to mechanical at the turbine and in analysis by the steady flow equation the kinetic energy of the wind must feature. It is however inconceivable that this energy would have found another application in engineering practice therefore it is not being 'invested' in the wind turbine. The argument in [4] that anything with an EROEI of less than unity is non-viable is also suspect. Biodiesels have low EROEI values, possibly below unity, but this is offset by the fact that being carbon-neutral they generate carbon credits [5]. The statement from reference [6] quoted in the table indicates that EROEI is being treated there as an energy balance.

Reference [7] in its title propagates the error that lower values of the EROEI and oil depletion are closely correlated. In the first of the sentences from [7] quoted the erroneous idea that EROEI is an energy balance is propagated. As for the link between EROEI and Enhanced Oil Recovery (EOR) the following points can be added. Whilst EOR by steam injection does reduce the EROEI [8], other means of EOR including reinjection of associated gas to increase the reservoir pressure or by burning some of the oil *in situ* hardly affect it at all. Where carbon dioxide is sequestrated at a producing oilfield as a means of EOR, as at the Sleipner field in the North Sea, the EROEI cannot be considered separately from the benefits of the sequestration. So the generalised statement about the effect of EOR on EROEI in [7] cannot be validated. The first of the statements quoted from [9] also suggests treatment of EROEI as an energy balance, but the second statement quoted leads to a highly important conclusion: EROEI is a ratio of energies to which finances are irrelevant. It is has been widely known for decades that unit amount of heat from crude oil costs about five times unit amount of heat from natural gas, and this alone would cause EROEI values to become erratic and quite meaningless were energy return, energy invested or both to have a monetary rather than energy basis. There has inevitably been confusion over this, with the result that inflated EROEI values for oil production have been reported simply because electricity for production has by concession been obtained at below-market prices [10]. Such EROEI values are not of course correct.

If the present author has emphasised negativity in his review of published work on EROEI it is to stress that three points central to a proper understanding of it are still not fully appreciated. These have been made in the preceding paragraph and will be repeated here.

- Lower values of the EROEI for oil cannot be identified with 'depletion'.
- An EROEI calculation is not an energy balance.
- A correctly formulated EROEI is a ratio of energies, not of energy prices.

Published work on EROEI which avoids these difficulties will be reviewed in what follows.

A Model Using Newtonian Mechanics and Field Data

From analysis of EROEI for crude oil performed at Aberdeen, for oil obtained from depth 'h' metres:

$EROEI = 5.5 \times 10^5/25h = 2 \times 10^4/h$ [11].

Where for an onshore well h = well depth and for an offshore well, h = (well depth + sea depth). The factor in the numerator of the expression in bold incorporates work done in raising the oil and was obtained from application of basic physics. The factor in the denominator was deduced by fitting the expression to EROEI-depth data pairs for particular fields. There are very few such data pairs in the public domain.

The factor of 25 arises from the energy requirements other than those of 'raising' the oil, such things as oil and gas separation, crude oil stabilisation, produced water removal (see below) and 'incidentals' such as vehicles travelling to and from an onshore field or vessels and helicopters travelling to and from an offshore one. An example of a field to which the approach developed at Aberdeen has been applied is the Qarn Alam field in Oman [8], an onshore field with well depths ≈ 1600 m, for which:

$EROEI = 2 \times 10^4/h$ gives:

EROEI = 13 to the nearest whole number.

There was consultation with the oil company who operate the field who had obtained a similar value for the EROEI from their own energy auditing. The operators of the field were unaware of the above approach until it was brought to their notice by the present author. A reader is encouraged to examine where this value would fall on Figure 1.

Use of a single numerical factor for energy additional to pumping is an approximation and, very importantly, its precise value could be reduced by improved practices. In [12] the question of whether it would be possible to express it as: f (infrastructure performance) is put. In [13] it is suggested that the value of the factor could be moved up or down from 25 according to the proportion of produced water and that field data from coal bed methane production, where produced water is often particularly abundant, could to this end be extended to oil fields.

A Means of Reducing the EROEI of Oil

We return to the expression:

$EROEI = 5.5 \times 10^5/25h = 2 \times 10^4/h$

and note that in the derivation [11] two efficiencies feature: that of the conversion of thermal energy to electrical at the scene of electricity generation and that of conversion of electrical energy to mechanical at the pump. A value of 0.35 is assigned to each of these in the treatment in [11]. In the hypothetical limit where each efficiency was unity the EROEI would be raised by a factor of $1/0.35^2 \approx 8$. It is shown in [14] and [15] that if the electricity was generated isothermally one of those efficiencies of 0.35 need not appear, raising the EROEI by a factor: $1/0.35 \approx 3$.

The most obvious example of isothermally generated electricity is a wind turbine. The point emphasized earlier that EROEI is not an energy balance is the basis of this analysis: if thermal generation involving a fuel is replaced by isothermal generation involving wind energy which occurs naturally without any previous human endeavour ('adventitious energy'), such energy does not feature in an EROEI calculation. A wind turbine does itself operate at an efficiency of about 0.35, but that is irrelevant to the calculation of EROEI.

There is therefore a conceptually simple way of increasing the EROEI from oil wells by a factor of three and it would involve interfacing oilfields with wind farms. There would be carbon credits as a significant bonus. Developments await acceptance of these ideas and estimates of the capital expenditure required.

Extension of the EROEI Concept to Refined Fractions

Once crude oil has been obtained at a particular EROEI, how is that affected by refining? Equivalently, how does the EROEI of a gasoline, kerosene or diesel fraction differ from that of the parent crude? This point is addressed in [16]. It was stated earlier in this paper that the monetary value of unit amount of energy from crude oil is about five times that of unit amount of energy from natural gas, and this is closely reasoned in [16], using benchmark prices for the two. Oil is often accompanied by associated gas and because of its low monetary value is sometimes flared although this practice has been reduced greatly over the last decade because of the carbon dioxide emissions which flaring entails. The term adventitious energy is used in parenthesis in the previous paragraph to describe wind energy. Natural gas otherwise destined for flaring is equally describable as being 'adventitious' even though, of course, its use is highly non-isothermal. On this basis on continuing to emphasise that EROEI is not an energy balance, it is reasoned in [16] that the EROEI of a petroleum fraction is not significantly different from that of the parent crude.

As wind farms are proliferating in response to Kyoto targets, countries are compiling 'wind atlases' which express wind speeds and patterns thereof in different regions of the respective countries. If on this basis wind farm operators were charged a government levy for installation of turbines at sites most favourable according to the information in the atlases that would not at first consideration seem at all unreasonable, and the kinetic energy of the wind would become a resource incurring a charge. On the other hand, natural gas is in some situations seen as being close to worthless (why else would it be flared?), and we could start to see wind energy attracting a higher price than natural gas!

Fuels Other Than Conventional Oil and Natural Gas

That biodiesels have low EROEIs has been noted. About a decade ago there was major interest in natural gas hydrates, which comprise methane trapped in ice in what is termed in structural chemistry a clathrate structure. They occur in vast quantities on continental shelves and on the sea floor and the amount of methane which could be obtained from them far exceed to present known reserves internationally of conventional gas. The view of the present author is that activity into obtaining natural gas from hydrates declined because of the discoveries in recent years of large amounts of tight gas, and also because of the increased production of coal bed methane. The matter of the EROEI of gas from hydrates has been looked into [17-19]. Liberation of methane from the ice structure is of course possible by melting the ice by heat exchange to it of low quality heat, that is, heat in a fluid itself at ordinary temperatures. The most obvious such fluid for the purpose is sea water. Again otherwise useless energy is applied. It is suggested in [20] that the soundest way of viewing this is not that the EROEI of natural gas hydrates is infinity but that the concept of EROEI does not apply to the process of releasing gas from the clathrate structure by melting the ice with sea water where everything is at low temperature.

As for oil from shale, it is widely held that this has a low EROEI partly because of the need to retort the shale to get syncrude. There are huge amounts of shale in places including the Rocky Mountains states of the US and recent development work there has been into producing the oil *in situ*, that is, without excavating the rock. Even if a reasonable EROEI, say in the region 5-10, is realised the viability of the operation is not assured as it will depend on prices of conventional crude oil with which it has to compete. It has happened twice in history – once about the time of WW1 and once about 30 years ago – that incipient shale oil production has crashed leaving investors out of pocket.

Concluding Remarks

This paper has brought together some of the current thinking on EROEI and has added one or two original thoughts including the possibility that energy in isothermal processes might, in the energy milieu of the second decade of the 21st Century, be charged for in the same way that energy from fuels is. The dual responsibility of supplying the world with energy and of doing so in a way consistent with Kyoto targets is a formidable one in which the EROEI of the various means of producing energy will be a key factor even though, as noted above, it has only started to enter the research literature over the last few years.

References

1. Bardi U, Lavacchi A, Yaxley L (2011) Modelling EROEI and net energy in the exploitation of non renewable resources. Ecol Model 223: 54-58.
2. van den Bergh JCJM (2013) Environmental and climate innovation: limitations, policies and prices. Technol Forecast Soc 80: 11-23.
3. Jones JC (2009) Effects of exploratory drilling on the EROEI for crude oil. Fuel 88: 2321.
4. Desideri U, Proietti S, Zepparelli F, Sdringola P, Bini S (2012) Life Cycle Assessment of a ground-mounted 1778 kWp photovoltaic plantand comparison with traditional energy production systems. Appl Energ 97: 930-943.
5. Jones JC (2010) Comments on recently reported flame temperature measurements. Fuel 89: 3607.
6. García-Olivares A, Ballabrera-Poy J, García-Ladona F, Turiol A (2012) A global renewable mix with proven technologies and common materials. Energ Policy 41: 561-574.
7. Verbruggen A, Al Marchohi M (2010) Views on peak oil and its relation to climate change policy. Energ Policy 38: 5572-5581.
8. Jones JC (2009) EOR by steam injection at the Qarn Alam Field. Oilfield Technology.
9. Roma A, Pirino D (2009) The extraction of natural resources: The role of thermodynamic efficiency. Ecol Econ 68: 2594-2606.
10. Jones JC (2007) Energy in, value out. Oil and Gas Middle East. 3: 46-47.
11. Jones JC, Cardno S, Service J, Udensi I (2012) Calculations and hypotheses concerning the energy return on energy invested, with reference to hydrocarbon fuels. International Journal of Mechanical Engineering Education 36: 176-181.

12. Jones JC (2013) Letter to the Editor: Towards refinement of the formulation for the EROEI of crude oil. Fuel 112: 591.

13. Jones JC (2013) Addendum to 'Towards refinement of the formulation for the EROEI of crude oil'. Fuel 112: 592.

14. Jones JC (2010) Wind turbines and the EROEI of crude oil. Fuel 89: 260.

15. Jones JC (2010) Addendum to 'Wind turbines and the EROEI of crude oil'. Fuel 89: 1745.

16. Jones JC (2013) On the possible extension of the EROEI concept to downstream operations. Fuel 112: 593.

17. Jones JC (2008) Addendum: Application of the phase rule to natural gas hydrates. International Journal of Mechanical Engineering Education 36: 182.

18. Jones JC (2009) Energy-return-on-energy-invested for natural gas hydrates. Fuel 88: 1531.

19. Jones JC (2009) Addendum to `Energy-return-on-energy-invested for natural gas hydrates. Fuel: 88: 1532.

20. Jones JC (2010) Energy Resources for the Past, Present and Future. The Open Thermodynamics Journal 4: 72-75.

Bacterial Degradation of the Saturate Fraction of Arabian Light Crude oil: Biosurfactant Production and the Effect of ZnO Nanoparticles

Wael Ismail[1]*, Noora A Alhamad[1], Wael S El-Sayed[2,4], Ashraf M El Nayal[1], Yin-Ru Chiang[3], Riyad Y Hamzah[1]

[1]*Biotechnology Program, College of Graduate Studies, Arabian Gulf University, Manama, Kingdom of Bahrain*
[2]*Taibah University, Faculty of Science, Biology Department, Al-Madinah Al-Monawarah, 344, Kingdom of Saudi Arabia*
[3]*Biodiversity Research Center, Academia Sinica, Taipei 115, Taiwan*
[4]*Microbiology Department, Faculty of Science, Ain Shams University, Cairo, Egypt*

Abstract

The aim of this research was to study the influence of nanoparticles on the biodegradation of crude oil. Production of biosurfactants was also assessed. Crude oil-utilizing bacteria were isolated from oil fields via enrichment in chemically defined medium with crude oil as a sole carbon source. The isolates could be affiliated to the genera *Bacillus*, *Pseudomonas*, *Achromobacter*, and *Microbacterium* by 16S rRNA gene sequencing and phylogenetic analysis. GC/FID analysis revealed 52 to 98% degradation of the oil saturate fraction within one month. Nanoparticles of ZnO inhibited growth and crude oil biodegradation by one isolate (NBHCO4) in a concentration-dependent manner. Growth of the cultures containing ZnO nanoparticles was relatively slower than that of the nanoparticles-free cultures. Moreover, the oil degradation extent in the nanoparticles-containing cultures (22-50%) was less than that achieved in cultures lacking ZnO nanoparticles (80%). Two strains, NBHCO2 and NCEOW, emulsified and utilized water-in-oil emulsions (chocolate mousse). Biosurfactant production in only one crude oil culture (I-19) could be confirmed by the observed reduction in surface tension. Some isolates produced biosurfactants from water-soluble substrates such as glucose. The NBHCO2 strain produced a lipopeptide biosurfactant which reduced the surface tension of the growth medium from 72 to 27 mN/m. A gene of catechol dioxygenase was detected in the I-19, NBHCO4, and NCEOW isolates indicating the potential catabolic capabilities of these strains for degradation of the aromatic oil fraction. In conclusion, metal oxide nanoparticles can interfere with crude oil biodegradation. Biosurfactants are not necessarily a prerequisite for crude oil biodegradation. The isolates can be applied for bioaugmentation of petroleum-polluted soil and biosurfactants production.

Keywords: Bioremediation; Biosurfactant; Emulsification; Nanoparticles; *Bacillus*

Introduction

Petroleum hydrocarbons are among the most common environmental contaminants [1,2]. Petroleum and its hazardous products can be released into the environment through various production, processing, and consumption operations. Consequently, environmental pollution with petroleum and its products has become an issue of global concern, particularly with the anticipated boost in the production and consumption of fossil fuels [3-5].

Pollution with petroleum hydrocarbons in the Arabian Gulf region has become a serious environmental issue. In deed the Arabian Gulf is one of the most heavily polluted aquatic environments worldwide. It was estimated that ca 60% of the oil transported via the sea is shipped in the Arabian Gulf [5,6]. Furthermore, the Arabian Gulf region has witnessed the largest oil spill to date in terms of the amount of crude oil discharged. During the 1991 Gulf war the release of about 11 million barrels of crude oil has lead to the pollution of all environmental components in the region [7].

In Kuwait, where the situation was the worst, about 190 km of border trenches were filled with 3.5 million barrels of crude oil. Moreover, 400 km^2 of land were polluted because of the fallout from 800 burning wells which released 50 million m^3 of crude oil [8]. The marine environment and ground water reservoirs were also heavily impacted. Another devastating consequence is the formation of more than 300 oil lakes that covered over 49 km^2 of land with more than 40 million tons of crude oil [9,10]. In addition to land and water pollution with petroleum hydrocarbons, the 1991 Gulf war contributed massively to air pollution [11].

Although more than two decades have elapsed since the catastrophic Gulf war oil spill occurred, the consequences are still apparent today [10]. Following the 1991 Gulf war, some studies have been conducted to evaluate bioremediation as a cleanup technology in the Arabian Gulf territory [9,12,13]. Nonetheless, the work achieved to date is not up to the scale of the massive pollution problem. Most of the research done has focused on the marine environment of the Arabian Gulf. In contrast, a few studies have addressed the crude oil-degrading microbial populations in petroleum-impacted soil.

It is well known that one of the factors which often impede the biodegradation of hydrophobic substrates like hydrocarbons and crude oil is their low bioavailability [14,15]. This is because these substrates have poor aqueous solubility and therefore, are not readily accessible for microbial attack. To circumvent this problem, microbes, which exist predominantly in the aqueous phase, have evolved different mechanisms to get access to and utilize hydrophobic compounds. Some

***Corresponding author:** Wael Ismail, Biotechnology Program, College of Graduate Studies, Arabian Gulf University, Manama, Kingdom of Bahrain, Tel: +97336146948; Fax: +97317239664

crude oil-degrading bacteria produce biosurfactants/bioemulsifiers that promote the aqueous solubility, and thus the bioavailability, of petroleum hydrocarbons by solubilization and emulsification [16,17]. Consequently, crude oil biodegradation can be enhanced by biosurfactants/bioemulsifiers production. However, this is not always the case because biosurfactants might also negatively impact the biodegradation process [18]. Therefore, further investigations are still needed to better understand biosurfactants production patterns and their role in crude oil biodegradation and action mechanisms. Noteworthy, studies on biosurfactants production by crude oil-degrading bacteria or bacteria inhabiting crude oil-polluted environments in the Arabian Gulf region are very scarce.

The presence of other environmental contaminants such as nanoparticles, can probably have an impact on crude oil biodegradation. Nanoparticles have gained tremendous interest due to their unique physicochemical properties compared to their bulk counterparts [19]. Potential impact of commercial nanoparticles on crude oil biodegradation has not been investigated so far. This is particularly demanding due to the increasing applications of engineered nanoparticles and nanotechnology which will inevitably lead to discharge of nanoparticles into the environment via various routes [19-21]. Accordingly, it is likely that nanoparticles may interfere somehow with the biodegradation of environmental pollutants since some of them are known to have antimicrobial activity [22].

This study was undertaken to isolate crude-oil degrading bacteria from petroleum-polluted soil collected from oil fields in Bahrain and Kuwait and to assess their biodegradation potential. Biosurfactants production by the isolated bacteria was also investigated. Furthermore, we tested the effect of commercially available ZnO nanoparticles on growth and oil biodegradability.

Materials and Methods

Chemicals, bacteria, soil, and oil samples

Chemicals and molecular biology reagents were obtained from Fluka (Switzerland), Sigma (USA) Promega and Qiagen (Germany), and the sterile filters (0.22 and 0.45 μm) were from Millipore (USA). Arab light crude oil was provided by Saudi Aramco (Kingdom of Saudi Arabia) and BAPCO (The Bahrain Petroleum Company). Nanoparticles (ZnO nanopowder < 100 nm) were purchased from Sigma-Aldrich (Germany). Petroleum-contaminated soil samples were collected from different locations in Bahrain (Awali oil field), and Kuwait (oil-impacted locations including those that date back to the 1991 Gulf War; Al Sabiriya field, and Al Moqawwa field). Moreover, crude oil samples mixed with water (production water) were collected from pipelines. All samples were collected in October 2010 in clean glass containers and kept at room temperature in the laboratory. These samples were used as a source of potential petroleum-degrading microorganisms. The I-19 strain is a *Pseudomonas* sp. isolated by a MSc student at the Arabian Gulf University [23]. The source of the I-19 strain is crude oil-contaminated soil collected from Al Sabiriya oil field in Kuwait. It was isolated from an enrichment culture in chemically defined medium with biphenyl as a sole carbon source. The I-19 strain could emulsify and grow on crude oil and it is included in the current study to assess its biodegradation potential. *Bacillus subtilis* DSM 3257 was purchased from Deutsche Sammlung von Mikroorganismen und Zellkulturen (DSMZ, Germany). Genomic DNA from *B. subtilis* was used as a positive control for detection of the biosurfactants-related genes as described below.

Culture media and growth conditions

Enrichment of crude oil-degrading-bacteria and the biodegradation experiments (shake-flask cultures) were performed in mineral salts medium containing crude oil as a sole carbon source [24]. Filter-sterilized (0.22 μm Stericup filters, Millipore) crude oil was added to sterile mineral salts medium (2 or 5%, v/v). Biosurfactants production was also tested in 50 ml mineral salts medium containing 10 mM glucose from filter-sterilized stock solution as a sole carbon source. Isolation and purification of the bacteria was done via spread and streak plate techniques on Luria-Bertani (LB) agar. Inocula for the biodegradation experiments were prepared in 50 ml LB-broth (in 100 ml Erlenmeyer flasks). All liquid cultures were routinely incubated at 30°C with shaking at 200 rpm. LB-agar plates were incubated at 30°C for 48 hours. Growth of bacteria in liquid cultures was monitored by measuring the optical density at 580 nm (OD_{580}) with a Perkin Elmer lambda EZ210 spectrophotometer (1 cm optical path).

Enrichment and isolation of bacteria

Soil samples were suspended in distilled water (20 g/100 ml) and the suspension was shaken for 30 minutes. The soil suspension was then kept at room temperature and allowed to settle and aliquots of the soil slurry (1 ml) were inoculated in 100 ml mineral salts medium (in 250 ml Erlenmeyer flasks) containing 2% crude oil as a sole carbon source. Non-sterilized crude oil was used as inoculum in mineral salts medium (10%) without the addition of soil slurry to check for the presence of bacteria in the oil. Production water was inoculated (2%) into mineral salts medium containing 2% filter-sterilized crude oil as a carbon source. Enrichment cultures were incubated for one week. Samples from the various enrichments that showed signs of turbidity, emulsification/dispersion and/or change in color were serially diluted in sterile saline (0.9% NaCl). Aliquots from the culture dilutions (100 μl) were spread over LB-agar plates and incubated for 48 hours. Morphologically distinct single colonies from the spread plates were streaked across LB-agar plates three times consecutively for purification.

Preparation of inocula for the biodegradation experiments

To screen the isolated bacteria for crude oil degradation capacity, the various isolates were grown in LB-broth for 16-18 hours. The OD_{580} of the cultures was then measured and the cells were harvested by centrifugation (3500 rpm, 20 minutes) at 4°C. The cell pellet was washed once with basal medium (phosphate buffer+NH_4Cl) [24] and the washed cells were resuspended in 5 ml of basal medium. These cell suspensions were used as inocula. All manipulations of the cells were done on ice. Cells were harvested from 2 ml of the cell suspension by centrifugation (14,000 rpm, 5 minutes) and heated at 105°C for 15 hours in a drying oven for measuring the dry weight.

Crude oil biodegradation assays

Routinely, 2 ml of each cell suspension (0.04-0.2 gram dry cell weight/l) were inoculated into mineral salts medium containing 2% crude oil as a sole carbon source and incubated for 4 weeks. Cultures were inspected visually at intervals for turbidity, color change, and oil emulsification/dispersion. Cultures prepared for the growth curve experiments contained 5% crude oil. In all experiments, crude oil biodegradation was assessed by GC-FID after *n*-hexane extraction from the various cultures as described later. Uninoculated controls were included to monitor sterility and to compensate for abiotic loss of oil. Change in biomass was monitored by measuring the optical density (OD_{580}) at hourly intervals during the initial two days of incubation then once weekly.

Biosurfactants production

To test selected isolates for biosurfactants production, the bacteria were grown in mineral salts medium with crude oil (2%) as described earlier. Biosurfactants production was also assessed in cultures growing on water soluble substrates such as glucose (10 mM). Uninoculated medium was included as a control. After different time intervals (24 hours and 24 days), samples from the crude oil cultures (25 ml from the aqueous phase) and the glucose cultures (25 ml, after 4 days) were withdrawn and the cells were removed by centrifugation for 20 minutes at 3500 rpm. The supernatants were then filtered under vacuum through 0.22 µm membrane filters to remove residual cells. The surface tension of cell-free culture supernatants was measured with a Kruss K100MK3 tensiometer (Kruss, Germany) equipped with a platinum plate at room temperature via the Wilhelmy plate method [25].

Effect of ZnO nanoparticles on growth and biodegradation capacity

This experiment was conducted to assess the impact of commercially available ZnO nanoparticls on growth and crude oil biodegradation. Nanoparticles of ZnO were chosen because they are widely applied [26]. The isolate NBHCO4 was cultured in mineral salts medium containing crude oil in the presence of ZnO nanoparticles. One set of cultures contained 1 µg/ml and another set contained 10 µg/ml of ZnO nanoparticles. Control cultures without nanoparticles and uninoculated flasks were also included. All cultures were incubated for 28 days at 30°C. Growth was monitored by measuring the OD_{580} after time intervals and oil biodegradation was assessed by GC-FID.

Gas Chromatography (GC)

Gas chromatography was adopted to monitor the degradation of saturated hydrocarbons in the *n*-hexane-soluble fraction. The crude oil from all cultures was extracted once in *n*-hexane. The culture contents were mixed with 15 ml of *n*-hexane by vigorous shaking in a separating funnel for 5 minutes. After the two phases were resolved, the aqueous phase was allowed to drain and the organic phase was collected in clean glass flasks. A portion (1 ml) of the organic phase was then centrifuged at 13,000 rpm for 10 minutes. The supernatant was transferred to a clean Eppendorf tube and diluted 50x in *n*-hexane. Samples (1 µl) from all dilutions were injected into an Agilent GC7890A-Markes Headspace System equipped with a FID (flam ionization detector). The column was a HP-5ms (5%-phenylmethylpolysiloxane, 30 m x 0.25 mm id, 0.25 µm film thickness) and the carrier gas helium was used at flow rate of 0.8 ml/min. The injector temperature was 250°C and the oven temperature was held at 70°C for 2 min then increased to 300°C at 20°C /min, thereafter held at 300°C for 8.5 min. The peak areas for some components of the *n*-hexane soluble fractions in the GC/FID chromatograms of the different experiments were compared. The peak area represents the % of the hydrocarbon constituent in the total amount of the hexane soluble fraction (relative abundance). To assess the extent of oil (hexane soluble fraction) biodegradation, the total area of all the peaks in each GC/FID chromatogram was taken as an indicator.

Polymerase Chain Reaction (PCR)

Polymerase chain reaction was performed to screen selected isolates for the presence of genes involved in the degradation of aromatic compounds and biosurfactants production. Genomic DNA was isolated from overnight LB-cultures of the different bacteria using the Wizard Genomic DNA Purification kit (Promga, USA) and was used as a template. The primers used are listed in Table 1. The PCR was run on a Biorad DYAD thermocycler (Biorad, USA) and the PCR mixture (20 µl) contained 0.2 µM from each of the forward and reverse primers, 7-200 ng of genomic DNA, and 2 µl of 20x Taq PCR master mix (Qiagen, Germany). The PCR conditions for each pair of primers were as mentioned in the respective reference.

Amplification, sequencing of 16S rRNA gene and phylogenetic analysis

A loopful of overnight grown cultures was transferred to 50 µl TE buffer (100 mM Tris-HCl–10 mM EDTA, pH 8.0) and boiled for 5 min. Then resulting cell extract was used as a template for PCR using Platinum Supermix (Invitrogen, USA). The 16S rRNA genes were amplified with eubacterial universal primers (Table 1). PCR mixture consisted of 25 µl Platinum Supermix, 1 µl of each primer (from 10 µM stock), 1 µl of cell extract, and was adjusted to 50 µl with distilled water. PCR was performed in Thermal Cycler (Applied Biosystem 2720, USA) and the PCR conditions were adjusted to 5 min for initial denaturation at 94°C followed by 30 cycles of 1 min at 94°C, 1 min at 54°C, and 1 min at 72°C, and finally 10 min at 72°C. The amplicons were subjected to electrophoresis using 1% agarose gel containing ethidium bromide (0.05%) with size marker (DNA ladder, Promega, USA).

Nucleotide sequences of amplified 16S rRNA were determined by automated fluorescent dye terminator sequencing method originally developed by Sanger et al. [32] using DYEynamic ET Terminator Cycle Sequencing Kit (Amersham Pharmacia Biotech) with a model ABI 310 genetic sequence analyzer (Applied Biosystems, CA, USA) according to user manual. Obtained sequences were analyzed by Genetyx-Win MFC application software version 4.0. The reference 16S rRNA gene sequences were retrieved from the GenBank database (National Center for Biotechnology Information NCBI, National Library of Medicine, USA) [33]. Sequences were compared with their closest matches in GenBank with nucleotide-nucleotide BLAST to obtain the nearest phylogenetic neighbors (www.ncbi.nlm.nih.gov/BLAST). Multisequence alignments were performed by Clustal W1.83 XP software and phylogenetic trees were constructed with MEGA (The Biodesign Institute) [34] using evolutionary distance and neighbor-joining method [35].

Primer name	Primer sequence 5'-3'	Amplified gene or DNA fragment	Expected product size (bp)	Reference
23cat-f 23cat-r	CGACCTGATCTCCATGACCCA TCAGGTCAGCACGGTCA	2,3 catechol dioxygenase	238	[27]
sfp0-F sfp0-R srfA-F srfA-R	CTAGAATTCAGATTTACCGAATTTATATG GGGGAATTCAGGGTGTGCGGCGCATAC TCCGTTTTTCCTTGTTCACC TCTTTCTGCCACTGCATCAC	the genes of lipopeptide biosurfactants from *Bacillus subtilis*	642 707	[28] [28]
Kpd1 Kpd2	GCCCACGACCAGTTCGAC CATCCCCCTCCCTATGAC	the *rhlB* gene for rhamnolipid biosynthesis	226	[29]
27F 1500R	AGAGTTTGATCCTGGCTCAG ACGGCTACCTTGTTACGACT	The 16S rRNA genes	1500	[30] [31]

Table 1: Primers used in this study.

Sequence accession numbers

The 16S rRNA gene sequences obtained in this study have been deposited in GenBank under accession numbers AB824289-AB824294 and AB836753 to AB836756

Results

Isolation and identification of bacteria

Various enrichment cultures revealed different patterns of color change, turbidity and dispersion of crude oil within one week of incubation. These changes indicated the ability of the organisms in the enrichments to utilize petroleum hydrocarbons as a carbon and energy source. Strong emulsification of the crude oil was also observed in the enrichment which contained only crude oil as an inoculum and carbon source. In total, 10 different bacterial strains could be isolated from the spread plates. Based on the results of the phylogenetic analysis, the isolated bacteria could be affiliated to the genera *Bacillus, Pseudomonas, Achromobacter,* and *Microbacterium* (Figure 1, Table 2). The majority of the isolates belong to the *Pseudomonas* and *Bacillus* genera. The phylogenetic analysis was based on full 16S rRNA gene sequence for the isolates NKCO2 and NKCO4 (isolated from Kuwaiti oil fields), NBHCO3 (isolated from Bahraini oil fields), and NCEOW (isolated from production water), whereas only partial sequence was considered for the remaining isolates (Table 2).

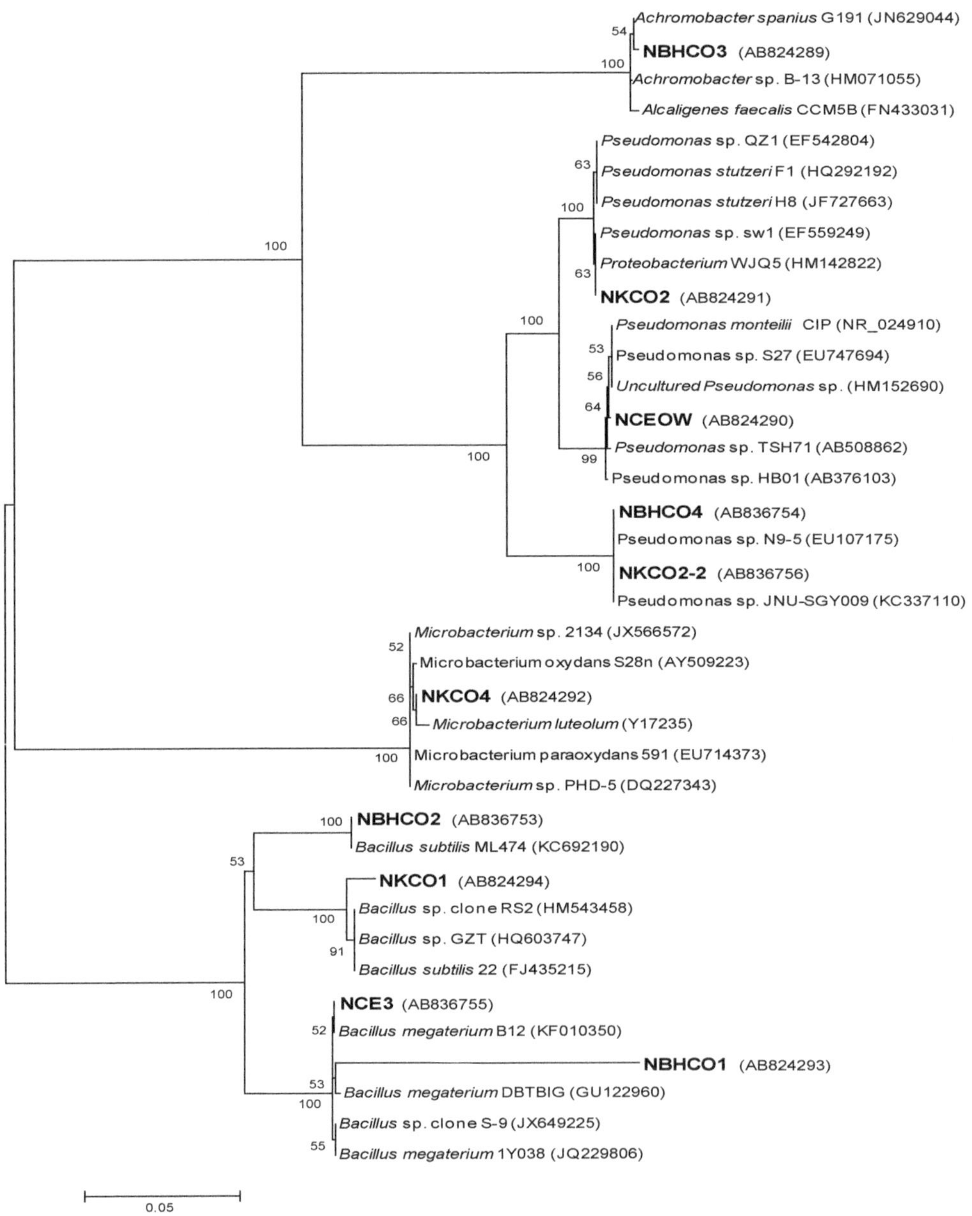

Figure 1: Consensus neighbour-joining phylogenetic tree based on 16S rRNA gene sequences showing the phylogenetic positions of isolated strains and related reference taxa. Numbers at nodes indicate bootstrap values of neighbor-joining analysis for 1000 replicates. The scale bar indicates substitutions per nucleotide. The GenBank accession numbers for the 16S rRNA gene sequences are given in parentheses after the strain name.

Strain	Source	Closest matches	Accession No.[a]	% of similarity	Accession No.[b]
NKCO2	Kuwait, Al Sabiriya Oil Field	*Pseudomonas* sp. sw1	EF559249	99	AB824291
NKCO4	Kuwait, Al Sabiriya Oil Field	*Microbacterium luteolum*	Y17235	99	AB824292
NBHCO1	Bahrain, Awali Oil Field	*Bacillus* sp. clone S-9	JX649225	98	AB824293
NKCO1	Kuwait, Al Sabiriya Oil Field	*Bacillus* sp. clone RS2	HM543458	99	AB824294
NBHCO3	Bahrain, Awali Oil Field	*Achromobacter spanius* G191	JN629044	99	AB824289
NCEOW	Bahrain, Production Water	*Pseudomonas* sp. TSH71	AB508862	99	AB824290
NBHCO2	Bahrain, Awali Oil Field	*Bacillus subtilis* ML474	KC692190	99	AB836753
NBHCO4	Bahrain, Awali Oil Field	*Pseudomonas* sp. N9-5	EU107175	99	AB836754
NCE3	Crude Oil	*Bacillus megaterium* B12	KF010350	99	AB836755
NKCO2-2	Kuwait, Al Sabiriya Oil Field	*Pseudomonas* sp. JNU-SGY009	KC337110	99	AB836756

[a]GeneBank sequence accession numbers of most closely related sequences
[b]GeneBank sequence accession numbers of isolated strain

Table 2: Bacteria isolated from oil fields.

Growth of the isolated bacteria on crude oil

Growth curves for the isolates NBHCO2 and NBHCO4 (isolated from Bahraini oil fields), NCE3 (isolated from crude oil), NCEOW (isolated from production water), NKCO2-2 and NKCO1 (isolated from Kuwaiti oil fields), during the initial 48 hours of incubation are shown in Figure 2. Fluctuations in the OD_{580} were observed in the NCE3 and NBHCO4 cultures. However, a net increase in turbidity was observed up to 26 hours of incubation. The OD_{580} then decreased and remained stable. The NBHCO2 culture was the only one which exhibited no fluctuations in the OD_{580}. After a 10-hour lag phase, the OD_{580} increased then became constant after 30 hours for 15 hours. The highest OD_{580} during the first two days was attained in the NBHCO4 and NCE3 cultures after 26 hours. After 28 days of incubation the OD_{580} of the NCE3 culture was 1.6 whereas that of the NBHCO4 was 2.4. Fluctuation in culture turbidity was also observed in the NKCO1 culture. An increase in the OD_{580} of the NKCO1 culture occurred up to 23 hours followed by sudden decrease which lasted for almost 8 hours. After 32 hours of incubation the OD_{580} increased again and remained almost constant. The culture turbidity of the NKCO2-2 strain increased steadily during the initial 48 hours. After 21 days of incubation the OD_{580} of the NKCO1 culture reached 5.5 whereas that of the NKCO2-2 was 1.25. The NCEOW culture turbidity decreased during the initial 6 hours then remained almost constant.

In some oil cultures, water-in-oil emulsions (the so called chocolate mousse) were produced within 24~72 hours of incubation. In those cultures no turbidity in the aqueous phase was observed even after prolonged incubation (4 weeks) except the NBHCO2 and NCEOW cultures. In both cultures, the chocolate mousse was dispersed and the aqueous phase became turbid after 3~4 weeks.

Biodegradation of crude oil by selected bacteria

As shown in Figure 3, the tested bacteria degraded saturated hydrocarbons in the *n*-hexane-soluble fraction of Arabian light crude oil within 28 days of incubation. This was confirmed by the detected decrease in the amounts of the saturate components as compared to uninoculated controls. To estimate the biodegradation extent achieved by the different isolates, the relative amount of the consumed saturate fraction was calculated in comparison with that remaining in the uninoculated controls. The NCE3 strain achieved the highest biodegradation extent (98%) followed by NBHCO4 (78%), then NKCO1 (75%) (Figure 4). Figure 5 displays the change in the relative amount of some saturated hydrocarbons after 28 days of incubation at 30°C. Both of the NKCO1 and NBHCO4 strains reduced the (%) of *n*-C12 (dodecane), *n*-C13 (tridecane), *n*-C18 (octadecane) and *n*-C19

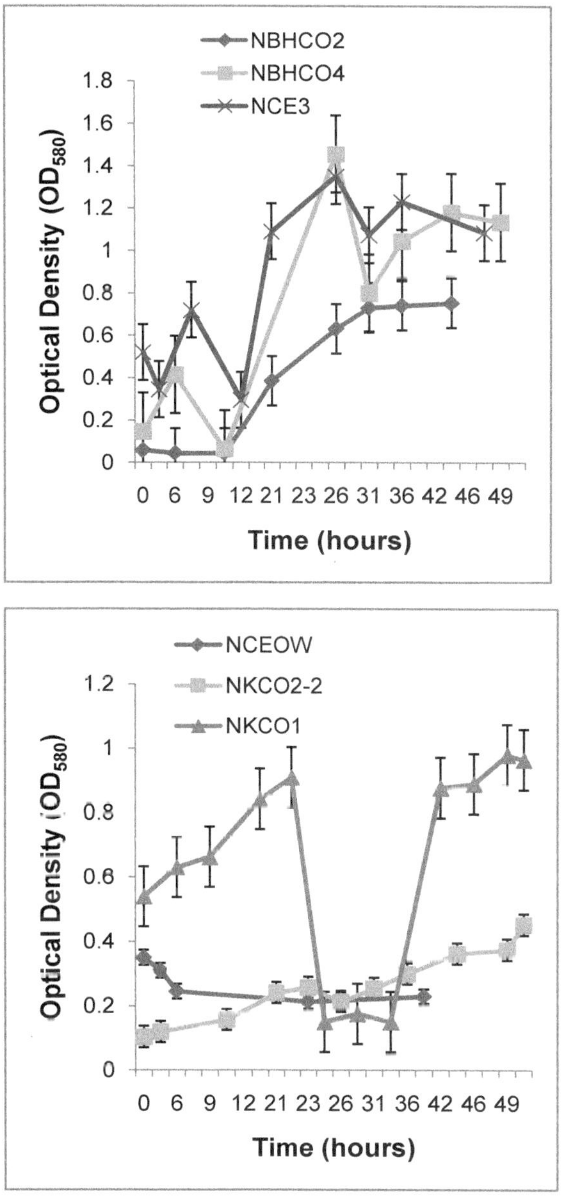

Figure 2: Growth of isolated bacteria in mineral salts medium containing crude oil (5% v/v) at 30°C during the initial 48 hours of incubation. Results are means of duplicate measurements from two independent experiments. Error bars represent standard error.

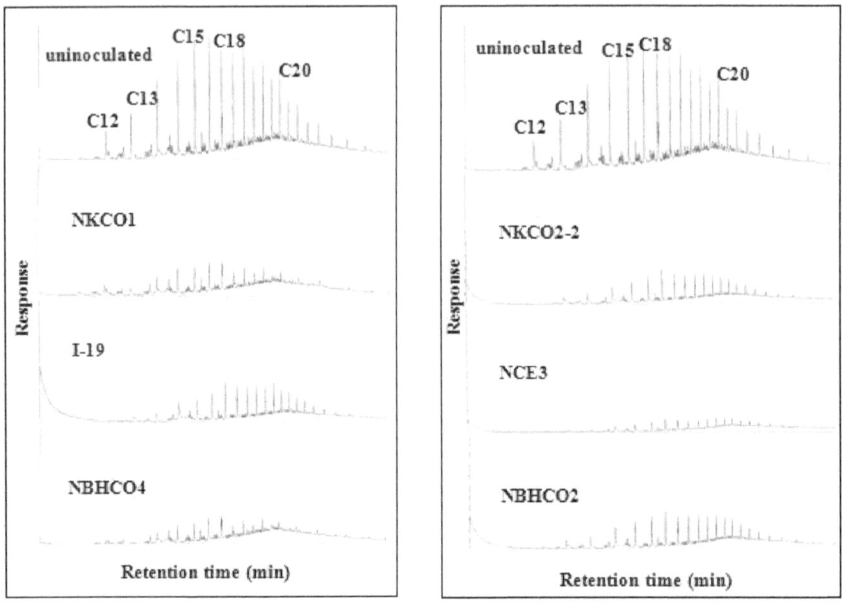

Figure 3: GC/FID chromatograms showing the biodegradation of the hexane-soluble fraction of Arabian light crude oil by some isolates at 30°C for 28 days. The tested bacterial isolates are NBHCO2 and NBHCO4 (isolated from Bahraini oil fields), NCE3 (isolated from crude oil), I-19 (isolated from Kuwaiti oil field), NKCO-2-2 and NKCO1 (isolated from Kuwaiti oil fields).

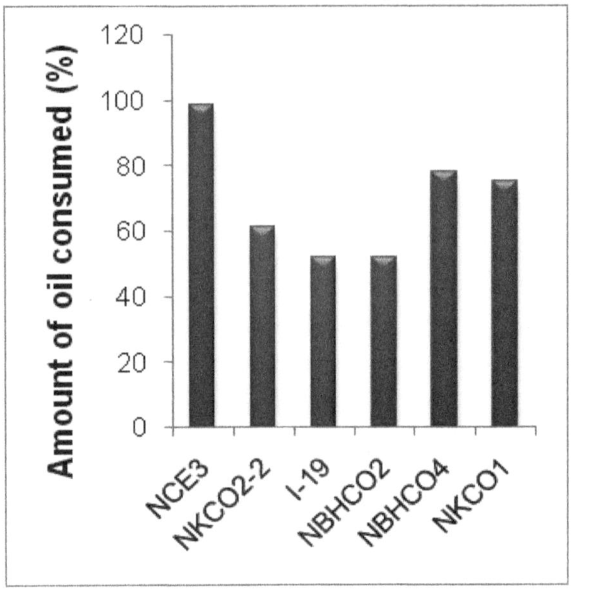

Figure 4: Amount of oil (%, hexane-soluble fraction) consumed after biodegradation of crude oil by the isolated bacteria at 30°C for 28 days.

(nonadecane) hydrocarbons. The other tested isolates (I-19, NCE3, NKCO2-2) drastically reduced the *n*-C12 hydrocarbons % as compared to the uninoculated control. However, little increase in the *n*-C12% was observed in the NBHCO2 culture. The NCE3, NBHCO2, I-19 and NKCO2-2 isolates also reduced hydrocarbons in the range *n*-C13 and *n*-C18. The largest % reduction was achieved by the NCE3 strain. The degradation preference changed with the *n*-C15 (pentadecane) and *n*-C20 (eicosane) hydrocarbons. A relative increase in the % of the *n*-C15 constituents was observed in the NBHCO2, I-19 and NKCO2-2 cultures. On the contrary, the NCE3 strain achieved almost 50% reduction in the relative proportion of *n*-C15 hydrocarbons. The % proportion of the *n*-C20 hydrocarbons did not reveal significant change in most cultures. Exceptionally, the NCE3 strain caused ca 35% increase in the *n*-C20 hydrocarbon relative amount.

Effect of ZnO nanoparticles on growth and crude oil biodegradation

The ZnO nanoparticles were added to the NBHCO4 culture (1 μg/ml and 10 μg/ml) to test their effect on bacterial growth and oil biodegradation. The NBHCO4 strain grew in crude oil cultures containing ZnO nanoparticles within 28 days at 30°C. The color of the cultures turned brownish and the oil got dispersed. No apparent differences in color or oil dispersion pattern were observed between cultures having and those lacking nanoparticles. However, the growth of the nanoparticles-containing cultures was slower than that of the cultures containing no nanoparticles. The maximum OD_{580} was higher in the culture containing nanoparticles at 1 μg/ml (Figure 6).

The GC/FID chromatograms (Figure 7) revealed degradation of the hexane-soluble oil fraction in the nanoparticles-containing cultures. Interestingly, the amount of oil consumed by NBHCO4 was much less in nanoparticles-containing cultures. Cultures containing 1 and 10 μg/ml of ZnO nanoparticles consumed 50 and 22% of the hexane-soluble fraction, respectively, after 28 days of incubation. The nanoparticles-free culture consumed 80% of the hexane soluble fraction.

Biosurfactants production by the isolated bacteria

Dispersion/emulsification of crude oil in the growth medium by most of the tested bacteria occurred within the initial 2-3 days of incubation. In some cultures such as I-19, NBHCO4, and NCE3, oil dispersion was observed even a few hours after inoculation. The strongest dispersion occurred in the I-19 and NBHCO4 cultures. The cultures revealed different sizes of the oil droplets. These data indicate the presence of some sort of tensioactive agents (biosurfactants/bioemulsifiers) in the crude oil cultures. There was no significant

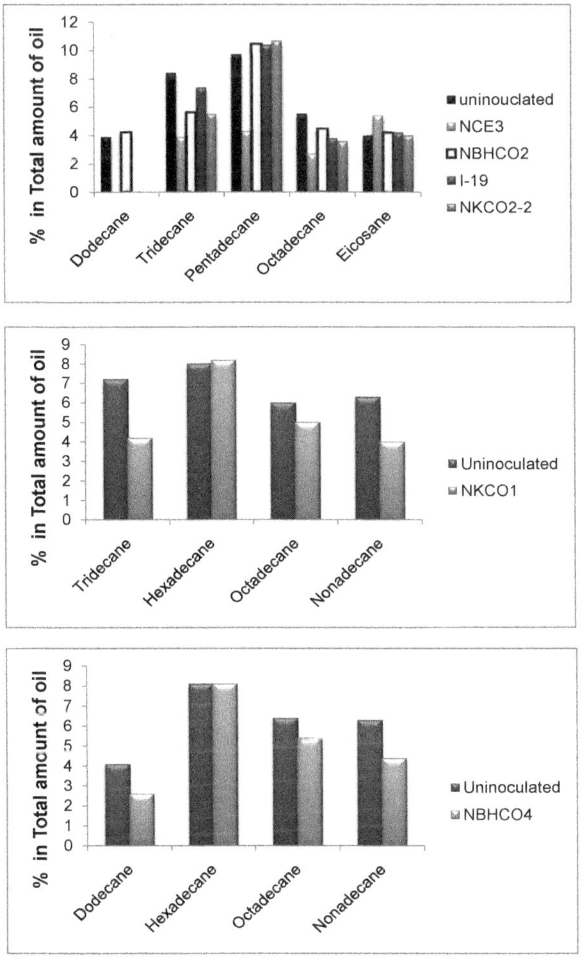

Figure 5: Changes in the relative proportions of hydrocarbon components in the hexane-soluble fraction of Arabian light crude oil after biodegradation by the isolated bacteria for 28 days at 30°C.

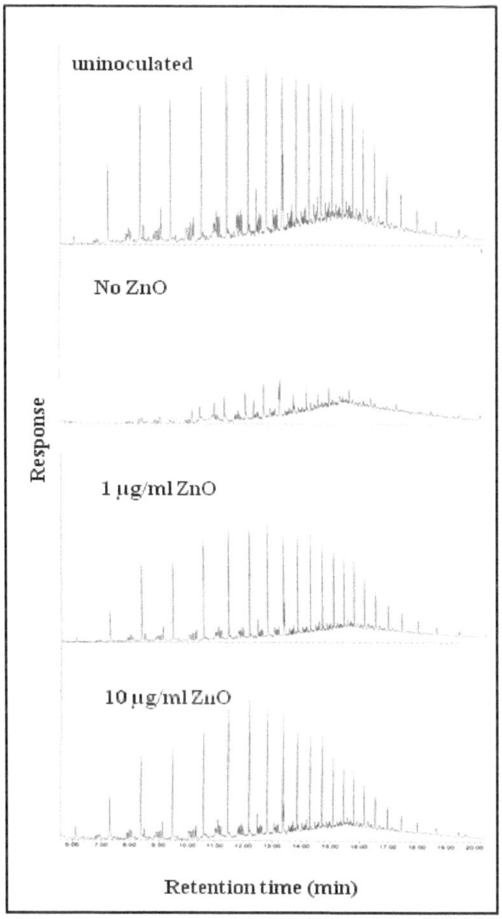

Figure 7: GC/FID chromatograms showing the influence of ZnO nanoparticles on the biodegradation of the hexane-soluble fraction of Arabian light crude oil by the NBHCO4 isolate at 30°C for 28 days.

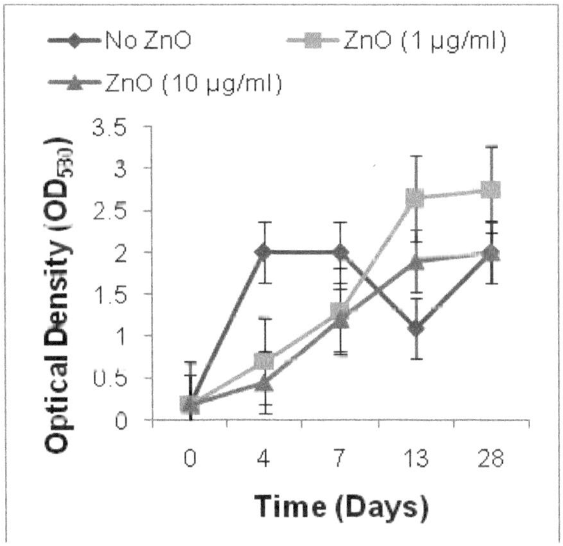

Figure 6: Growth of the NBHCO4 strain in mineral salts medium containing crude oil (5% v/v) at 30°C in the presence or absence of ZnO nanoparticles.

difference in surface tension between most of the crude oil cultures and the uninoculated medium (Figure 8). After 24 hours of incubation, the lowest surface tension was measured in the NKCO1 culture (54 mN/m). However, after 24 days, the surface tension in the I-19 culture decreased from 59.5 (after 24 hours) to 52.7 to record the lowest surface tension among the tested cultures. After two weeks of incubation, most of the oil droplets in the I-19 and NCEOW cultures coalesced. The tested bacteria produced biosurfactants using glucose as a sole carbon source. This was confirmed by measurement of surface tension (Figure 8). The differences in surface tension between the cultures and the uninoculated medium were much larger than those observed in the crude oil cultures. The lowest surface tension was measured in the NBHCO2 glucose culture (27 mN/m) after 4 days of incubation.

Detection of genes involved in aromatic hydrocarbons biodegradation and biosurfactants production

Gene of 2,3-catechol dioxygenase was detected only in NBHCO4, I-19, NCEOW. Genes of lipopeptide biosurfactants biosynthesis in *Bacillus subtilis* were detected only in the NBHCO2 strain (Figure 9). A rhamnosyl transferase-encoding gene involved in biosynthesis of rhamnolipids biosurfactants in *Pseudomonas aeruginosa* could not be detected in any of the tested isolates.

Discussion

As shown by phylogenetic analysis, the majority of the isolated bacteria belong to *Bacillus* and *Pseudomonas* genera. Literature search revealed that various species of *Pseudomonas* and *Bacillus* are common inhabitants of petroleum-polluted ecosystems. Furthermore, these organisms are well known for their capacity to degrade a range of petroleum hydrocarbons [36-38]. In contrast *Achromobacter* and *Microbacterium* spp. have been much less frequently isolated from crude oil-polluted niches [39,40]. This also agrees with the results of the phylogenetic analysis reported in this study. In the Arabian Gulf region, hydrocarbon-utilizing *Microbacterium* spp. were isolated from legume leaves [41] and along the Arabian Gulf coast [42,43]. Recently, Tanase et al. [44] isolated hydrocarbon-degrading *Achromobacter xylosoxidans* from oil-polluted soil. Also in line with our results, Minf et al. [16] isolated hydrocarbon-degrading *Achromobacter* sp. from Tunisian oil fields. To our knowledge, crude oil-degrading and emulsifying *Achromobacter* and *Microbacterium* species have not been reported before as isolates from oil-polluted soil in Kuwait.

Changes in culture turbidity, color, and oil dispersion/emulsification indicate the ability of the tested isolates to utilize petroleum hydrocarbons for growth and energy production. The fluctuation of the OD_{580} in most of the cultures during the initial two days could be attributed to attachment and detachment of the bacterial cells to and from the oil droplets. Bacterial adhesion to hydrocarbons is one strategy by which hydrocarbonoclastic bacteria overcome the extremely low aqueous solubility of hydrocarbons [45]. Similar results were reported by Bredholt et al. [46]. Moreover, Obuekwe et al. [47] investigated hydrocarbon degradation as a function of cell surface hydrophobicity and found that higher degradation extent is correlated with higher cell-surface hydrophobicity. Another plausible reason for the observed fluctuations in culture density could be the structural complexity and heterogeneous nature of the crude oil. Different types of interactions between the cells and the myriad of compounds in crude oil make the biodegradation process very dynamic. Petroleum hydrocarbons differ in their toxicity, accessibility, and recalcitrance. The turbidity profile of the NKCO1 culture mimics a diauxic growth curve. Transient decline in the turbidity of the NKCO1which remained for several hours could be due to growth inhibition by toxic oil components or oxygen deficiency. The recovery from this temporary stationary phase resulted in increase in culture turbidity again. However, other factors like the attachment to /detachment from the oil droplets may be involved.

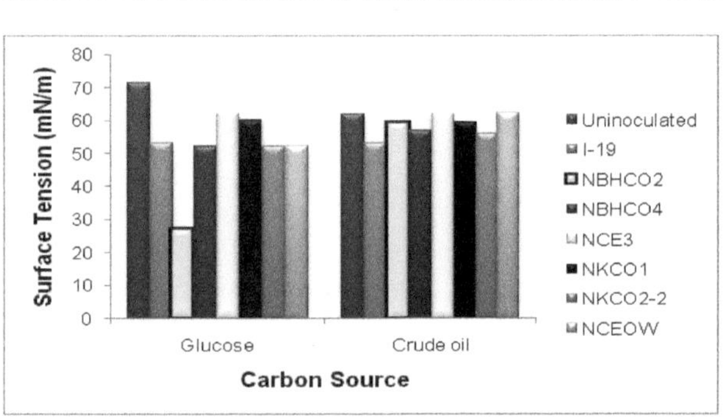

Figure 8: Change in surface tension of cell-free culture supernatants. Bacteria were grown at 30°C in mineral salts medium containing either glucose (10 mM) or crude oil (2%, v/v).

The isolated bacteria caused substantial degradation of the hexane-soluble (saturate) fraction of Arabian light crude oil within one month. Accordingly, these bacteria possess the catabolic machinery that enables them to utilize various aliphatic hydrocarbons of crude oil. This is common among bacteria isolated from crude oil-impacted ecosystems [44,48]. However the tested bacterial isolates achieved different degradation extents. Crude oil-utilizing microorganisms differ in their degradation capacities [47,49]. Although we do not have a direct evidence for the degradation of the aromatic fraction, the presence of a gene of catechol dioxygenase in some isolates suggests the ability of these bacteria to utilize aromatic substrates. Differences in biodegradation extent may be attributed to various factors such as cell surface hydrophobicity, enzyme activity and specificity, genes arrangement and regulation of gene expression, as well as biosurfactants/bioemulsifiers production. The highest degradation extent was recorded for NCE3 and NBHCO4. These two strains are among the isolates that caused strong dispersion of crude oil in the culture medium. This could be a reason for their higher biodegradation capacities [17].

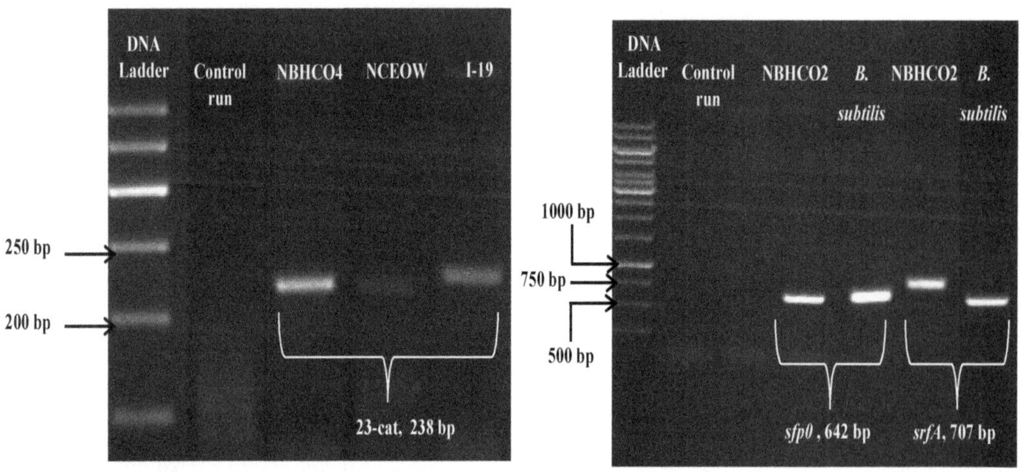

Figure 9: Images of agarose gels showing results of PCR for the detection of genes related to lipopeptide biosurfactants production (*sfp0* and *srfA*) and catechol dioxygenase in some isolates. Genomic DNA isolated from different bacteria was used as a template. Genomic DNA from B. subtilis was used a positive control for the *sfp0* and *srfA* genes. Control runs are PCR runs without template DNA.

It is worth mentioning that stronger oil dispersion does not always correlate with enhanced biodegradation efficiency [50]. The stability (decay constant) of the crude oil emulsion is another important factor. In line with this, NKCO1, which caused slight dispersion of the oil, achieved degradation extent comparable to that achieved by the stronger emulsifier NBHCO4. Interestingly, although the NCE3 isolate achieved degradation extent higher than that achieved by the NKCO1 strain, the NKCO1 culture reached OD_{580} more than 3 fold (5.5) higher than that of the NCE3 culture (1.6) after 3 weeks of incubation. It can be proposed that the NKCO1 strain utilized hydrocarbons from the other fractions such as aromatics, asphaltenes and resins and attained higher OD_{580}. Another explanation for this observation is that NKCO1 can achieve complete degradation of the hydrocarbons whereas NCE3 can only affect partial or incomplete degradation. This also might explain why the NKCO1 culture attained much higher biomass yield than that of the NBHCO4 culture although both achieved similar degradation extents for the saturate fraction. However, this needs further investigations.

The changes observed in the relative proportions of the various aliphatic hydrocarbons indicated that the tested bacteria did not utilize them equally. Preferential degradation of hydrocarbons by various bacteria has been reported by many authors [51,52]. The differential degradation of hydrocarbons could be due to toxicity and recalcitrance effects. Petroleum hydrocarbons are usually degraded in the following order *n*-alkanes > cycloalkanes and branched alkanes > aromatic hydrocarbons > resins and asphaltenes [51,53]. The differential utilization of the aliphatic hydrocarbons in the *n*-hexane-soluble fractions may reflect the different degradation extents.

The isolated bacteria appear not to be capable of utilizing the *n*-C16 and *n*-C20 hydrocarbons because no significant changes could be observed in the relative amounts of these compounds. The observed increase in the *n*-C20 hydrocarbons in the NCE3 culture can be attributed to recalcitrance or biotransformation of long-chain alkanes to *n*-C20. The same might apply for the observed increase in the *n*-C15 alkanes in the NBHCO2, I-19, and NKCO2-2 cultures. The reduction in the relative amounts of the *n*-C12, *n*-C13, and *n*-C18 hydrocarbons indicates utilization of the compounds by the tested bacterial isolates. The results reported by some authors are in a good agreement with the results reported here. Recently, Wang et al. [48] studied the degradation of crude oil by a novel *Dietzia* strain isolated from a deep oil reservoir. After 5 days of cultivation on crude oil, they observed a decrease in hydrocarbons with the chain length ≤ *n*-C25 and increase in the relative abundance of the ≥ *n*-C28 hydrocarbons. After 22 days of incubations, the ratios of ≤ *n*-C22 and ≥ *n*-C28 hydrocarbons decreased and there was an accumulation of hydrocarbons ranging from *n*-C23 to *n*-C27.

Palittapongarnpim et al. [54] reported that degradation of crude oil by some yeasts revealed decrease in the amount of long chain alkanes and accumulation of short chain fractions around *n*-C10. They explained these results by assuming that the long chain alkanes are degraded and transformed to short chain ones. Biosurfactants/bioemulsifiers interact variably with different hydrocarbons [48,55]. This differential behavior should preferentially enhance the aqueous solubility of certain hydrocarbons and consequently, promote their degradation over the less bio-available compounds.

Nanoparticles are gaining increasing interest due to their unique physicochemical properties and potential applications in the environmental and biological fields [19]. Very few studies have addressed the impact of nanoparticles on bioremediation and biodegradation of environmental pollutants. In this study, ZnO nanoparticles were chosen to test their effect on the growth of NBHCO4 and crude oil biodegradation. The results showed a negative impact of ZnO nanoparticles on the growth and crude oil degradation extent. The slower growth and lower degradation percentages of the ZnO–containing culture is probably due to some sort of inhibitory effect of ZnO nanoparticles. The observation that the effect of ZnO nanoparticles on the NBHCO4 culture was concentration-dependent further corroborates the inhibitory action. Some types of nanoparticles including ZnO, exhibit antimicrobial activity against several bacteria [22]. Recently, Tilston et al. [56] reported that nanoscale zerovalent iron inhibited microbial degradation of chloroaromatic compounds in soil. In contrast Fe/Ni nanoparticles promoted the biodegradation of phenol [57]. Obviously, the effect of nanoparticles on the biological degradation of environmental contaminants awaits further investigations.

Dispersion/emulsification of crude oil in the culture medium suggests the presence of tensioactive agents. Production of biosurfactants is known amongst crude oil and hydrocarbon-degrading bacteria [14,16]. This is another way by which hydrocarbonoclastic microbes circumvent the hydrophobicity and low aqueous solubility of the hydrocarbon substrates. In contrast to the proposed bacterial adherence to the oil droplets which was inferred from the observed fluctuations in culture density, the production of biosurfactants was confirmed by the observed reduction in surface tension in the I-19 crude oil culture. For the other crude oil cultures, the extent of surface tension reduction, as compared to the uninoculated control, was so small to conclude biosurfactants production. Probably, crude oil inhibited biosurfactants production or the tensioactive agent may be a bioemulsifier-type biosurfactant [58]. Alternatively, the bacterial cells themselves can act as biosurfactants/bioemulsifiers [59] or the produced biosurfactants are not excreted to the extracellular medium. This could be corroborated by the fact that the surface tension measurements were performed in cell-free culture supernatants. Moreover, dispersion of crude oil in most of the cultures was observed a few hours after the incubation. It is also worth mentioning that the presence of oils in the growth medium usually interferes with surface tension measurements [25]. Finally, the role of biosurfactants in biodegradation of hydrophobic compounds is still controversial [14].

The large reduction in surface tension observed in most of the glucose cultures may be due to biosurfactants that are qualitatively and/or quantitatively different from those produced in the crude oil cultures. Nayak et al. [60] reported that an efficient biosurfactant reduces the surface tension of the culture medium by more than 20 mN/m. This agrees with the surface tension measurements performed in the NBHCO2, I-19, NBHCO4, NBHCO2, NKCO2-2, and NCEOW cultures. The disability of these bacteria to reduce the surface tension in the crude oil cultures, similar to what they did in the glucose cultures, suggests that biosurfactants production might be prohibited. The carbon source is a significant determinant of the type, yield, and activity of the produced biosurfactants [61]. Glucose is widely used for biosurfactant production by many bacteria [61,62]. The biosurfactant produced by the NBHCO2 strain is a lipopeptide as confirmed by the detection of the relevant genes. Neither genes of lipopeptide nor those of rhamnolipid biosurfactants production could be detected in the other tested bacteria. This suggests that these bacteria may produce other types of biosurfactants/bioemulsifiers.

The absence of turbidity in most of the cultures containing water-in-oil emulsions (chocolate mousse) shows the inability of the tested bacteria to utilize these emulsions. Water-in-oil emulsions are known to be microbiologically recalcitrant [63]. This explains why most of the tested bacteria could not grow on them. Interestingly, two of the isolated bacteria, NBHCO2 and NCEOW, transformed chocolate mousse into oil-in-water emulsions after prolonged incubation. This was concomitant with the appearance of turbidity in cultures of these two isolates. It can be concluded that both of these isolates are able to emulsify chocolate mousse and utilize the resulting oil-in-water emulsions after prolonged incubation. Similar results were reported for *Pseudomonas* sp. and *Rhodococcus* sp. [46,64].

Conclusions

From the results it can be concluded that oilfields in Kuwait and Bahrain represent a good source of petroleum-degrading bacteria. The multiple phenotypes (crude oil degradation and biosurfactants production) of the isolated bacteria make them promising for bioaugmentation of petroleum-polluted soil in the Arabian Gulf region as well as biosurfactant production. We showed that commercially available ZnO nanoparticles can interfere with the biodegradation process. However, it is important to conduct in-depth investigations of the impact of nanoparticles on the growth and biodegradation capacity of environmental microorganisms. Biodegradation of crude oil by some bacteria may occur without production of biosurfactants. However, the role of biosurfactants, in the bioremediation/biodegradation processes needs further investigations.

Acknowledgements

The authors gratefully acknowledge the financial support by the Arabian Gulf University. We also thank Kuhnikrishnan Thottathan and Nada Obaid (General Directorate of Forensic Science Evidence-Public Prosecution-Kingdom of Bahrain) for technical support with the GC analysis.

References

1. Philp CJ, Bamforth SM, Singleton I, Atlas RM (2005) Environmental pollution and restoration: A role for bioremediation. Washington DC, ASM Press 1-48.
2. Margesin R, Moertelmaier C, Mair J (2013) Low-temperature biodegradation of petroleum hydrocarbons (n-alkanes, phenol, anthracene, pyrene) by four actinobacterial strains. Int Biodeter Biodegr 84: 185-191.
3. http://www.iea.org/publications/freepublications/publication/fs_GlobalEnergyTrends.pdf
4. Dehghani A, Sattarin M, Bridjanian H, Mohamadbeigy KH (2009) Investigation on effectiveness parameters in residue upgrading method. Petroleum and Coal 51: 229-23.
5. Mahmoud HM, Al-Hasan RH, Sorkhoh NA, Eliyas M, Radwan SS (2009) Attenuation of oil pollutants in the Arabian Gulf water by bacteria naturally associated with live fish. Int Biodeter Biodegr 63: 615-620.
6. Vazquez MA, Allen KW, Kattan YM (2000) Long-Term Effects of the 1991 Gulf War on the Hydrocarbon Levels in Clams at Selected Areas of the Saudi Arabian Gulf Coastline. Mar Pollut Bull 40: 440-448.
7. Michel J (2011) 1991 Gulf war oil spill. Fingas M (ed), Oil Spill Science and Technology. Elsevier, USA, p. 1127-1132.
8. Vandecasteele JP (2008) Petroleum Microbiology: Concepts, Environmental Implications, Industrial Applications. Editions Technip France.
9. Balba MT, Al-Daher R, Al-Awadhi N, Chino H, Tsuji H (1998) Bioremediation of oil-contaminated desert soil: The Kuwaiti experience. Environ Int 24: 163-173.
10. Alfailakawi A (2013) Monitoring changes and their causes in places with hydrocarbon pollution in the State of Kuwait using remote sensing and geographic information system techniques. Arabian Gulf University, Kingdom of Bahrain.
11. Ahmed MH, El-Raey M, Nasr SM, Frihy OE (1998) Socioeconomic impact of pollution on ecosystems of the Arabian Gulf. Environ Int 24: 229-237.
12. Al-Bader D, Eliyas M, Rayan R, Radwan S (2013) Subsurface associations of Acaryochloris-related picocyanobacteria with oil-utilizing bacteria in the Arabian Gulf water body: promising consortia in oil sediment bioremediation. Microb Ecol 65: 555-565.
13. Al-Mailem DM, Eliyas M, Radwan SS (2013) Oil-bioremediation potential of two hydrocarbonoclastic, diazotrophic Marinobacter strains from hypersaline areas along the Arabian Gulf coasts. Extremophiles 17: 463-470.
14. Ward OP (2010) Microbial biosurfactants and biodegradation. Adv Exp Med Biol 672: 65-74.
15. Franzetti A, Tamburini E, Banat IM (2010) Applications of biological surface active compounds in remediation technologies. Adv Exp Med Biol 672: 121-134.
16. Mnif S, Chamkha M, Labat M, Sayadi S (2011) Simultaneous hydrocarbon biodegradation and biosurfactant production by oilfield-selected bacteria. J Appl Microbiol 111: 525-536.
17. Chen Q, Bao M, Fan X, Liang S, Sun P (2013) Rhamnolipids enhance marine oil spill bioremediation in laboratory system. Mar Pollut Bull 71: 269-275.
18. Mohanty S, Mukherji S (2013) Surfactant aided biodegradation of NAPLs by Burkholderia multivorans: comparison between Triton X-100 and rhamnolipid JBR-515. Colloids Surf B Biointerfaces 102: 644-652.
19. Liu WT (2006) Nanoparticles and their biological and environmental applications. J Biosci Bioeng 102: 1-7.
20. Fabrega J, Luoma SN, Tyler CR, Galloway TS, Lead JR (2011) Silver nanoparticles: behaviour and effects in the aquatic environment. Environ Int 37: 517-531.
21. Navarro E, Baun A, Behra R, Hartmann NB, Filser J, et al. (2008) Environmental behavior and ecotoxicity of engineered nanoparticles to algae, plants, and fungi. Ecotoxicology 17: 372-386.
22. Azam A, Ahmed AS, Oves M, Khan MS, Habib SS, et al. (2012) Antimicrobial activity of metal oxide nanoparticles against Gram-positive and Gram-negative bacteria: a comparative study. Int J Nanomedicine 7: 6003-6009.
23. Rowaihi AS (2012) Isolation and characterization of biosurfactants-producing microorganisms. Arabian Gulf University, Kingdom of Bahrain Bahrain.
24. Mohamed Mel-S, Ismail W, Heider J, Fuchs G (2002) Aerobic metabolism of phenylacetic acids in Azoarcus evansii. Arch Microbiol 178: 180-192.
25. Walter V, Syldatk C, Hausmann R (2010) Screening concepts for the isolation of biosurfactant producing microorganisms. Adv Exp Med Biol 672: 1-13.
26. Chen P, Powell BA, Mortimer M, Ke PC (2012) Adaptive interactions between zinc oxide nanoparticles and Chlorella sp. Environ Sci Technol 46: 12178-12185.
27. Baek KH, Yoon BD, Cho DH, Kim BH, Oh HM, et al. (2009) Monitoring bacterial population dynamics using real-time PCR during the bioremediation of crude-oil-contaminated soil. J Microbiol Biotechnol 19: 339-345.
28. Sekhon KK, Khanna S, Cameotra SS (2011) Enhanced biosurfactant production through cloning of three genes and role of esterase in biosurfactant release. Microb Cell Fact 10: 49.
29. Bodour AA, Drees KP, Maier RM (2003) Distribution of biosurfactant-producing bacteria in undisturbed and contaminated arid Southwestern soils. Appl Environ Microbiol 69: 3280-3287.
30. Stackebrandt E, Goodfellow M (1991) Nucleic acid techniques in bacterial systematics. Wiley USA.
31. Massol-Deya AA, Odelson DA, Hickey RF, Tiedje JM (1995) Bacterial community fingerprinting of amplified 16S and 16–23S ribosomal DNA gene sequences and restriction endonuclease analysis (ARDRA). Molecular Microbial Ecology Manual 289-296.
32. Sanger F, Nicklen S, Coulson AR (1977) DNA sequencing with chain-terminating inhibitors. Proc Natl Acad Sci U S A 74: 5463-5467.
33. Altschul SF, Madden TL, Schäffer AA, Zhang J, Zhang Z, et al. (1997) Gapped BLAST and PSI-BLAST: a new generation of protein database search programs. Nucleic Acids Res 25: 3389-3402.
34. Kumar S, Tamura K, Nei M (2004) MEGA3: Integrated software for Molecular Evolutionary Genetics Analysis and sequence alignment. Brief Bioinform 5: 150-163.
35. Saitou N, Nei M (1987) The neighbor-joining method: a new method for reconstructing phylogenetic trees. Mol Biol Evol 4: 406-425.
36. Rahman KS, Rahman T, Lakshmanaperumalsamy P, Banat IM (2002) Occurrence of crude oil degrading bacteria in gasoline and diesel station soils. J Basic Microbiol 42: 284-291.

37. Radwan SS, Al-Hasan RH, Ali N, Salamah S, Khanafer M (2005) Oil-consuming microbial consortia floating in the Arabian Gulf. Int Biodeter Biodegr 56: 28-33.

38. Zhang Z, Gai L, Hou Z, Yang C, Ma C, et al. (2010) Characterization and biotechnological potential of petroleum-degrading bacteria isolated from oil-contaminated soils. Bioresour Technol 101: 8452-8456.

39. Hassanshahian M, Ahmadinejad M, Tebyanian H, Kariminik A (2013) Isolation and characterization of alkane degrading bacteria from petroleum reservoir waste water in Iran (Kerman and Tehran provenances). Mar Pollut Bull 73: 300-305.

40. Mahjoubi M, Jaouani A, Guesmi A, Ben Amor S, Jouini A, et al. (2013) Hydrocarbonoclastic bacteria isolated from petroleum contaminated sites in Tunisia: isolation, identification and characterization of the biotechnological potential. N Biotechnol 30: 723-733.

41. Ali N, Dashti N, Al-Mailem D, Eliyas M, Radwan S (2012) Indigenous soil bacteria with the combined potential for hydrocarbon consumption and heavy metal resistance. Environ Sci Pollut Res Int 19: 812-820.

42. Al-Awadhi H, Sulaiman RH, Mahmoud HM, Radwan SS (2007) Alkaliphilic and halophilic hydrocarbon-utilizing bacteria from Kuwaiti coasts of the Arabian Gulf. Appl Microbiol Biotechnol 77: 183-186.

43. Radwan S, Mahmoud H, Khanafer M, Al-Habib A, Al-Hasan R (2010) Identities of epilithic hydrocarbon-utilizing diazotrophic bacteria from the Arabian Gulf Coasts, and their potential for oil bioremediation without nitrogen supplementation. Microb Ecol 60: 354-363.

44. Tanase AM, Ionescu R, Chiciudean L, Vassu T, Stoica I (2013) Characterization of hydrocarbon-degrading bacterial strains isolated from oil-polluted soil. Int Biodeter Biodegr 84: 150-154.

45. Warne Zoueki C, Ghoshal S, Tufenkji N (2010) Bacterial adhesion to hydrocarbons: role of asphaltenes and resins. Colloids Surf B Biointerfaces 79: 219-226.

46. Bredholt H, Josefsen K, Vatland A, Bruheim P, Eimhjellen K (1998) Emulsification of crude oil by an alkane-oxidizing Rhodococcus species isolated from seawater. Can J Microbiol 44: 330-340.

47. Obuekwe CO, Al-Jadi ZK, Al-Saleh ES (2009) Hydrocarbon degradation in relation to cell-surface hydrophobicity among bacterial hydrocarbon degraders from petroleum-contaminated Kuwait desert environment. Int Biodeter Biodegr 63: 273-279.

48. Wang XB, Chi CQ, Nie Y, Tang YQ, Tan Y, et al. (2011) Degradation of petroleum hydrocarbons (C6-C40) and crude oil by a novel Dietzia strain. Bioresour Technol 102: 7755-7761.

49. Al-Mailem DM, Sorkhoh NA, Al-Awadhi H, Eliyas M, Radwan SS (2010) Biodegradation of crude oil and pure hydrocarbons by extreme halophilic archaea from hypersaline coasts of the Arabian Gulf. Extremophiles 14: 321-328.

50. Giedraityte G, Kalediene L, Bubinas A (2001) Correlation between biosurfactant synthesis and microbial degradation of crude oil hydrocarbons. Ekologija (Vilnius) 3: 38-41.

51. Trindade PV, Sobral LG, Rizzo AC, Leite SG, Soriano AU (2005) Bioremediation of a weathered and a recently oil-contaminated soils from Brazil: a comparison study. Chemosphere 58: 515-522.

52. Zheng C, He J, Wang Y, Wang M, Huang Z (2011) Hydrocarbon degradation and bioemulsifier production by thermophilic Geobacillus pallidus strains. Bioresour Technol 102: 9155-9161.

53. Atlas RM (1991) Microbial hydrocarbon degradation—bioremediation of oil spills. J Chem Technol Biot 52: 149-156.

54. Palittapongarnpim M, Pokethitiyook P, Upatham ES, Tangbanluekal L (1998) Biodegradation of crude oil by soil microorganisms in the tropic. Biodegradation 9: 83-90.

55. Urum K, Grigson S, Pekdemir T, McMenamy S (2006) A comparison of the efficiency of different surfactants for removal of crude oil from contaminated soils. Chemosphere 62: 1403-1410.

56. Tilston EL, Collins CD, Mitchell GR, Princivalle J, Shaw LJ (2013) Nanoscale zerovalent iron alters soil bacterial community structure and inhibits chloroaromatic biodegradation potential in Aroclor 1242-contaminated soil. Environ Pollut 173: 38-46.

57. Kuang Y, Zhou Y, Wang QP, Chen ZL (2012) [Effect of Fe and Fe/Ni nanoparticles on the biodegradation of phenol by BFN at different pH values]. Huan Jing Ke Xue 33: 3160-3166.

58. Karanth N, Deo P, Veenanadig N (1999) Microbial production of biosurfactants and their importance. Ferment Sci Technol 77: 116-126.

59. Rosenberg M, Kjellberg S (1985) Hydrophobic interactions in bacterial adhesion. Adv Microb Ecol 9: 353-393.

60. Nayak AS, Vijaykumar MH, Karegouder TB (2009) Characterization of biosurfactant produced by Pseudomonas sp. PNK-04 and its application in bioremediation. Int Biodeterior Biodegradation 63: 73-79.

61. Mulligan CN, Gibbs BF (1993) Factors influencing the economics of biosurfactants. Marcel Dekker Inc. New York 329-371.

62. Ismail W, Al-Rowaihi I, Al-Humam AA, Hamza R, El-Nayal A, et al. (2013) Characterization of a lipopeptide biosurfactant produced by a crude oil-emulsifying Bacillus sp. Int Biodeterior and Biodegradation 84: 168-178.

63. Floodgate GD (1984) The fate of petroleum in marine ecosystems. Macmillan Publishing Company New York USA 355-398.

64. Zajic JE, Supplisson B, Volesky B (1974) Bacterial degradation and emulsification of No. 6 fuel oil. Environ Sci Technol 8: 664-668.

Evaluation of Screening Methods for the Isolation of Biosurfactant Producing Marine Bacteria

Rengathavasi Thavasi[1]*, Shilpy Sharma[1] and Singaram Jayalakshmi[2]

[1]*Department of Chemical and Biological Sciences, Polytechnic Institute of New York University, 6 Metrotech Center, Brooklyn, New York 11201, USA*
[2]*CAS in Marine Biology, Annamalai University, Parangipettai 608502, TN, Indi*

Abstract

Hemolytic assay, bacterial adhesion to hydrocarbons (BATH) assay, drop collapse assay, oil spreading assay, emulsification assay and surface tension measurement were compared for their reliability and ease of use to screen biosurfactant producing bacteria. 105 bacterial strains were screened for their biosurfactant production. Results from screening methods revealed that hemolytic and BATH assays are not reliable methods to check biosurfactant production because hemolytic assay relies on lysis of blood cells, which can be caused by compounds produced by microbes other than the biosurfactants. In that case hemolytic assay may include strains that do not produce biosurfactant. BATH assay relies on hydrophobicity or adherence of cells with hydrophobic compounds. Results obtained in BATH assay for three bacterial isolates, Branhamella catarrhalis, Citrobacter intermedius and Klebsiella ozaenae were positive, but when the same isolates screened for drop collapse, oil spreading, and surface tension measurement the results were negative. This indicates that these bacterial cells are either acting as biosurfactant themselves (cell surface) or they don't have the ability to produce biosurfactants. In conclusion, drop collapse and oil spreading assays are reliable methods to screen large number of samples. However, for confirmation surface tension measurement can be used at the end. Another interesting finding in this study was the use of crude oil as hydrophobic substrate for the isolation of biosurfactant producing bacterial strains; this approach may be useful during the initial isolation of biosurfactant producing bacteria to reduce the number of strains to be screened.

Keywords: Biosurfactants; Emulsification; Drop collapse; Oil spreading; Hemolysis

Introduction

Bacteria produce a wide range of extracellular products with many properties and applications. Biosurfactants are one such extracellular amphiphilic compounds produced by bacteria, yeast and fungi, especially when they are grown on hydrophobic substrates. Biosurfactants are surface active compounds produced by microorganisms. They reduce surface and interfacial tension by accumulating at the interface of immiscible fluids and thus increase the solubility, bioavailability and subsequent biodegradation of the hydrophobic or insoluble organic compounds [1]. There are many types of biosurfactants based on their chemical nature such as glycolipids, lipopolysaccharides, oligosaccharides and lipopeptides that have been reported to be produced by diverse bacterial genera [2-4]. Biosurfactants have received considerable attention in the field of environmental remediation processes such as biodegradation, soil washing and soil flushing. Biosurfactants influence these processes because of their efficacy as dispersion and remediation agents and their environmentally friendly characteristics such as low toxicity and high biodegradability [5-9]. There are many reports describing the effect of exogenously added microbial biosurfactants in enhancing the bioremediation of crude oil-polluted soils by indigenous microbes [10-12] and how to purify and detect these biosurfactants [13-15]. Biosurfactants are potential antimicrobial agents to specific microbes and thus can be exploited for its antimicrobial applications against plant, animal and human microbial pathogens [1].

Due to their unique properties and vast array of application, identification of new biosurfactant producing microbes is in great demand. There are nine different screening methods that have been reported as criteria to screen biosurfactant producing microbes such as hemolytic assay [16,17], bacterial adhesion to hydrocarbons (BATH) assay [18,19], drop collapse assay, oil spreading assay [20,21], emulsification assay [22,23], surface tension measurement, titled glass slide test, blue agar plate and hydrocarbon overlay agar assay [23]. Methods discussed above have both advantages and disadvantages. Methods like hemolytic assays are not reliable and sensitive, because this method will categorize microbes in two groups as hemolytic and non-hemolytic. Strains that are hemolytic are believed to be biosurfactant producers, but there are other products such as virulence factors that can lyse the blood cells and also biosurfactants with poor diffusion in agar may not be able to lyse the blood cells. Thus, the results from hemolytic assay on blood agar plate are not so reliable and sensitive. So there is a need to find robust, sensitive and easy to perform, collection of methods to screen biosurfactant producing microbes. This study was conducted to find sensitive, reproducible and easy to perform assay methods to screen biosurfactant producing bacteria. Screening criteria used in this study include: hemolytic assay, BATH assay, drop collapse assay, oil spreading assay, emulsification assay and surface tension measurement. Total of 105 different bacterial strains were screened to check their biosurfactant production using aforementioned six methods. From the results, it was found that the drop collapse and oil spreading assays directly correlated to each other and easy to perform without any specialized equipments. However, surface

***Corresponding author:** Rengathavasi Thavasi, 1Department of Chemical and Biological Sciences, Polytechnic Institute of New York University, 6 Metrotech Center, Brooklyn, New York 11201, USA, Tel: +1 718-260-3960; Fax: +1 718-260-3075

tension measurement results also had a direct correlation with drop collapse and oil spreading assays, but it needs a surface tensiometer to perform the assay, which may not be readily available in a microbiology laboratory. Apart from above mentioned screening methods, this study also analyzes the credibility of the use of hydrophobic substrates like crude oil for the isolation of biosurfactant producing bacteria from the environment.

Material and Methods

Microorganism

Bacterial cultures used in this study were isolated from sea water samples collected at Tuticorin harbor (08°45' N; 78°13' E), South East coast of India using Bushnell Haas agar enriched with 0.1% crude oil. Total of 105 bacterial strains were isolated and identified to their species level and reported elsewhere [24]. In above study, Thavasi and Jayalakshmi [24] isolated the strains to check their crude oil degradation, whereas in this study the same isolates were screened for their biosurfactant production and to prove the concept of use of hydrophobic substrates as an effective screening tool for the isolation of biosurfactant producing bacteria.

Screening for biosurfactant production

Culture medium and bacterial growth: Bacteria were grown aerobically in 500 ml Erlenmeyer flask with 100 ml of mineral salt medium containing (g l^{-1}) 1.0 K_2HPO_4, 0.2 $MgSO_4.7H_2O$, 0.05 $FeSO_4.7H_2O$, 0.1 $CaCl_2.2H_2O$, 0.001 $Na_2MoO_4.2H_2O$, 30 NaCl and crude oil (1.0%, w/v). Crude oil used in this study was obtained from Chennai Refineries Limited, Chennai, India with a specific gravity of 0.844 at 25°C. Flasks containing sterilized mineral salt medium were inoculated with a loopful of bacterial culture grown in crude oil containing nutrient agar plates and the culture flasks were maintained in a shaker for 7 days at 200 rpm and 30°C. After 7 days of incubation, culture broth from each flask was centrifuged at 6000 rpm and 4°C for 15 minutes and the supernatant was filtered through 0.45μm pore size filter paper (Millipore). This cell free culture broth was used for drop collapse assay, oil spreading assay, emulsification assay and surface tension measurement and the bacterial cells were used for hemolytic and BATH assay. All the screening experiments were performed in triplicates (until otherwise mentioned) and the mean values were used as results.

Hemolytic activity

Hemolytic assay was performed in 5% sheep blood agar plates [25]. 50μl of bacterial culture grown in mineral salt medium was spot inoculated on to blood agar plates and incubated for 48 h at 37°C. The plates were visually inspected for clear zone (hemolysis) around the colony. The diameter of the clear zone is a qualitative method used as an indicator of biosurfactant production [25,26].

BATH assay

Cell hydrophobicity was measured by bacterial adherence to hydrocarbons according to a method described by Rosenberg et al. [27]. The cell pellets collected in the above culture medium and bacterial growth section (in materials and methods) were washed twice and suspended in a buffer salt solution (g l^{-1}, 16.9 K_2HPO_4 and 7.3 KH_2PO_4) and diluted using the same buffer solution to an optical density (OD) of ~ 0.5 at 610 nm. To the cell suspension (2 ml) in test tubes (10 ml volume with 10 x 100 mm dimension) 100 μl of crude oil was added and vortex-shaken for 3 min. After shaking, crude oil and aqueous phases were allowed to separate for 1 h. OD of the aqueous phase was then measured at 610 nm in a spectrophotometer (Varian, Cary Eclipse Spectrophotometer). From the OD values, percentage of cells attached to crude oil was calculated using the following formula:

$$\% \text{ of bacterial cell adherence} = (1-(OD_{\text{shaken with oil}}/OD_{\text{original}})) \times 100$$

Where: $OD_{\text{shaken with oil}}$ - OD of the mixture containing cells and crude oil

OD_{original} - OD of the cell suspension in the buffer solution (before mixing with crude oil)

A few drops of 2-(4-iodophenyl)-3-(4-nitrophenyl)-5-phenyltetrazolium chloride (INT) solution was added to the above BATH assay solution and observed under a light microscope. The INT turned red if it was reduced inside the cells, indicating the viability of the cells adhered to the crude oil droplets [28].

Drop collapse test

Drop collapse test was performed by following the procedure described by Jain et al. [29] and modified by Bodour and Miller-Maier [20]. 2 μl of crude oil was applied to the well regions delimited on the covers of 96-well micro-plate lids and allowed to equilibrate for 24 h. Five microliters of the cell free culture broth was transferred to the oil-coated regions and the drop size was observed 1 min later with the aid of a magnifying glass. A result was considered positive for biosurfactant production when the drop diameter was at least 1 mm larger than that produced by de-ionized water (negative control). Triton X-100 (a chemical surfactant) solution prepared at 1mg/ml concentration was used as a standard.

Oil spreading assay

Oil spreading experiment was performed using the method described by Morikawa et al. [30]. In brief, 20 ml of distilled water was added to a plastic Petri dish followed by addition of 20 μl of crude oil to the surface of the water. 10 μl of cell free culture broth was then added to the oil surface. If biosurfactant is present in the cell free culture broth, the oil will be displaced with an oil free clearing zone and diameter of this clearing zone indicates the surfactant activity, also called oil displacement activity. A negative control was maintained with distilled water (without surfactant), in which no oil displacement or clear zone was observed and Triton X-100 was used as the positive control.

Emulsification assay

Cell free culture broth was used as the biosurfactant source to check the emulsification of crude oil. 1 ml of cell free culture broth was added to 5 ml of 50mM Tris buffer (pH 8.0) in a 30 ml screw-capped test tube. Five milligram of hydrocarbon was added to the above solution and vortex-shaken for 1 minute and the emulsion mixture was allowed to stand for 20 minutes. OD of the emulsion mixture was measured at 610 nm in a spectrophotometer and the results were expressed as D_{610} [31]. A negative control was maintained only with buffer solution and crude oil and Triton X-100 was used as the positive control.

Surface tension measurement

Surface tension measurement of cell-free culture broth from each strain was determined in a K6 tensiometer (Krüss GmbH, Hamburg, Germany), using the du Nouy ring method. Triton X-100 solution prepared at 1mg/ml concentration was used as a standard.

Results and Discussion

Microorganisms

Oil degrading and total heterotrophic bacteria isolated from the sea water sample were 1.05x10^3 and 2.13x10^5 CFU/ml, respectively. Percentage calculation showed that oil degrading bacteria contributed 0.5% of the total heterotrophic bacterial (99.5%) population in the water sample. All the 105 oil degrading bacterial strains isolated in Bushnell Haas agar were subjected to screening for their biosurfactant production. Results on identification of 105 bacterial strains revealed that out of 105 isolates, 34 strains belong to gram positive and 71 strains to gram negative group represented by 18 bacterial species from 11 genera. Scientific names of the isolated bacterial species and number of isolates from each species (values in bracket) were as follows: Bacillus cereus (3), B. megaterium (3), B. subtilis (4), Branhamella catarrhalis (2), Citrobacter intermedius (5), Corynebacterium kutscheri (15), C. xerosis (2), Enterobacter aerogenes (2), Escherichia coli (6), Klebsiella ozaenae (5), Lactobacillus casei (2), L. delbrueckii (2), Proteus inconstans (6), Pseudomonas aeruginosa (23), P. fluorescens (1), P. diminuta (12), P. mallei (9) and Staphylococcus aureus (3). Species dominance results showed that among the 105 strains isolated, 45 strains belong to the genus Pseudomonas (42.8%) in which Pseudomonas aeruginosa was found as the dominant species [24]. Dominant existence and biosurfactant producing Pseudomonas strains in hydrocarbon polluted environment was reported by many researchers [32-34] supporting the results obtained in this study.

Screening for biosurfactant production

Hemolytic assay: From 105 bacterial strains tested, 101(96.1%) strains were positive for hemolysis and P. aeruginosa showed the maximum hemolytic activity of 3.05 cm and E. aerogenes and L. casei showed the least activity of 1.15 cm (Table 1). Reason for using hemolytic assay in this study as a criterion for biosurfactant production was because it is a widely used method to screen biosurfactant production and in some reports it is the sole method used to screen biosurfactant production [16,17]. The hemolytic activity of biosurfactants was first discovered when Bernheimer and Avigad [35] reported that the biosurfactant produced by B. subtilis, surfactin, lysed red blood cells. Carrillo et al. [36] found an association between hemolytic activity and surfactant production and they recommended the use of blood agar lysis as a primary method to screen biosurfactant production. None of the studies reported in the literature [21-23,36-38] mention the possibility of biosurfactant production without a hemolytic activity. However, in some studies hemolytic assay excluded many good biosurfactant producers and in some reports strains with positive hemolytic activity were found negative for biosurfactant production [21]. In addition, not all biosurfactants have hemolytic activity and compounds other than biosurfactants may cause hemolysis. Hence, in the present investigation BATH assay, drop collapse test, oil spreading assay, emulsification assay and surface tension measurements were included to confirm biosurfactant production.

BATH assay

Rosenberg et al. [27] developed a procedure to estimate the cell hydrophobicity. Cell adherence with hydrophobic compounds like crude oil is considered as an indirect method to screen bacteria for biosurfactant production, because cells attach themselves with oil droplets by producing surface active compounds called biosurfactants. BATH assay conducted in this study with 105 isolates revealed that

Name of bacteria and total number of strains	Hemolytic assay[a]	BATH assay[b]
Bacillus cereus (3)	+(2) ++(1)	+(2) ++(1)
B. megaterium (3)	++(2) +++(1)	++(2) +++(1)
B. subtilis (4)	++(1) +++(3)	++(2) +++(2)
Branhamella catarrhalis (2)	+(2)	+(2)
Citrobacter intermedius (5)	+(2) ++(3)	+(2) ++(2)
Corynebacterium kutscheri (15)	++(5) +++(10)	++(10) +++(2)
C. xerosis (2)	+(2)	+(1) ++(1)
Enterobacter aerogenes (2)	+(1) ++(1)	+(2)
Escherichia coli (6)	+(4) ++(2)	+(3) ++(1)
Klebsiella ozaenae (5)	++(4) +++(1)	+++(5)
Lactobacillus casei (2)	+(1)	+(2)
L. delbrueckii (2)	++(2)	+++(2)
Proteus inconstans (6)	+(5)	+(4)
Pseudomonas aeruginosa (23)	+(5) ++(7) +++(11)	++(19) +++(4)
P. fluorescens (1)	++(1)	++(1)
P. diminuta (12)	+(10)	+(11)
P. mallei (9)	+(5) ++(4)	+(5)
Staphylococcus aureus (3)	++(3)	+(2)

Hemolytic assay[a]: '+' - hemolysis with a clear zone of 1-1.5 cm, '++' - hemolysis with a clear zone of 1.6-2.5 cm, '+++ - hemolysis with a clear zone of 2.6-3.5 cm.
BATH assay[b]: '+++' - cell adhesion > 90%, '++' - 60 to 89% cell adhesion, '+' - 40 to 59% cell adhesion.
Values in () near '+' sign are the number of positive strains in that category. Values in () near bacterial name are the total number of strains isolated.

Table 1: Hemolytic and BATH assay.

91 (86.6%) bacterial strains were positive for the BATH assay, which indicated the affinity of the bacterial cells towards hydrophobic substrate. Cell adherence found for above positive strains with crude oil was in the range of 34±1.41 to 95.15±0.21%. Maximum cell attachment with crude oil was found with P. aeruginosa (95.15±0.21%) followed by L. delbrueckii (92.6±0.84%) (Table 1). Strains of Pseudomonas genus showed highest cell adherence with crude oil than other bacterial strains screened, which is complemented by other earlier reports on cell hydrophobicity and biosurfactant production by Pseudomonas strains [19,39,40]. Bacterial strains with high cell hydrophobicity are reported as potential biosurfactant producers [18,19] and in some reports BATH assay was used as a principle method to screen biosurfactant production [18]. While correlating the BATH assay results with hemolytic assay, strains with hemolytic activity of 1.75 ±0.07 to 3.05±0.07 cm showed high cell hydrophobicity and their cell adhesion was in the range of 41.1±1.27 to 95.15±0.21%, which indicated a positive correlation between hemolytic assay and BATH assay. Positive cell hydrophobicity was reported as an indication of biosurfactant production [41]. Visualization of bacterial cells adhered to crude oil confirmed the affinity of cells towards crude oil droplets.

Emulsification assay

Emulsification assay is an indirect method used to screen biosurfactant production. It was assumed that if the cell free culture broth used in this assay contains biosurfactant then it will emulsify the hydrocarbons present in the test solution. In this study, crude oil was used as the hydrophobic substrate. Results observed in this study revealed that from 105 strains screened, 85(80.9%) strains showed positive emulsification activity. No emulsification activity was found with the following 20 (19.1%) strains namely Citrobacter intermedius (1), Corynebacterium kutscheri (3), Escherichia coli (3), Klebsiella ozaenae (all 5 strains), Proteus inconstans (2), Pseudomonas diminuta (2) and Staphylococcus aureus (2). 5 strains from K. ozaenae species were positive for both hemolytic and BATH assay where as all of them were negative for emulsification assay. Thus, hemolytic and BATH assays are not very reliable methods to test the biosurfactant production and which also inferred that extracellular products other than biosurfactants are responsible for the positive hemolytic and BATH activity observed with the strains showing negative emulsification activity. Three P. aeruginosa strains showed the highest emulsification activity of D_{610} 1.95±0.5 (Table 2).

Drop collapse test

Drop collapse method is a sensitive and easy to perform method which requires small volume (~5μl) of culture broth or biosurfactant solution to test the surfactant property. Among the 105 strains screened, 82 (78.1 %) strains were positive for drop collapse activity (Table 3). All 12 strains from three species namely, Branhamella catarrhalis (2), Citrobacter intermedius (5) and Klebsiella ozaenae (5) were negative for drop collapse activity, but they were all positive for hemolytic and BATH assays. The reason behind the negative drop collapse and positive hemolytic and BATH results obtained with above 12 strains might be that some bacterial cells act as biosurfactant themselves [42] and have high cell hydrophobicity, but do not produce extracellular biosurfactants. In this experiment cell free culture broth was used as the biosurfactant source. For strains which produce extracellular biosurfactant there was a drop collapse activity and for strains which do not produce biosurfactant the results were negative, which also inferred that to check the biosurfactant production of any microbial strain, cell free culture broth should be used instead of using culture broth with cells. This criterion will exclude microbial strains having high cell hydrophobicity and hemolytic activity but no biosurfactant production. Accuracy and reliability of results obtained in drop collapse assay in this study was similar to the results reported by Bodour and Miller-Maier [20].

Name of bacteria and total number of strains	Emulsification assay D_{160}[a]
Bacillus cereus (3)	+(1)
	++(2)
B. megaterium (3)	++(2)
	+++(1)
B. subtilis (4)	++(2)
	+++(2)
Branhamella catarrhalis (2)	NA
Citrobacter intermedius (5)	+(2)
	++(2)
Corynebacterium kutscheri (15)	++(10)
	+++(2)
C. xerosis (2)	+(1)
	++(1)
Enterobacter aerogenes (2)	+(2)
Escherichia coli (6)	+(3)
Klebsiella ozaenae (5)	NA
Lactobacillus casei (2)	+(2)
L. delbrueckii (2)	++(2)
Proteus inconstans (6)	+(4)
Pseudomonas aeruginosa (23)	++(20)
	+++(3)
P. fluorescens (1)	+(1)
P. diminuta (12)	+(10)
P. mallei (9)	+(9)
Staphylococcus aureus (3)	+(1)
Triton X-100	+++

Emulsification assay[a]: 'NA' - no emulsification, '+++'- emulsification activity D_{610} >3, '++' - D_{610} 2 to 2.9, and '+' - D_{610} 0.5 to 2.
Values in () near '+' sign are the number of positive strains in that category.
Values in () near bacterial name are the total number of strains isolated.

Table 2: Emulsification assay.

Name of bacteria and total number of strains	Drop collapse assay[a]	Oil spreading assay[b]	Surface tension (mN/m)[c]
Bacillus cereus (3)	+(2)	+(2)	+(1)
			++(2)
B. megaterium (3)	++(1)	++(1)	+(1)
	+++(2)	+++(2)	+++(2)
B. subtilis (4)	++(1)	++(2)	+(2)
	+++(3)	+++(2)	++(2)
Branhamella catarrhalis (2)	NA	NA	NA
Citrobacter intermedius (5)	NA	NA	NA
			NA
Corynebacterium kutscheri	++(4)	++(3)	++(13)
(15)	+++(11)	+++(12)	+++(2)
	+(1)	+(1)	+(1)
C. xerosis (2)			++(1)
	+(2)	+(2)	+(2)
Enterobacter aerogenes (2)	+(5)	+(5)	+(3)
Escherichia coli (6)	NA	NA	NA
Klebsiella ozaenae (5)	+(1)	+(1)	+(2)
Lactobacillus casei (2)	++(2)	++(2)	++(2)
L. delbrueckii (2)	+(5)	+(5)	+(4)
Proteus inconstans (6)	++(5)	++(4)	++(3)
Pseudomonas aeruginosa	+++(18)	+++(19)	+++(20)
(23)	++(1)	++(1)	+(1)
	+(10)	+(10)	+(10)
P. fluorescens (1)	+(6)	+(5)	+(4)
P. diminuta (12)	+(2)	+(2)	+(1)
P. mallei (9)	+++	+++	+++
Staphylococcus aureus (3)			
Triton X-100			

Drop collapse assay[a]: 'NA'- no drop collapse activity, '+++'- drop collapse within 1 minute, '++'- drop collapse after 1minute and '+' - drop collapse after 3 minutes of biosurfactant addition.
Oil spreading assay[b]: 'NA' - no oil spreading activity, '+' - oil spreading with a clear zone of 0.5-1.5 cm, '++' - oil spreading with a clear zone of 1.6 to 2.5 cm, '+++' - oil spreading with a clear zone of 2.6 to 3.5 cm.
Surface tension[c]: 'NA'- no surfactant activity, '+++'- surface tension <40 mN/m '++'- surface tension 40 to 50,'+'- surface tension 51 to 70.
Values in () near '+' sign are the number of positive strains in that category. Values in () near bacterial name are the total number of strains isolated.

Table 3: Drop collapse, oil spreading assays and surface tension measurement.

Another merit associated with drop collapse assay is the sample volume required to check the drop collapse. 5 to 10μl of surfactant solution is enough to conduct a duplicate measurement. To further confirm the biosurfactant production of above strains with positive and negative results, cell free culture broth from all 105 bacterial strains was subjected to oil spreading and surface tension measurement experiments.

Oil spreading assay

Oil spreading assay results were in corroboration with drop collapse assay results. Strains found with positive drop collapse results were positive for oil spreading assay also (Table 3). These results confirmed the presence (for strains with positive results) and absence (for strains with negative results) of surface active compound (biosurfactant) in the cell free culture broth. Morikawa et al. [30] reported that the area of oil displacement in oil spreading assay is directly proportional to the concentration of the biosurfactant in the solution. However, in this study there was no quantitative study conducted on biosurfactant concentration versus oil spreading activity, but a qualitative study to check the presence of biosurfactant in the cell free culture broth was in concurrence with the above mentioned earlier report. As found in drop collapse assay, 12 strains from Branhamella catarrhalis, Citrobacter intermedius, Klebsiella ozaenae showed no oil spreading activity and in total out of 105 strains, 81 (77.1%) strains were positive for the oil spreading assay. 5 strains namely Bacillus megaterium, B. subtilis, Corynebacterium kutscheri, Lactobacillus delbrueckii, Pseudomonas aeruginosa and P. fluorescens showed highest oil spreading activity. Similar results with drop collapse and oil spreading assay was reported by Youssef et al. [21] while screening bacteria for biosurfactant production and also recommended that both drop collapse and oil spreading assay methods as reliable techniques for testing biosurfactant production.

Surface tension measurement

Surface tension measurement of cell free culture broth revealed that out of 105 strains screened, 79 (75.2%) strains showed reduction in surface tension and highest surface tension reduction was observed with 5 strains namely Bacillus megaterium (30.8±1.13mN/m), B. subtilis (38.75±0.3mN/m), Corynebacterium kutscheri (36.9±0.77mN/m), C. xerosis (37.8±0.42mN/m), Lactobacillus delbrueckii (32.5±0.70mN/m), Pseudomonas aeruginosa (28.7±0.98mN/m) and P. fluorescens (34.7±0.35mN/m) (Table 3). There was a direct correlation found between drop collapse, oil spreading and surface tension assays. Strains highly active in any one of these methods were active in other two methods. Similar direct correlation between drop collapse method and surface tension was reported by Bodour and Miller-Maier [20]; and direct correlation between drop collapse and oil spreading method by Youssef et al. [21]. Above earlier reports and results from this study indicated that drop collapse and oil spreading assays are easy, reliable and sensitive methods to check the biosurfactant production.

As a confirmation, 4 strains from above 5 potential strains (i.e., Bacillus megaterium, Corynebacterium kutscheri, Lactobacillus delbrueckii and Pseudomonas aeruginosa) showing positive activity in drop collapse, oil spreading and surface tension assays were already reported as potential biosurfactant producers in a different study reported by Thavasi et al. [43,44].

Conclusion

Earlier reports on screening of biosurfactant producing microbes discussed either a single method or combination of two or three methods like hemolytic assay, BATH assay, drop collapse method, oil spreading assay and surface tension measurement [20-22]. Few reports have 6 to 8 screening methods but with no surface tension measurement. Whereas, in this study six different methods were used to screen bacterial isolates for biosurfactant production and it was found that drop collapse and oil spreading methods are reliable methods to screen large number of samples for biosurfactant production because these methods require only 5-10 μl of biosurfactant solution and have direct correlation with each other as reposted in this study and in other earlier studies. After screening the strains for biosurfactant production using oil spreading and drop collapse assay, for confirmation surface tension method can be included in the screening criteria. Results from this study show that the order of suitable methods to screen biosurfactant production is surface tension measurement > drop collapse assay = oil spreading > emulsification assay.

Compared to other earlier reports the percentage of biosurfactant producing strains isolated in this study was high (75.2, 81.77, and 82.7%, for surface tension, oil spreading and drop collapse assays, respectively). The reason for this higher percentage may be that the bacterial strains used in this study were isolated using crude oil enrichment medium which might have excluded all other bacterial strains that are not able to use crude oil for growth. Other earlier reports mentioned in this manuscript used strains isolated with nutrient rich culture medium e.g., Nutrient agar for the isolation. Thus, use of crude oil or hydrophobic carbon substrate enrichment can be used as a criterion to isolate new biosurfactant producing bacteria. In many reports crude oil degradation is directly correlated to surfactant production [23,45]. However, a comparative study with hydrocarbon enrichment and without hydrocarbon enrichment is needed to justify the above statement on the use of hydrophobic substrate as a criterion to isolate new biosurfactant producing bacteria.

Acknowledgement

Drs. Thavasi and Jayalakshmi thank the authorities of Annamalai University for providing the facilities and Department of Ocean Development (DOD) and Council of Scientific and Industrial Research (CSIR), Government of India for providing financial support.

References

1. Van Hamme JD, Singh A, Ward OP (2006) Physiological aspects: Part 1 in a series of papers devoted to surfactants in microbiology and biotechnology. Biotechnol Adv 24: 604-620.
2. Banat IM, Makkar RS, Cameotra SS (2000) Potential commercial applications of microbial surfactants. Appl Microbiol Biotechnol 53: 495-508.
3. Banat IM, Franzetti A, Gandolfi I, Bestetti G, Martinotti MG, et al. (2010) Microbial biosurfactants production, applications and future potential. Appl Microbiol Biotechnol 87: 427-444.
4. Franzetti A, Tamburini E, Banat IM (2010) Application of biological surface active compounds in remediation technologies. Adv Exp Med Biol 672: 121-134.
5. Mulligan CN (2005) Environmental applications for biosurfactants. Environ Pollut 133: 183-198.
6. Das P, Mukherjee S, Sen R (2009) Biosurfactant of marine origin exhibiting heavy metal remediation properties. Bioresour Technol 100: 4887-4890.
7. Sivapathasekaran C, Mukherjee S, Ray A, Gupta A, Sen R (2010) Artificial neural network modeling and genetic algorithm based medium optimization for the improved production of marine biosurfactant. Bioresour Technol 101: 2884-2887.
8. Kiran SG, Thomas TA, Selvin J, Sabarathnam B, Lipton AP (2010) Optimization and characterization of a new lipopeptide biosurfactant produced by marine Brevibacterium aureum MSA13 in solid state culture. Bioresour Technol 101: 2389-2396.
9. Satpute SK, Banat IM, Dhakephalkar PK, Banpurkar AG, Chopade BA (2010) Biosurfactants, bioemulsifiers and exopolysaccharides from marine microorganisms. Biotechnol Adv 28: 436-450.
10. Abalos A, Vinas M, Sabate J, Manresa MA, Solanas AM (2004) Enhanced biodegradation of Casablanca crude oil by a microbial consortium in presence of a rhamnolipid produced by Pseudomonas aeruginosa AT 10. Biodegradation 15: 249-260.
11. Cubitto MA, Moran AC, Commendatore M, Chiarello MN, Baldini MD, et al. (2004) Effects of Bacillus subtilis O9 biosurfactant on the bioremediation of crude oil-polluted soil. Biodegradation 15: 281-287.
12. Owsianiak M, Chrzanowski Ł, Szulc A, Staniewski J, Olszanowski A, et al. (2009). Biodegradation of diesel/biodiesel blends by a consortium of hydrocarbon degraders: Effect of the type of blend and the addition of biosurfactants. Bioresour Technol 100: 1497-1500.
13. Satpute SK, Banpurkar AG, Dhakephalkar PK, Banat IM, Chopade BA (2009) Methods for investigating biosurfactants and bioemulsifiers: a review. Crit Rev Biotechnol 30: 127-144.
14. Smyth TJP, Perfumo A, Marchant R, Banat IM (2010) Isolation and analysis of low molecular weight microbial Glycolipids. In: Handbook of hydrocarbon

and lipid microbiology (Timmis KN, Ed.), Springer-Verlag, Heidelberg pp 3705-3723.

15. Smyth TJP, Perfumo A, McClean S, Marchant R, Banat IM (2010) Isolation and analysis of lipopeptides and high molecular weight biosurfactants. In: Handbook of hydrocarbon and lipid microbiology (Timmis KN, Ed.), Springer-Verlag, Heidelberg pp 3689-3704.

16. Banat IM (1993) The isolation of a thermophilic biosurfactant producing *Bacillus* sp. Biotechnol Lett 15: 591-594.

17. Yonebayashi H, Yoshida S, Ono K, Enomoto H (2000) Screening of microorganisms for microbial enhanced oil recovery process. Sekiyu Gakkaishi 43: 59-69.

18. Volchenko NN, Karasev SG, Nimchenko DV, Karaseva EV (2007) Cell hydrophobicity as a criterion of selection of bacterial producers of biosurfactants. Microbiology 76: 112-114.

19. Tuleva BK, Ivanov RG, Christova NE (2002) Biosurfactant production by an new Pseudomonas putida strain. Z Naturforsch 57: 356-360.

20. Bodour AA, Miller-Maier R (1998) Application of a modified dropcollapse technique for surfactant quantification and screening of biosurfactant-producing microorganisms. J Microbiol Methods 32: 273-280.

21. Youssef NH, Duncan KE, Nagle DP, Savage KN, Knapp RM, et al. (2004) Comparison of methods to detect biosurfactant production by diverse microorganisms. J Microbiol Methods 56: 339-347.

22. Afshar S, Lotfabad TB, Roostaazad R, Najafabadi AR, Noghabi AK (2008) Comparative approach for detection of biosurfactant-producing bacteria isolated from Ahvaz petroleum excavation areas in south of Iran. Ann Microbiol 58: 555-560.

23. Satpute SK, Bhawsar BD, Dhakephalkar PK, Chopade BA (2008) Assessment of different screening methods for selecting biosurfactant producing marine bacteria. Indian J Mar Sci 37: 243-250.

24. Thavasi R, Jayalakshmi S (2003) Bioremediation potential of hydrocarbonoclastic bacteria in Cuddalore harbour waters (India). Res J Chem Environ 7: 17-22.

25. Mulligan CN, Cooper DG, Neufeld RJ (1984) Selection of microbes producing biosurfactants in media without hydrocarbons. J Ferment Technol 62: 311-314.

26. Rodrigues LR, Teixeira JA, van der Mei HC, Oliveira R (2006) Isolation and partial characterization of a biosurfactant produced by Streptococcus thermophilus A. Colloids Surf B Biointerfaces 53: 105-112.

27. Rosenberg M, Gutnick D, Rosenberg E (1980) Adherence to bacteria to hydrocarbons: a simple method for measuring cell-surface hydrophobicity. FEMS Microbiol Lett 9: 29-33.

28. Betts RP, Bankers P, Banks JG (1989) Rapid enumeration of viable microorganisms by staining and direct microscopy. Lett Appl Microbiol 9: 199-202.

29. Jain DK, Collins-Thompson DL, Lee H, Trevors JT (1991) A dropcollapsing test for screening surfactant-producing microorganisms. J Microbiol Methods 13: 271-279

30. Morikawa M, Hirata Y, Imanaka, TA (2000) A study on the structure-function relationship of lipopeptide biosurfactants. Biochim Biophys Acta 1488: 211-218.

31. Rosenberg E, Zuckerberg A, Rubinovitz C, Gutnick DL (1979) Emulsifier of Arthrobacter RAG-I: isolation and emulsifying properties. Appl Environ Microbiol 37: 402-408.

32. Yateem A, Balba MT, Al Shayji Y, Al-Awadhi N (2002) Isolation and characterization of biosurfactant-producing bacteria from oil-contaminated soil. Soil Sediment Contam 11: 41-55.

33. Bodour AA, Drees KP, Miller-Maier R (2003) Distribution of biosurfactant-producing bacteria in undisturbed and contaminated arid southwestern soils. Appl Environ Microbiol 69: 3280-3287.

34. Das K, Mukherjee AK (2005) Characterization of biochemical properties and biological activities of biosurfactants produced by Pseudomonas aeruginosa mucoid and non-mucoid strains isolated from hydrocarbon-contaminated soil samples. Appl Microbiol Biotechnol 69: 192-199.

35. Bernheimer AW, Avigad LS (1970) Nature and properties of a cytolytic agent produced by Bacillus subtilis. J Gen Microbiol 61: 361-369.

36. Carrillo PG, Mardaraz C, Pitta-Alvarez SJ, Giulietti AM (1996) Isolation and selection of biosurfactant-producing bacteria. World J Microbiol Biotechnol 12: 82- 84.

37. Moran AC, Martinez AM, Sineriz F (2002) Quantification of surfactin in culture supernatant by hemolytic activity. Biotechnol Lett 24: 177- 180.

38. Walter V, Syldatk C, Hausmann R (2010) Screening concepts for the isolation of biosurfactant producing microorganisms. In: Biosurfactants: advances in Experimental Medicine and Biology (Sen R, Ed.), Springer-Verlag, Heidelberg, Vol. 672 pp 1-13.

39. Zhang Y, Miller RM (1994) Effect of Pseudomonas rhamnolipid biosurfactant on cell hydrophobicity and biodegradation of octadecane. Appl Environ Microbiol 60: 2101-2106

40. Deziel E, Paquette G, Villemur R, Lepine F, Sailton JB, (1999) Biosurfactant production by a soil Pseudomonas strains growing on poly aromatic hydrocarbons. Appl Environ Microbiol 62: 1908-1912.

41. Franzetti A, Caredda P, Colla PL, Pintus M, Tamburini E, et al. (2009) Cultural factors affecting biosurfactant production by Gordonia sp. BS29. Int Biodeterioration Biodegrad 63: 943-947.

42. Hommel RK (1994) Formation and function of biosurfactants for degradation of water-insoluble substrates. In: Biochemistry of Microbial Degradation (Ratledge C, Ed.), Kluwer Academic Publishers, Boston p 63-87.

43. Thavasi R, Nambaru VRMS, Jayalakshmi S, Balasubramanian, T, Banat IM (2011) Biosurfactant production by Pseudomonas aeruginosa from renewable resources. Indian J Microbiol 51: 30-36.

44. Thavasi R, Jayalakshmi S, Banat IM (2011) Effect of biosurfactant and fertilizer on biodegradation of crude oil by marine isolates of Bacillus megaterium, Corynebacterium kutscheri and Pseudomonas aeruginosa. Bioresour Technol 102: 772-778.

45. Rahman KSM, Rahman TJ, Lakshmanaperumalsamy P, Marchant R, Banat IM (2003) The Potential of Bacterial Isolates for Emulsification with a Range of Hydrocarbons. Acta Biotechnol 23: 335-345.

Experimental Study of Use of Ionic Liquids in Enhanced Oil Recovery

Mabkhot S Bin-Dahbag[1]*, Abdulrahman A Al Quraishi[2], Mohammed S Benzagouta[1], Mustafa M Kinawy[1], Inas M Al Nashef[3] and Emad Al Mushaegeh[2]

[1]Department of Petroleum and Natural Gas Engineering, King Saud University, Saudi Arabia
[2]Oil and Gas Research Institute, King Abdulaziz City for Sciences and Technology, Saudi Arabia
[3]Department of Chemical Engineering, King Saud University, Saudi Arabia

Abstract

Chemical flooding process has shown great potential in Enhanced Oil Recovery (EOR). Unfortunately, chemicals used have some disadvantages such as high cost, high toxicity and high adsorption tendency. In this study, we aim at using Ionic Liquids (ILs) as alternatives for traditional chemicals. Ionic liquids are salts having melting point below 100°C and they found as a liquid at room temperature.

Nine Ammonium and Phosphonium based ILs were screened. The screening was based on their solubility in brines of different compositions, thermal stability and ability to reduce the aqueous-oleic phase's Interfacial Tension (IFT). The screening process flagged Ammoeng 102 as the favored ionic liquid. More investigations of Ammoeng 102 solutions indicated a sharp exponential decrease of IFT values with increasing concentration. On contrary to surfactant solutions, Lower IFT values were obtained with increasing brine salinity indicating the ILs superiority in high salinity reservoirs.

Two tertiary flooding experiments were conducted using 500 ppm Ammoeng 102 diluted in 10% and 20% (w/w) brine salinity to investigate its recovery efficiency. Lower salinity secondary brine flooding provided higher recovery. The opposite trend occurred in tertiary ionic solution flooding where recovery is higher for high salinity ionic solution indicating the effectiveness of ILs in recovering oil in high salinity, high temperature environment. In addition, the low cost and low toxicity are more advantages to promote the use of Ionic liquids in future EOR processes.

Keywords: Ammonium; Phosphonium; Surfactant solutions; Enhanced oil recovery

Introduction

Ionic liquids, known as salts having a melting point below 100°C, have recently attracted considerable attention as potential alternatives to conventional organic solvents. ILs consist typically of organic cations such as imidazolium, or pyridinium cation and inorganic anions such as halide, $[AlCl4]^-$, $[BF4]^-$, or $[PF6]^-$ and are very often found as liquids at room temperature [1,2]. There are many types of ILs but the most common types are cation organic ILs, anion organic ILs and anion inorganic ILs.

ILs are characterized by properties such as negligible vapor pressure and non-flammability under ambient conditions, high thermal conductivity, high polarity, high heat capacity, high density, extremely low volatility and high thermal stability [1,3,4]. Itis known for itlow cost and commercial availability, water stability, low toxicity, low environmental impact, noncorrosive properties, and recyclability. However, the most important attribute of ILs is the possibility of designing their properties based on the number of cations and anions combination [1,5].

Owing to difficulties in producing large amount of crude oil by traditional methods, enhanced oil recovery techniques have been developed. Surfactant flooding is one of the methods used in this regard. This process is intended to decrease Interfacial Tension (IFT) between injected brine and resident crude oil and/or alter rock wetting properties. Some disadvantages that limit the implementation of surfactant flooding are the chemicals high cost, high toxicity, and reservoir damage due to chemicals adsorption on rock grain surfaces. Several studies were conducted to investigate the effect of ILs on interfacial tension and its favorable impact on production [6-8]. ILs can lower the interfacial tension effectively; hence, it can replace the conventional chemicals such as solvents or co-solvents that are known for their toxicity in addition to their high cost compared to ILs.

The anticipated potential of ILs in Enhanced Oil Recovery (EOR) process and the lack of deep research work propped us to investigate ILs as future EOR chemicals. The work was started by screening different ILs to determine the most effective one for our application. Different factors were investigated and among these factors are aqueous phase composition and salinity, ILs concentration, and testing conditions of temperature and pressure. The most efficient IL was tested further and its feasibility as flooding chemical was investigated on Berea sandstone core samples at reservoir conditions of pressure and temperature.

Material

Many commercialAmmonium and Phosphonium based ionic liquids were screened. Distilled water and three types of brines composed of sodium chloride (NaCl) and calcium chloride ($CaCl_2$) dissolved in distilled water at 10% salinity (w/w) were used as aqueous phases in screening process. These brines are: Brine I (100% NaCl),

***Corresponding author:** Mabkhot S Bin-Dahbag, Department of Petroleum and Natural Gas Engineering, King Saud University, Saudi Arabia, Tel: 9665-56019372

Brine II (95% NaCl and 5% $CaCl_2$), Brine III (83% NaCl and 17% $CaCl_2$). The oleic phase was Saudi medium crude oil. Brine III at 10% and 20% salinity (w/w) was used as aqueous phase in further investigation of the chosen ionic liquid. Table 1 lists the physical properties of oleic phase and aqueous phase (Brine III) at 10% and 20% (w/w) salinity measured at reservoir conditions of 2000 psig and 60°C. Three homogenous Berea sandstone rock samples from same block were used as porous medium in the flooding runs. XRD analysis (Figure 1) showed that samples are mostly composed of Quartz with small amount of Kaolinite as cementing material. Table 2 summarizes rock samples petrophysical properties.

Experimental Work

Interfacial Tension (IFT) between oleic and aqueous phases was measured using pendant drop tensiometer (Figure 2). This device measures IFT based on Young-Laplace equation. The high pressure cell is first filled with the aqueous phase and pressure is increased to the desired level. Temperature is set to the desired value and we wait for temperature to stabilize. A needle is raised in the pressure cell and a sufficient drop size of oleic phase is injected. A digital image of the drop is made and transferred to the software where IFT is measured. Several measurements were made and an average IFT value is taken.

Flooding runs were performed using automated core flooding unit shown in Figure 3. Experiments were conducted at reservoir conditions of 5000 psig confining pressure, 2000 psig pore pressure and 60°C temperature. Each test started by saturating core samples under vacuum with free air brine solution. Saturated sample was placed in core holder and reservoir conditions were applied. Synthetic

Phase	Fluid Type	Viscosity(cp)	Density (gm/cm³)
Aqueous	10% salinity Bine III (83% NaCl and 17% $CaCl_2$)	0.8104	1.07101
	20% salinity brine III (83% NaCl and 17% $CaCl_2$)	1.0125	1.15096
Oleic	Medium Crude Oil	5.5	0.88289

Table 1: Fluid Physical Propertiesat 2000 psig and 60°C.

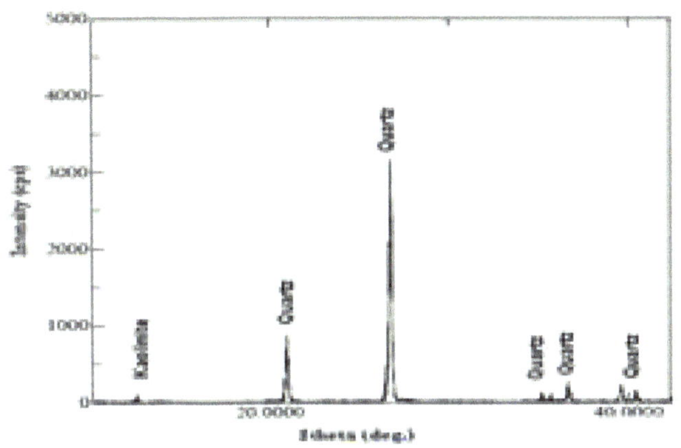

Figure 1: XRD Analysis for Berea Sandstone Core Sample.

Run No.	Diameter (cm)	Length (cm)	Porosity (%)	Absolute Permeability (md)
1	3.7	7.46	22	244
2	3.72	7.29	22	263
3	3.72	7.11	21.5	231

Table 2: Rock Physical Properties.

Figure 2: Pendant Drop Tensiometer.

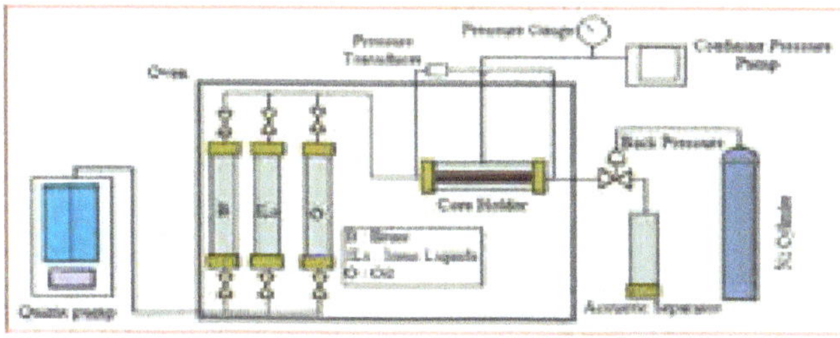

Figure 3: Schematic of Core Flooding Unit.

formation brine was injected through the core sample at different flow rates and absolute permeability was calculated using Darcy law. The fully saturated core was flooded with crude oil at 1 cm³/min. Flooding continued until water production ceases. Irreducible water saturation (Sw_{irr}) and initial oil saturation were calculated using material balance. Steady state pressure drop was measured to calculate crude oil effective permeability using Darcy law. The core sample was then flooded with brine at 1cm3/min. The flooding process continued until oil production ceases. Effluents were collected in acoustic separator and oil recovery curve and core fluids saturations were calculated using material balance equation. Tertiary flooding was started at residual oil saturation with selected ionic solution and extra oil recovery was measured.

Results and Discussion

Screening process

Nine Ammonium and Phosphonium based ILs were screened. Screening process was based on ILs solubility in brines (I, II, III) and distilled water, thermal stability (no turbidity) and their effectiveness in lowering aqueous-oleic phases IFT at different temperatures ranging from 22 to 90°C and atmospheric pressure. IFT measurements were conducted at ILs concentrations ranging from 100 to 1000 ppm diluted in distilled water and brines I, II, III. All brine solutions were prepared at 10% w/w salinity. Table 3 summarizes the ILs screened and the observations obtained at the corresponding conditions. All investigated ILs are soluble in brine and thermally stable at temperature below 80°C except the Ammoeng 130 and IL7. However, in contrast to other ILs, Ammoeng 102 was the only IL soluble in all aqueous phases tested with significant ability to lower IFT at increasing temperatures. Figure 4 illustrates the thermal stability of Ammoeng 102 up to 80°C. Turbidity starts at 82°C and at 86°C tensiometer was not able to do the measurement. Figure 5 plots the IFT measurements at of all ILs at

No.	Name	Base Type	Solubility	IFT Trend With Temperature	Temperature Stability
1	Ammoeng 102	Ammonium	Soluble in distilled water, brine I, II, III	Decreasing	Stable until 80°C in all brines and distillated water. IFT is very low.
2	Ammoeng 112	Ammonium	Not soluble in brine III	Increasing	Stable until 90° in brine (I).However, IFT is very high
3	Ammoeng 120	Ammonium	Soluble in distilled water, brine I	Increasing	Stable until 90° in distillatedwater and brine (I). However, IFT is very high
4	Ammoeng 130	Ammonium	Not soluble in distilled water and brines	Unable to measure because IL is not soluble	Unable to measure because IL is not soluble
5	IL6	Phosphonium	Soluble in distilled water, brine III	Decreasing	Stable until 90°C in brine (III).However, IFT is very high
6	IL7	Phosphonium	Not soluble in all brines	Unable to measure because IL is not soluble	Unable to measure because IL is not soluble
7	IL25	Phosphonium	Soluble in distilled water	Increasing	Stable until 90°C in brine (III).However, IFT is Very high.
8	IL34	Phosphonium	Soluble in distilled water, brine III	Increasing	Stable until 90°C in brine (III).However, IFT is very high
9	IL37	Phosphonium	Soluble in Brine I	Increasing	Stable until 80°C in brine (I).However, IFT is very high.

Table 3: Summary of Ionic Liquids Screening Results.

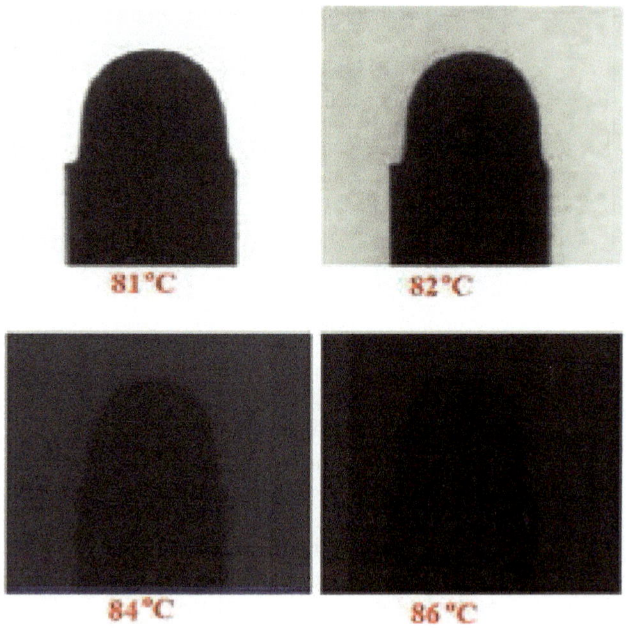

Figure 4: Torbidity of Ammoeng 102 Diluted in Brine III at 10% (w/w) Salinity.

different temperatures, and IL concentrations. Based on the screening process, Ammoeng 102 was chosen for further investigation.

Critical Micelle Concentration (CMC) is the concentration at which IL solutions begin to form micelles at large amount and the IFT is not reduced anymore as the IL concentrations increases. In order to find the CMC of the Ammoeng 102 solution, the IFT of crude oil-ionic solution at different concentrations were measured at reservoir conditions of 2000 psig and 60°C.Ammoeng 102 solutions were prepared with brine III at 10% and 20% salinity. The brine composition and salinities were chosen to resemble that of Saudi formation brine. Figure 6 plots the obtained results indicating exponential IFT decline with the increase of IL concentration. More IFT reduction was noticed with increasing brine salinity and that can be attributed to the higher salt content that can alter the distribution of ionic liquids molecules on crude oil-brine interfacial surface [9]. Critical micelle concentrations were determined around 250 ppm for both salinities with an IFT of 1.65mN/m for the 20% salinity compared to 3.36 mN/m for10% salinity. The obtained results prove that Ammoeng 102 possesses high capability to lower surface tensions and this can be related to its cationic nature with positive charges that are neutralized by the brine negatively charged ions leading to easier accumulation of IL molecules at the oil-brine interface, causing more IFT reduction.

Ammoeng 102 solution at 500 ppm concentration was chosen for the later tests and IFT of crude oil-Ammoeng 102 solution diluted in10% brine salinity was measured at various pressures and temperatures. Figures 7 and 8 present the results obtained indicating a slight IFT decreases with increasing temperature and slight increase with increasing pressure. The IFT trends obtained agrees with that presented by previous investigators in their work with conventional surfactants [10,11]. Similar trends are expected for Ammoeng 102 diluted in 20% brine salinity but with lower IFT values as proved previously. The thermal stability and high salinity tolerance prove that Ammoeng 102 is a good candidate for the EOR process in harsh reservoir conditions such as those characterizing Saudi reservoir.

Flooding process

Two flooding runs were conducted to investigate the efficiency of Ammoeng 102 as enhanced oil recovery chemical. Brine III at 10% and 20% salinity and Saudi medium crude oil were used as aqueous and oleic phases respectively. The ionic solution was prepared by diluting 500 ppm of Ammoeng 102 in previous aqueous phase. First run was conducted by flooding core sample with 10% brine salinity in secondary flooding mode followed by the tertiary flooding of ionic solution at the same salinity. The second run was similar to the first one in scenario but at 20% salinity. Figure 9 presents the recovery and pressure drop outcomes of the flooding experiments indicating the superiority of 10% salinity secondary brine flooding with an additional recovery of 10% Original Oil in Place (OOIP). This finding agrees well with the findings obtained by previous investigators on the efficiency of low salinity flooding [12-14]. Low Salinity fluid promote desorption of Ca^{2+} from the clay surface which consequently creates a local increase in pH close to the brine-clay interface. A fast reaction between OH- and the absorbed acidic and protonated basic material will cause desorption of organic material from the clay surface, and as the results the improves rock water wetness and hence enhance the oil recovery. Tertiary flooding of high salinity ionic solutions seems more effective providing additional recovery of 4.5 % OOIP compared to 3 OOIP for low salinity solution flooding. These values correspond to 8% and 5% Residual Oil in Place (ROIP) for high and low salinity ionic solution flooding respectively. This can be attributed to the effectiveness of high salinity ionic solution in lowering the interfacial tension and altering rock wettability [15,16].

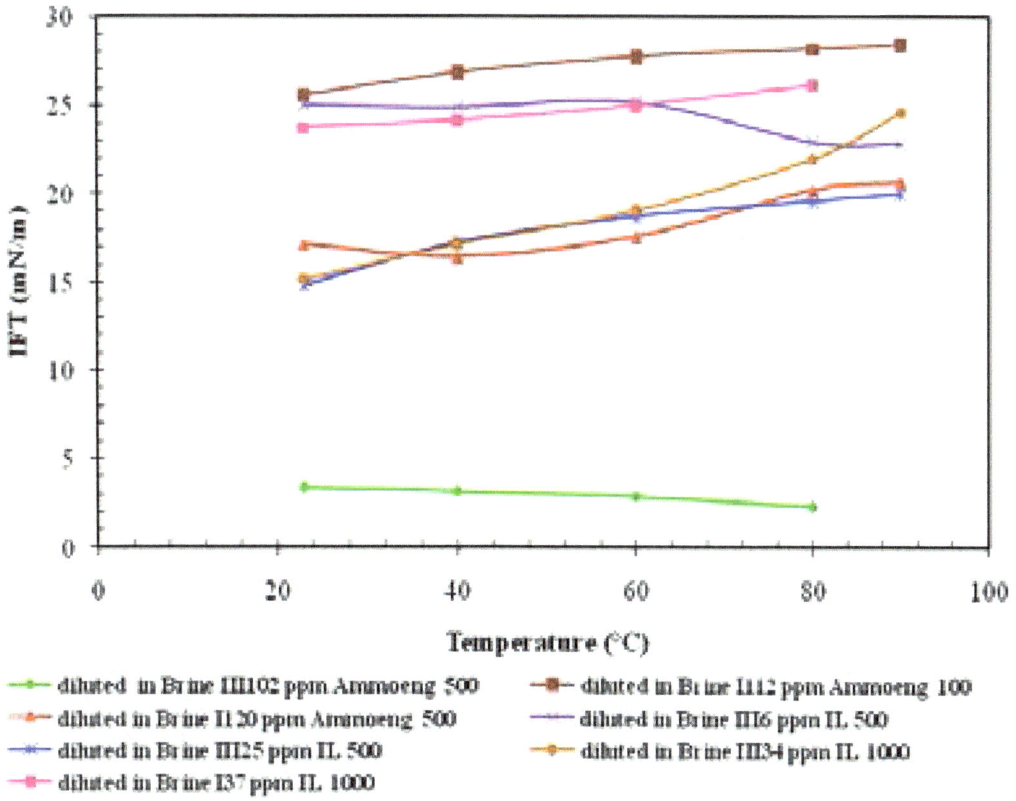

Figure 5: IFT Measurements of Ionic Liquids Diluted in Different Concentrations ofBrinesI and III at 10% (w/w) Salinity.

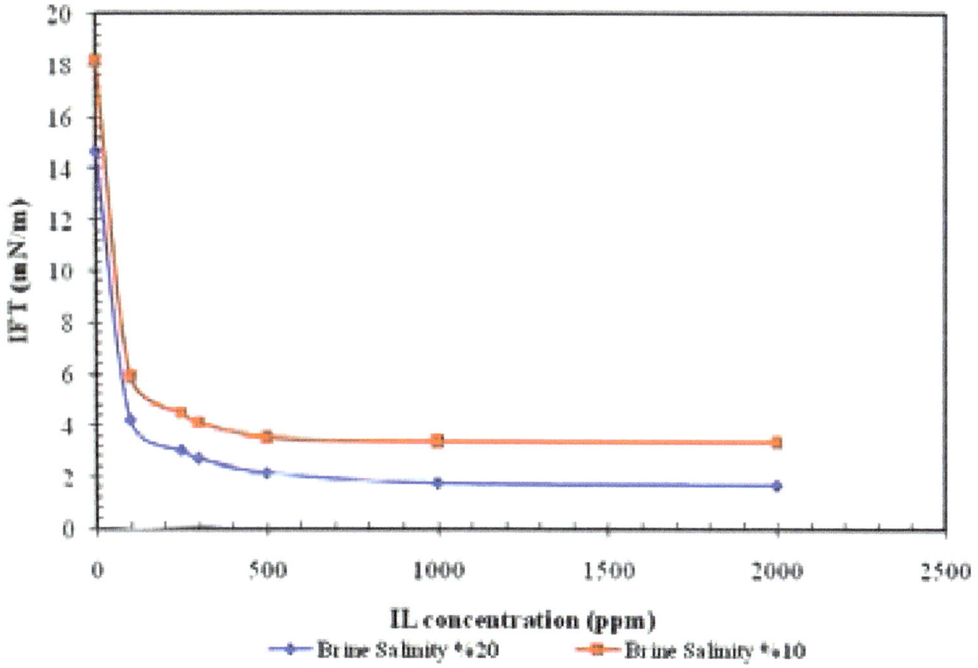

Figure 6: IFT of Ammoeng 102 Solution-Crude Oil at Different Concentrations and Two Brine Salinities.

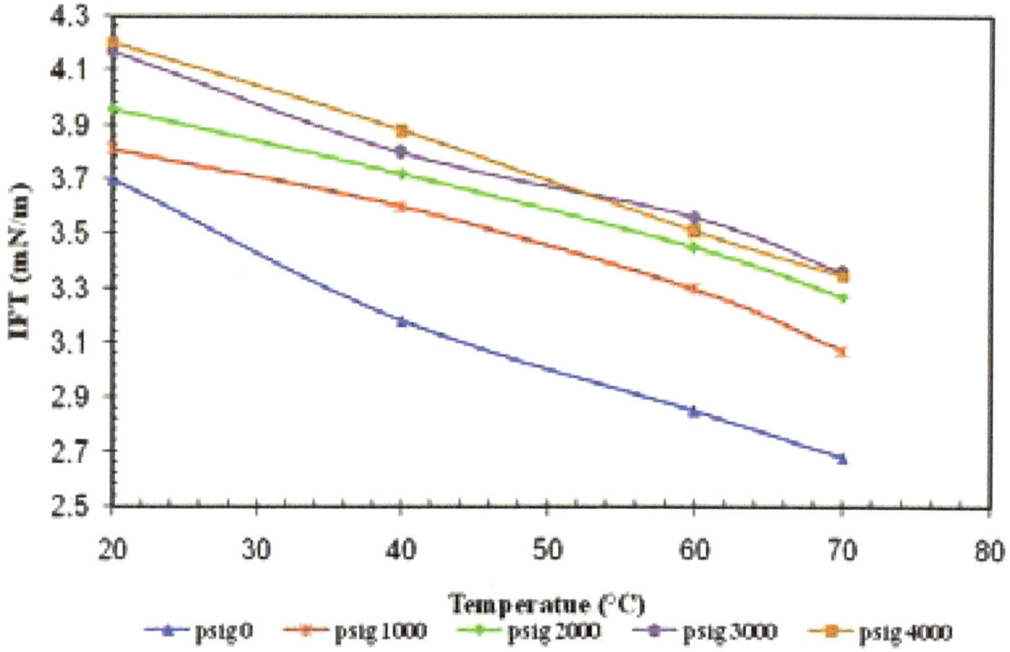

Figure 7: Effect of Temperature on 500 ppm Ammoeng 102 Solution-Crude Oil IFT.

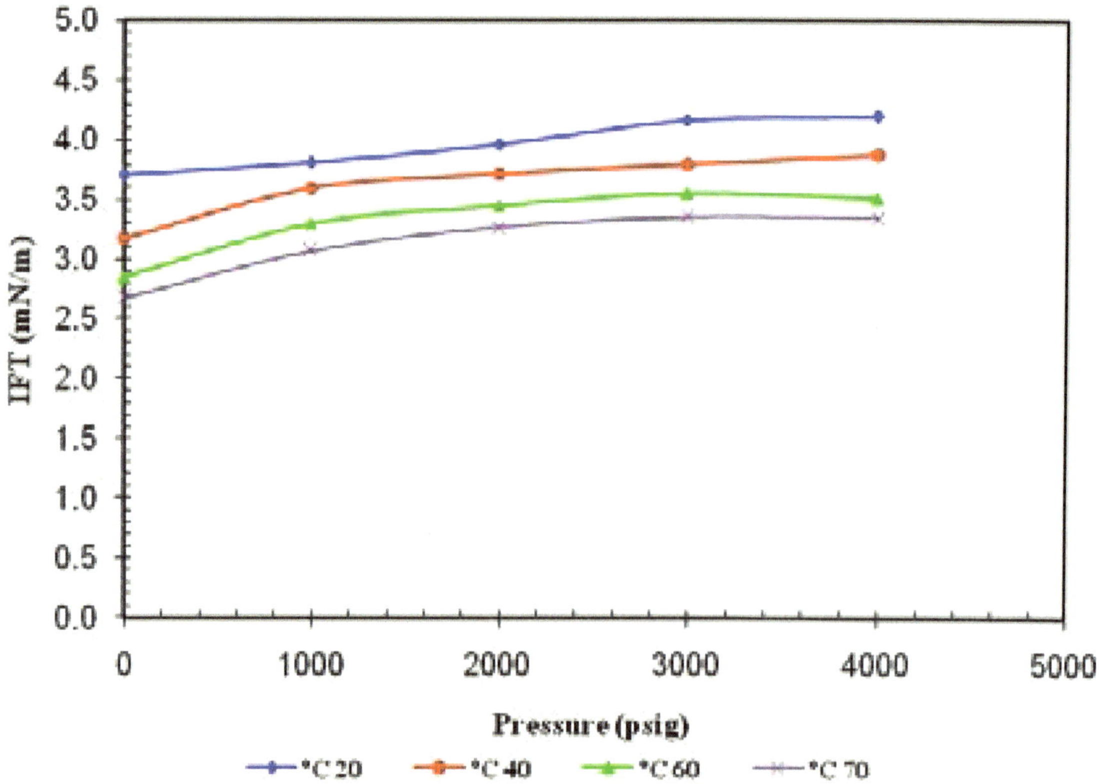

Figure 8: Effect of Pressure on 500 ppm Ammoeng 102 Solution-Crude Oil IFT.

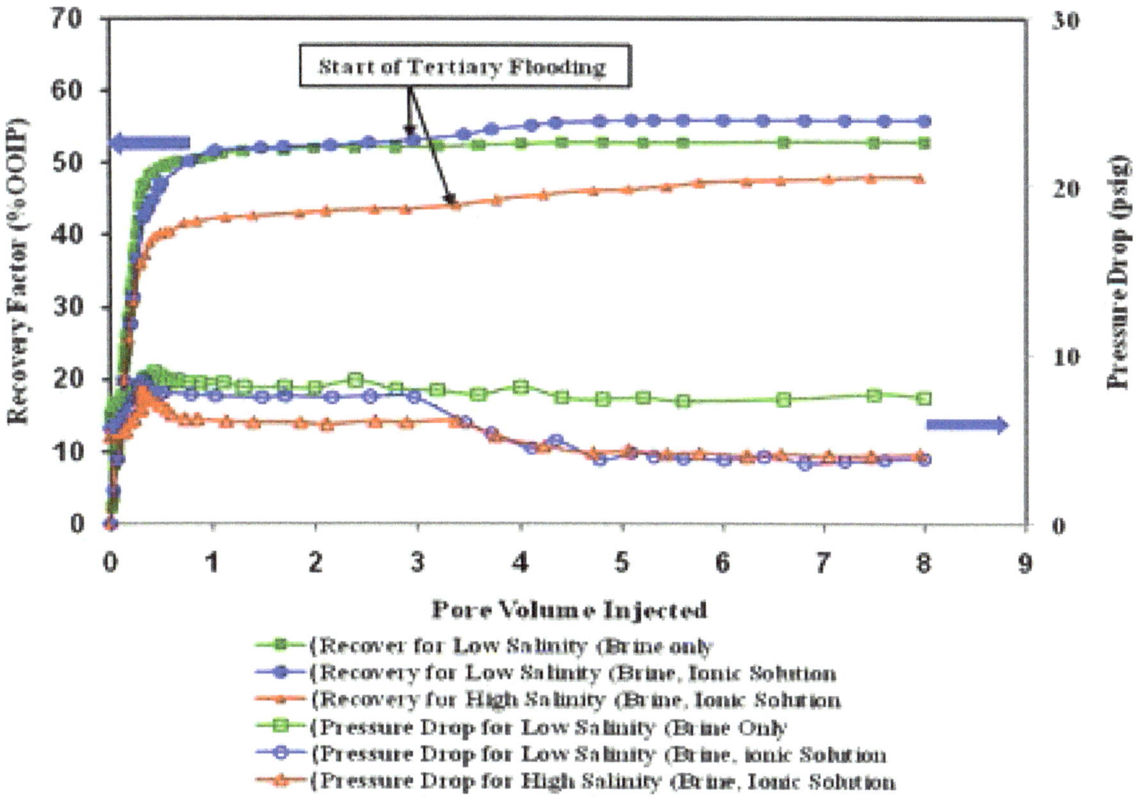

Figure 9: Recovery and Pressure Drop Curves of High and Low SalinitiesFlooding.

To check the reproducibility of our work, third run was conducted with 10% brine in secondary flooding mode on a fresh Berea core sample from the same block. The recovery was compared to that obtained previously in secondary flooding portion of run 1. Recovery curves plotted in Figure 9 was closely identical to that obtained in the first run indicating that salinity variation is the sole effect on flooding results.

The trend of pressure drop for all runs is presented in Figure 9. There exists a small difference in pressure during the secondary mode of runs 1 and 2 with higher pressure for low salinity secondary flooding (10% brine flooding). This difference diminishes as we start the tertiary flooding with ionic solutions with decreasing trend for both salinities. The slight drop in pressure is attributed to the lower IFT values between displacing and displaced phases. Degree of salinity has minor effect on IFT as seen on Figure 6, therefore, pressure profiles during tertiary flooding is overlapped.

Conclusion

This study aimed at investigating the efficiency of ILs for chemical EOR. The work started by screening several ILs based on the solubility in brines, thermal stability and ability to significantly reduce the oleic-aqueous phases IFT. Based on the screening process, Ammoeng 102 was the favorite IL. Further investigation of IFT between Ammoeng 102 diluted in brine III at 10% and 20% (w/w) salinity and medium Saudi crude oil was conducted. IFT measurement for different IL concentrations at 2000 psig and 60oC indicates IFT values exponentially decrease with increasing IL concentration. The CMC value for both solution salinities was 250 ppm with lower IFT values for high salinity ionic solutions. Ammoeng 102 solution- crude oil IFT was insignificantly affected by temperature and pressures within the ranges investigated.

Two tertiary flooding runs were conducted on Berea sandstone samples using 500 ppm of Ammoeng 102 diluted in Brine III at 10% and 20% salinity at reservoir condition. The obtained results indicated that flooding with10% brine salinity is more efficient than 20% in secondary mode, while the opposite state is taking place in tertiary flooding by the use of ionic solutions. Therefore, we suggest starting with low salinity brine secondary flooding followed by high salinity ionic solution for optimum recovery.

Acknowledgments

The authors express their sincere thanks and appreciations to the Department of Petroleum and Natural Gas Engineering in King Saud University and the oil and gas Research Institute at King Abdulaziz City for Sciences and Technology (KACST, Riyadh, KSA).

References

1. Domanska U (2005) Solubilities and thermophysical properties of ionic liquids. Pure Appl Chem 77: 543-557.
2. Simoni LD, Lin Y, Brennecke JF, Stadtherr MA (2008) Modeling Liquid–Liquid Equilibrium of Ionic Liquid Systems with NRTL, Electrolyte-NRTL, and UNIQUAC. Ind Eng Chem Res 47: 256-272.
3. Johnson KE (2007) What's an Ionic Liquid? Interface-Electrochemical Society 16: 38-41.

4. Bermúdez MD, Jiménez AE, Sanes J, Carrión FJ (2009) Ionic liquids as advanced lubricant fluids. Molecules 14: 2888-2908.

5. Murillo H, Jose A, Aburto J (2011) Current Knowledge and Potential Applications of Ionic Liquids in The Petroleum Industry. Ionic Liquids: Applications and Perspectives 18: 439-458.

6. Rodríguez H, Arce A, Soto A (2012) Liquid-liquid interfacial tension of equilibrated mixtures of ionic liquids and hydrocarbons. Science China Chemistry 55: 1519-1524.

7. Fröba AP, Kremer H, Leipertz A (2008) Density, refractive index, interfacial tension, and viscosity of ionic liquids [EMIM][EtSO4], [EMIM][NTf2], [EMIM][N(CN)2], and [OMA][NTf2] in dependence on temperature at atmospheric pressure. J Phys Chem B 112: 12420-12430.

8. Yan Y, Qiwei Y, Huabin X, Baogen S, Qilong R (2012) Surface/Interfacial Properties and Structures of Ionic Liquids. Progress in Chemistry 24: 659-673.

9. Bai JM, Fan WY, Nan GZ, Li SP, Yu BS (2010) Influence of Interaction between Heavy Oil Components and Petroleum Sulfonate on the Oil–Water Interfacial Tension. J Disper Sci Technol 31: 551-556.

10. Al-Sahhaf T, Elkamel A, Ahmed AS, Khan AR (2007) The Influence of Temperature, Pressure, Salinity, and Surfactant Concentration on the Interfacial Tension of the N-Octane-Water System. Chem Eng Commun 192: 667-684.

11. Amir B, Philip B (1990) The Influence of Temperature and Surfactant Concentration on Interfacial Tension of Saline Water and Hydrocarbon Systems in Relation to Enhanced Oil Recovery by Chemical Flooding. Society of Petroleum Engineers.

12. Yousef AA, Al-Saleh S, Al-Kaabi A, Al-Jawfi M, Aramco S (2010) Laboratory Investigation of Novel Oil Recovery Method for Carbonate Reservoirs. Society of Petroleum Engineers.

13. Yousef AA, Al-Saleh S, Al-Kaabi A, Al-Jawfi M, Aramco S (2011) Laboratory Investigation of the Impact of Injection-Water Salinity and Ionic Content on Oil Recovery From Carbonate Reservoirs. SPE Reserv Eval Eng 14: 578-593.

14. Tang GQ, Morrow NR (1997) Salinity, Temperature, Oil Composition, and Oil Recovery by Waterflooding. SPE Reservoir Eng 12: 269-276.

15. Wu S, Firoozabadi A (2009) Effect of Salinity on Wettability Alteration of Porous Media From Liquid Wetting to Intermediate Gas Wetting. Society of Petroleum Engineers.

16. Wu S, Firoozabadi A (2010) Effect of Salinity on Wettability Alteration to Intermediate Gas-Wetting. SPE Reserv Eval Eng 13: 228-245.

Geochemistry of Evaporites in Lacustrine Basin, Dongying Depression, Bohai Bay Basin, China

Zhonghong Chen[1]*, Jiqing Zhang[2], Li Wang[2] and Ming Zha[1]

[1]*School of Geosciences, China University of Petroleum, Qingdao, Shandong, China*

[2]*Petroleum Developing Center in Shengli Oilfield, SINOPEC, Dongying, Shandong, China*

Abstract

The middle and lower parts of the Fourth Member of the Shahejie Formation in the Dongying Depression are representatives of evaporites and mud rocks in Eastern China. To investigate hydrocarbon potential, some deep wells, such as the Hk1 well and Fs2 well, were intensively cored and evaluated by geochemistry.

The total salinity of the formation waters is very high and is dominated by $CaCl_2$, suggesting favorable seal development and preservation of petroleum. The content of Al, Mn, and Ba and the ratios of V/Ni and (Fe+Al)/(Ca+Mg) were low, whereas the content of Na, Ca, and Mg and the ratios of Sr/Ba, Fe/Mn, and Sr/Ca were high in the evaporites and mudstones of the Fourth Member compared with those in the mudstones of the Third Member. Sr content was remarkably high. The co-sedimentary system of evaporites and mudstones displays a heavier isotope of carbon. The values of $\delta^{13}C$ for different group compositions and n-alkanes ranged from -28‰ to -21‰ and -18‰ to -32‰, the fractionation of $\delta^{13}C$ between group compositions was no greater than -2‰, and $\delta^{13}C$ of C_{17} and C_{18} significantly fluctuated. $\delta^{13}C$ for high-maturity evaporite source rocks was significantly increased by thermal evolution and thermochemical sulfate reduction. High content of gammacerane and low Pr/Ph was exhibited in the evaporite system compared with that in the non-evaporite system. Different responses of biomarker parameters for the different sedimentary systems were exhibited, such as $C_{19}/(C_{19}+C_{23})$ terpanes, $C_{29}/(C_{27}+C_{28}+C_{29})$ steranes, and C_{24}/C_{23} and C_{22}/C_{21} tricyclic terpanes. The reverse trend of the maturation parameters at depths corresponding to abundant carbonate minerals indicates that carbonate retards the thermal evolution of organic matter. Evaporites and mudstones can generate and expel hydrocarbons. The tested samples were mostly types I and II_1 of organic matter, and their original generating capacity can reach 40 and 20 mg/g rock, respectively. The efficiency of hydrocarbon expulsion reached 60%, but the distribution of organic matter and its generative potential were highly variable. In general, the mudstones show greater generative potential than the evaporites.

Keywords: Mineral element; Carbon isotope; Evaporite; Hydrocarbon generation; Hydrocarbon expulsion; Dongying depression

Introduction

The formation, physical properties [1-4] and hydrocarbon potential of evaporites and mudstones has always been of interest to the petroleum industry [5,6]. Evaporite rocks are widely developed in petroliferous basins around the world and closely relate to the formation and distribution of the global giant oil and gas fields. Specifically in China evaporites are widespread in some Tertiary rifted lacustrine basins (Figure 1). For instance, the Jiyang, Jizhong, Dongpu, and Qianjiang Depressions are all piled with organic-rich gypsum-salt rocks and mudstones and form typical co-sedimentary systems. High bio-productivity in evaporite sedimentary environments that contain carbonate, sulfate, or chloride salt accumulates organic-rich sediment that may serve as source rocks [7-9].

The Dongying depression is a typical lacustrine basin containing evaporites and mudstones in eastern China. Dark mudstone is mainly deposited in the upper part of Member 4 (Es_4) and the lower and middle parts of Member 3(Es_3) of the Shahejie Formation in the Paleogene system; these parts are regarded as main hydrocarbon source rocks in the Depression [10-13]. However, exploration for many wells in the northern slope of the Depression has generated high yield oil and gas from the evaporites and mudstones in the middle part of Member 4. This yield is a favorable prospect for oil and gas exploration and thus arouses considerable concern regarding evaporites and mudstones [9, 13]. However, the geochemistry of evaporites and mudstones has rarely been investigated for fewer wells drilled through deeply buried source rocks. Exploration has carried out few deep drillings, such as the Hao-1 Well and Fs2 well. On this basis, more intensive coring and calculating was conducted to study the geochemistry of evaporites and mudstones.

The study primarily aims to (1) better understand the response of mineral elements and related ratios in a saline lacustrine compared with those in a fresh lacustrine, (2) define the response of $\delta^{13}C$ in a saline lacustrine, (3) improve understanding of biomarkers in a saline lacustrine, and (4) determine the capacity of evaporites to generate and expel hydrocarbon.

Geological Setting and Evaporite Distribution

The Dongying Depression is a typical negative rifted tectonic unit located in the southern part of the Bohai Bay Basin and characterized as northern faulted, southern overlapped, and opening-shaped. The depression developed from ancient Paleozoic bedrock with various tectonic movements, covering an area of 5700 km^2.

A set of clastic materials is deposited in the Paleogene system and is dominated by mudstone intermixed with sandstone, carbonate, and gypsum-salt rocks. Under the impact of tectonic movement, the sediment has undergone a sedimentary process in three different stages: salted lake, deep lacustrine, and river delta. In the process, the main source rocks deposited in this depression are Es_4, deposited at a hypersaline environment, and Es_3, deposited at a fresh-brackish environment.

***Corresponding author:** Chen Zhonghong, School of Geosciences, China University of Petroleum, Qingdao, 266580, Shandong, China

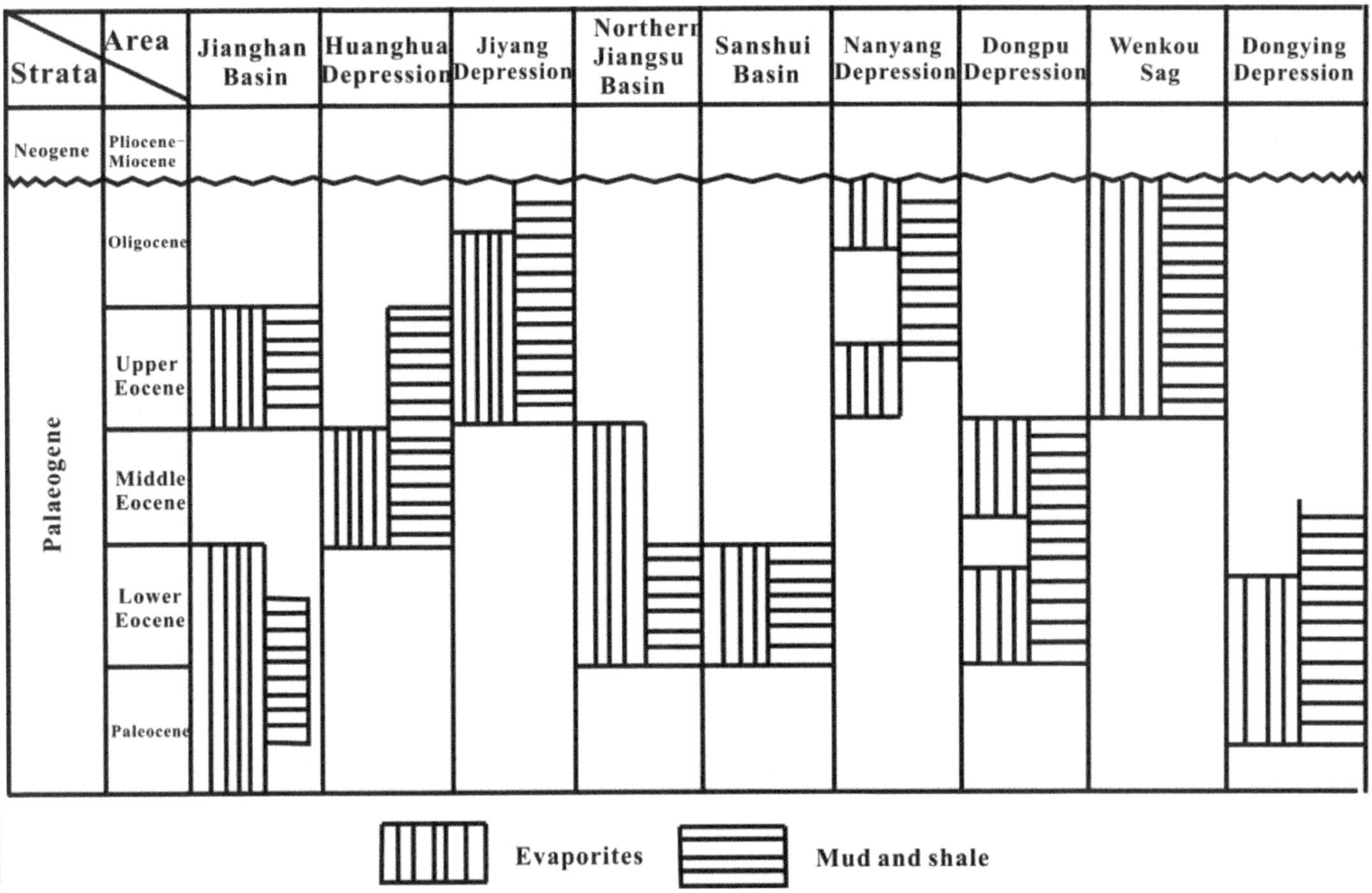

Figure 1: The distribution of the evaporates and mudstone in the Tertiary lacustrine basins in China.

In the lower part of Es_4, some brine-lake sediment, including gray shale, oil shale, salt rock, gypsum, and alluvial fan, is deposited in an evaporiting environment under weak tectonic activity and extremely hot and dry climate (Figures 2 and 3). In the central part of the lake during this period can be found about 400 m to 600 m of thick gray mudstone with salt rock and gypsum layers. In mid-Es_4 deposition, tectonic subsidence significantly increased, and a large rifted lake formed between high and continuous mountains. A large amount of lake water evaporated with result of high salinity. The highly saline center (>100 g/l) is distributed in the Minfeng sag of the northern basin, coinciding with the evaporite distribution (Figure 4). The salinity increased from the edge to the center of the salt lake, and some shales, marls, plasters, salts, and mudstones were deposited in sequence. In the alternation of oxygen environments with hypoxic environments, co-sediment of evaporites and mudstones was formed.

Experiment Methods

The contents of inorganic elements were measured by conventional ICP-AES and atomic absorption spectra in well drilling. Formation pressure and temperature were obtained by related testers in the drilling.

Sediment samples were crushed in a rotary mill, and aliquots were subjected to rock-eval pyrolysis to determine total organic carbon content, residual soluble hydrocarbon (S_1), pyrolysed hydrocarbon (S_2), production index, and hydrogen index (I_H), T_{max} (temperature of maximum kerogen pyrolysate yield). Vitrinite reflectance (R_o) was analyzed on randomly oriented grains by conventional microphotometric methods.

Chloroform is used as a solvent, and the samples are continuously extracted for 72 h at a constant temperature of 70°C in a Soxhlet extraction device. Ten to 15 milligrams of the samples were extracted, dissolved by petroleum ether, repeatedly washed, and moved in a flask, standing for 24 h (about 100 ml of petroleum ether in flask). The solution was finally filtered. The insoluble part was asphaltene, and the soluble part was a mixture of saturated hydrocarbons, aromatic hydrocarbons, and non-hydrocarbons.

A prepared filling column of silica-alumina and alumina was activated for 5 h at a constant temperature of 450°C, and silica was activated for 8 h at a constant temperature of 150°C. Filled column chromatography was conducted with the silica and alumina at a ratio of 3:l (silica in the up of alumina). After the column was wet with petroleum ether, the sample filtrate was poured into the column and washed with petroleum ether, dichloromethane, and diethyl ether. Saturated hydrocarbons, aromatic hydrocarbons, and non-hydrocarbons were then obtained.

HP-Chemistation was used as a GC-MS analytical instrument (model: HP-5890 II type for GC, HP-5890A for MS; column: DB5-MS; column length: 60 m; internal diameter: 0.25 mm). The initial temperature was 100°C, the heating rate was 4°C/min, the final temperature was 320°C, and the final time was 20 min. The temperature of gasification and transmission pipeline was 310°C, the ion source

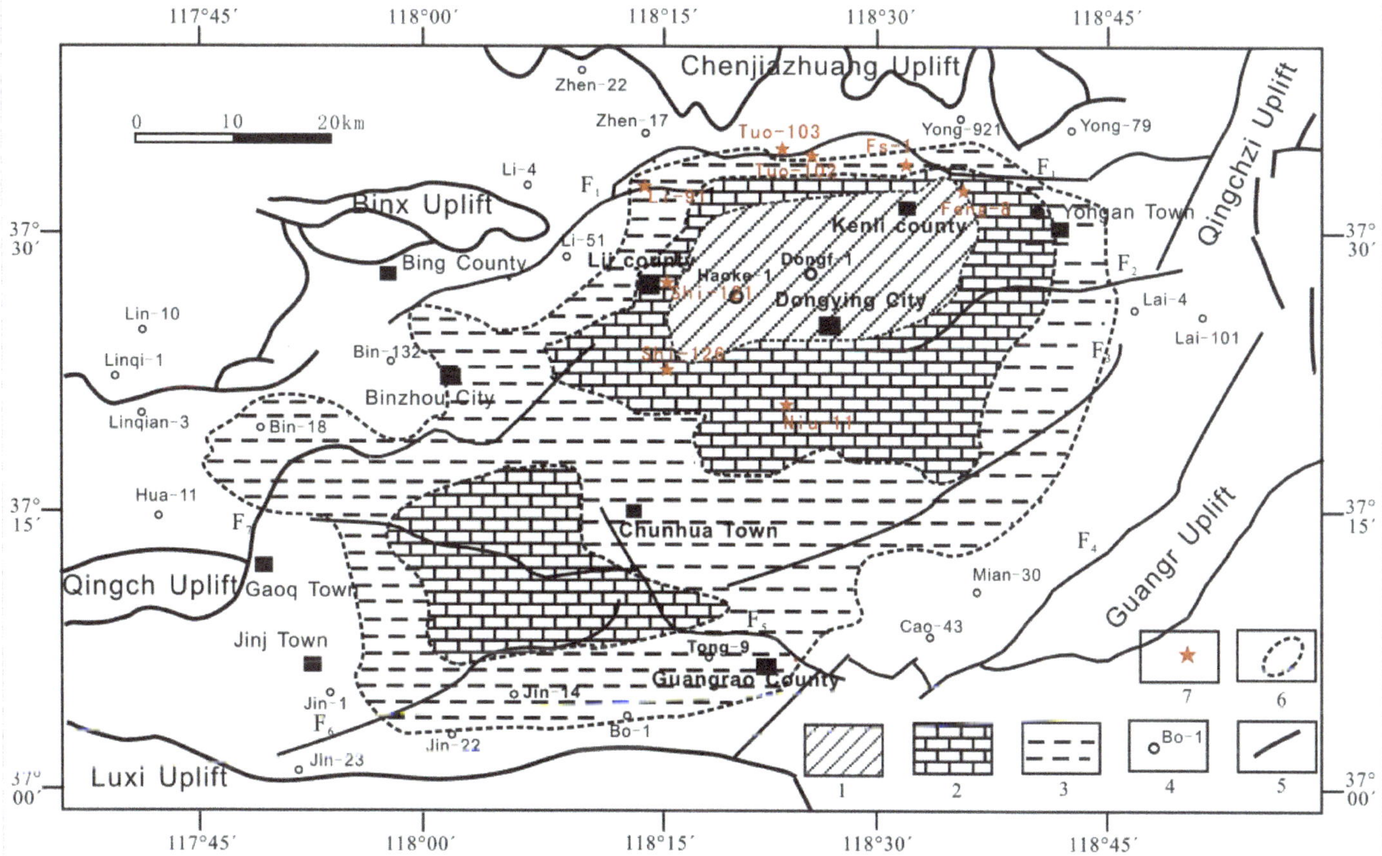

Figure 2: Sketch map showing the distribution of the co-sedimentary system of evaporites and source rocks in the middle part of Member 4 of the Shahejie Formation in the Dongying depression.
1. halite, gypsum; 2. limestone, dolomite; 3. mudstone; 4. wells; 5. faults; 6. boundary of different sedimentary systems; 7. the Wells for the samples coring.

temperature was 250°C, and the pre-column pressure was 170 kPa. The split ratio was 20:1, and the carrier gas was helium. Electron energy was 70 eV, the scan rate was 0.46 scans/s, and ionization was conducted by electronic impact.

Biomarkers were identified on the basis of mass fragment graphic responses and relative retention times, compared with a standard sample. Quantification was conducted by peak height responses in specific mass chromatograms, such as the m/z 191 and m/z 217 mass chromatograms.

The $\delta^{13}C$ values for n-alkanes were obtained by a Finnigan Mat GC-IRMS with a delta-S mass spectrometer. In a typical analysis, 1 μl of n-alkane concentrate was injected into a DB 5 chemically bonded fused silica capillary column (30 m × 0.32 mm i.d.) with splitless mode. Helium was used as a carrier gas. The GC oven temperature was programmed from 60°C to 120°C at 30°C min^{-1} and then to 310°C at 5°C min^{-1}. Injector temperature was 300°C. The measurement was conducted three or four times in most cases.

Results and Discussion

Inorganic elements in evaporites

In evaporate system the hydrocarbon generation is often associated with certain metals. For example, sulfate reducing agent can extract metal ions from salt water, and some metal ions can be selectively absorbed by the algae and higher aquatic plants around, which are released from decomposition of organic matter later. Amino acids generated from organic matter may improve the solubility of heavy metals. Therefore, the inorganic elements can reflect the sedimentary environment, water-rock interaction and hydrocarbon fluid flow.

The cores in the Paleogene system from the Hk1 Well were tested intensively and analyzed to investigate the inorganic elements of the evaporites and mudstone and to contrast the distribution rules for different sedimentary environments. The evaporites in Es_4 have low Al content (average=3.49%) but high Na, Ca, and Mg content (average=7.03%, 10.11%, 2.68%, respectively) compared with the mudstone in Es_3. Among the trace elements, Mn and Ba in the evaporites have significantly low content (average=697.9 × 10^{-6} and 473.6 × 10^{-6}, respectively), whereas Sr content is remarkably high (average=1135.35 × 10^{-6}). The ratios V/Ni and (Fe+Al)/(Ca+Mg) are low (average=1.99 and 0.60, respectively), whereas the ratios Sr/Ba, Fe/Mn, and Sr/Ca are high (average=3.99, 85.59, and 0.019, respectively). Particularly, the Sr/Ba ratio in the Es_4 evaporite system is five times that in the Es_3 non-evaporite system.

The chemical field of formation water also reflects the strong response of some major rock-forming elements in the evaporite system. Rock-forming elements sodium (Na), calcium (Ca), and magnesium (Mg) are the main cations in formation water, and their contents are well correlated with salinity. The salinity in the evaporite system is remarkably high, ranging from 1.63 g/l to 335.50 g/l and averaging 63.3 g/l.

The distribution of inorganic elements in the evaporites and

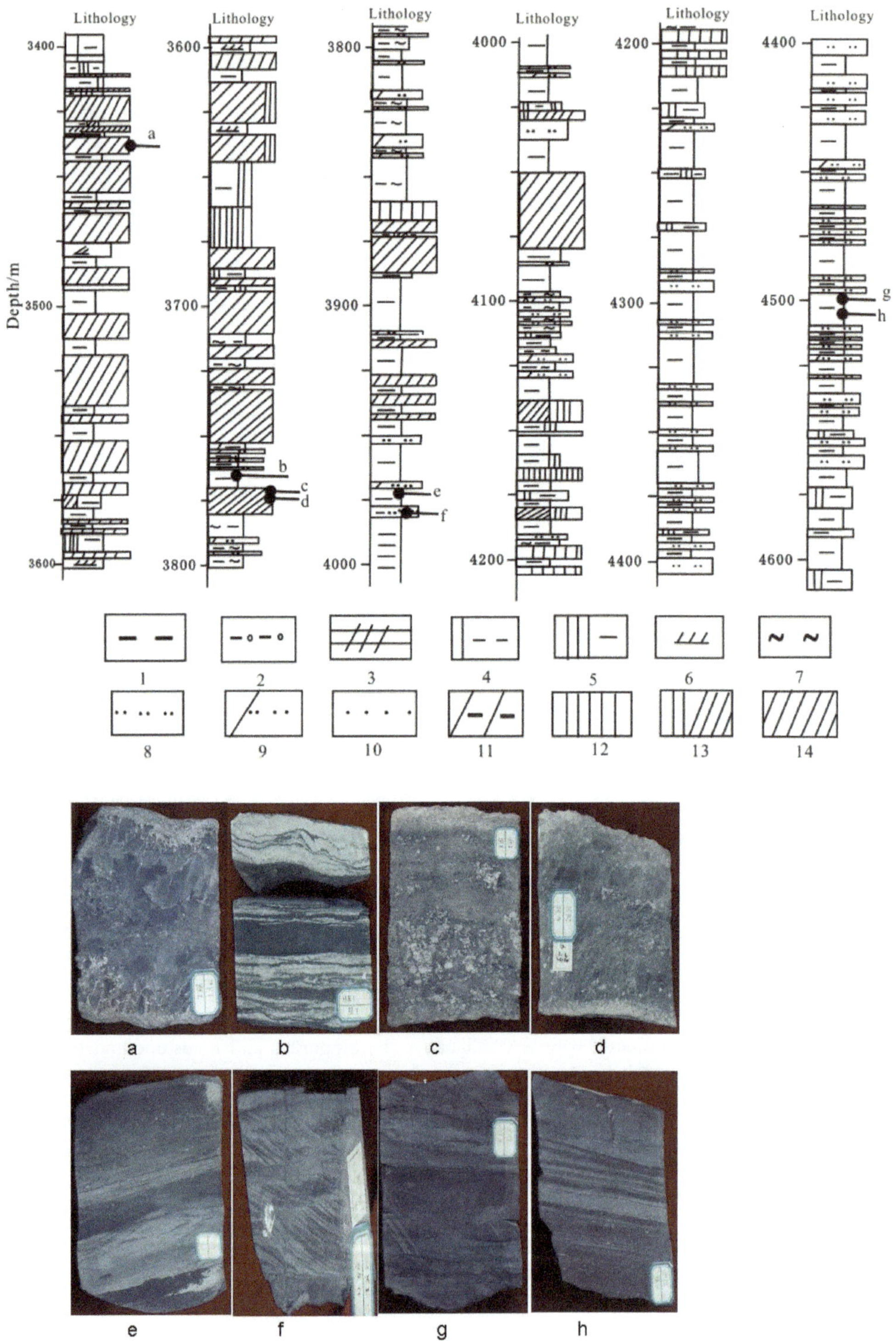

Figure 3: The lithologic log of the co-sedimentary system of evaporites and mudstone and their representative photos in the Es4 from Hk1 Well in Dongying depression.
a:3438.48m, salt rock, salt grain in the middle is larger than the edge of the salt rock.
b:3769.65m, dark gray mudstone with thin layers of salt and a deformation of bedding.
c: 3771.45m, salt rock with two rhythmic.
d: 3772.2m, salt rock with three rhythmics
e: 3972.1m, purple mudstone with white light-colored and rhythm with deformation
f: 3975.65m, purple argillaceous siltstone with bedding inclined plate
g: 4499.3m, dark purple-red mudstone with mud-gravel
h: 4506.6m, dark-purple, light gray rhythm bedding with calcite particles and bedding deformation
1, mudstone; 2,sandy mudstone; 3, oil shale; 4, calcareous clay; 5, gypsum mudstone; 6,saline mudstone; 7, soft mudstone; 8, siltstone; 9, dolomitic siltstone; 10, fine sandstone; 11, argillaceous dolomite; 12, gypsum; 13, gypsum salt rock; 14. Salt rock

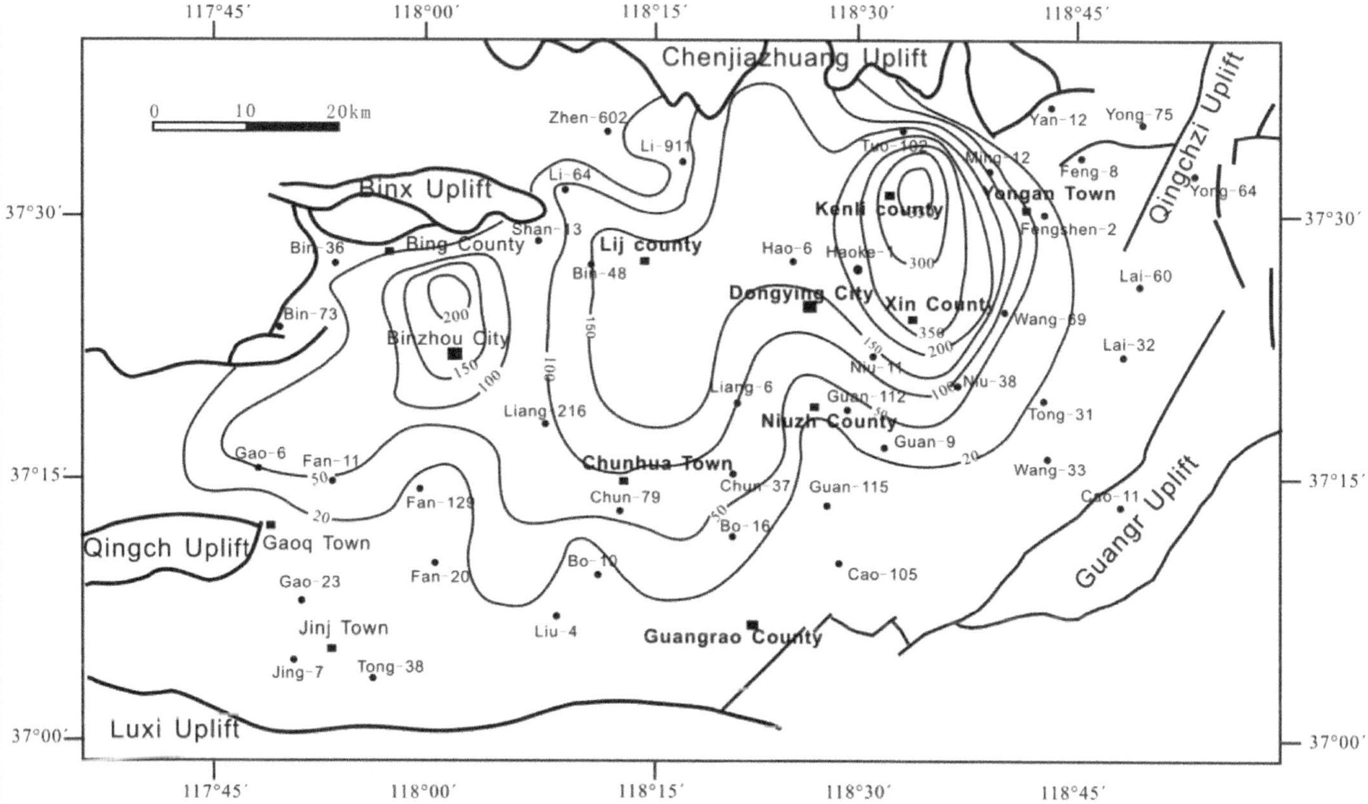

Figure 4: The distribution of total salinity (g/L) in the lower part of 4th Member of Shahejie Formation from the Paleogene System in Dongying depression.

mudstone of Es_4 shows distinct discreteness and fluctuation (Figure 5). The content of most inorganic elements is high in the mudstone but low in adjacent evaporites, except Na (Table 1). Such distribution is supposed to be associated with the special nature of the dissolution and adsorption of the inorganic element. When organic matter generates hydrocarbon, the organic acids generated can dissolve soluble inorganic salts in evaporites, and dissolved inorganic elements can easily migrate to adjacent mudstones and can accumulate related elements in clayed mudstones. The loss of elements in gypsum-salt for the clayed mudstones has a comparatively large surface area and strong adsorption. The evaporites and mudstone containing many catalytic transitional elements benefit the generation of oil and gas [14-16].

Water type in the evaporate system

The formation water chemical characteristics are reflection of different sedimentary and diagenetic environment, and ionic substances are the result of water-rock interaction and hydrocarbon accumulation. The ions Cl^-, SO_4^{2-}, HCO_3^-, Na^+, K^+, Ca^{2+} and Mg^{2+} were known as the main ions in formation water in oil fields. In a closed hydrological system the composition of concentrated lake brines largely depends on labile lithologies that are leached and altered by inflow waters in the drainage basin surrounding a salt lake [17,18]. For example, flow across and through limestone aquifers produces inflow waters rich in Ca and HCO_3, dolomite dissolution generates Mg-rich waters, igneous and metamorphic matrices tend to yield silica-rich Ca–Na–HCO_3 waters, and pyritic shales and other sulfur-rich sediments, such as tuffaceous ashfalls, will contribute inflows rich in sulfate, whereas basic and ultrabasic rocks tend to produce alkaline Mg–HCO_3 waters [19].

Due to the groundwater evaporation, concentration and salt precipitation chloride predominate in the groundwater with main NaCl salt. The long-term interaction between Water and rock make the greatly cation exchange effect for the absorbed Ca^{2+} in the rock with Na^+ in groundwater, and the exchange reduced the dissolution of the Na^+, increase the dissolved Ca^{2+}, resulting in the formation of $CaCl_2$ type water. Generally, $CaCl_2$ type water originates from the deep earth, and reflects a reductive environment with water stagnated and salt highly concentrated. Here, the formation water is dominated by $CaCl_2$, and some water shows $NaHCO_3$. The Na_2SO_4 and $MgCl_2$ types of formation water are rare. The development of $CaCl_2$ type water in the evaporites suggested favorable seal development and petroleum preservation.

Carbon isotope response to evaporites

The value of the carbon isotopes of crude oil ($\delta^{13}C$) mainly depends on organic matter. High $\delta^{13}C$ is displayed in many hypersaline environments, including some basins in China, such as Qaidam Basin [20], Qianjiang Depression in Jianghan Basin, [21] and Bohai Bay Basin [22]. The heavier carbon isotope is usually associated with the organic matter source mainly dominated by algae and aquatic bacteria [23,24].

High $\delta^{13}C$ is also shown in the evaporites and mudstones. The $\delta^{13}C$ of alkanes, aromatics, and non-hydrocarbons is greater than -27.6‰, -25.5‰, and -25.7‰, respectively, in the intervals of 3963 m to 3973 m and 4024 m to 4501 m of the Fs2 Well in the Dongying Depression (Table 2 and Figure 6). At the same time, the fractionation of $\delta^{13}C$ between different group compositions is relatively small, and the difference is less than 2‰ mostly. The highest fractionation occurs between the saturated hydrocarbons and aromatic hydrocarbons, reflecting the characteristics of $\delta^{13}C$ fractionation in salted lakes [25].

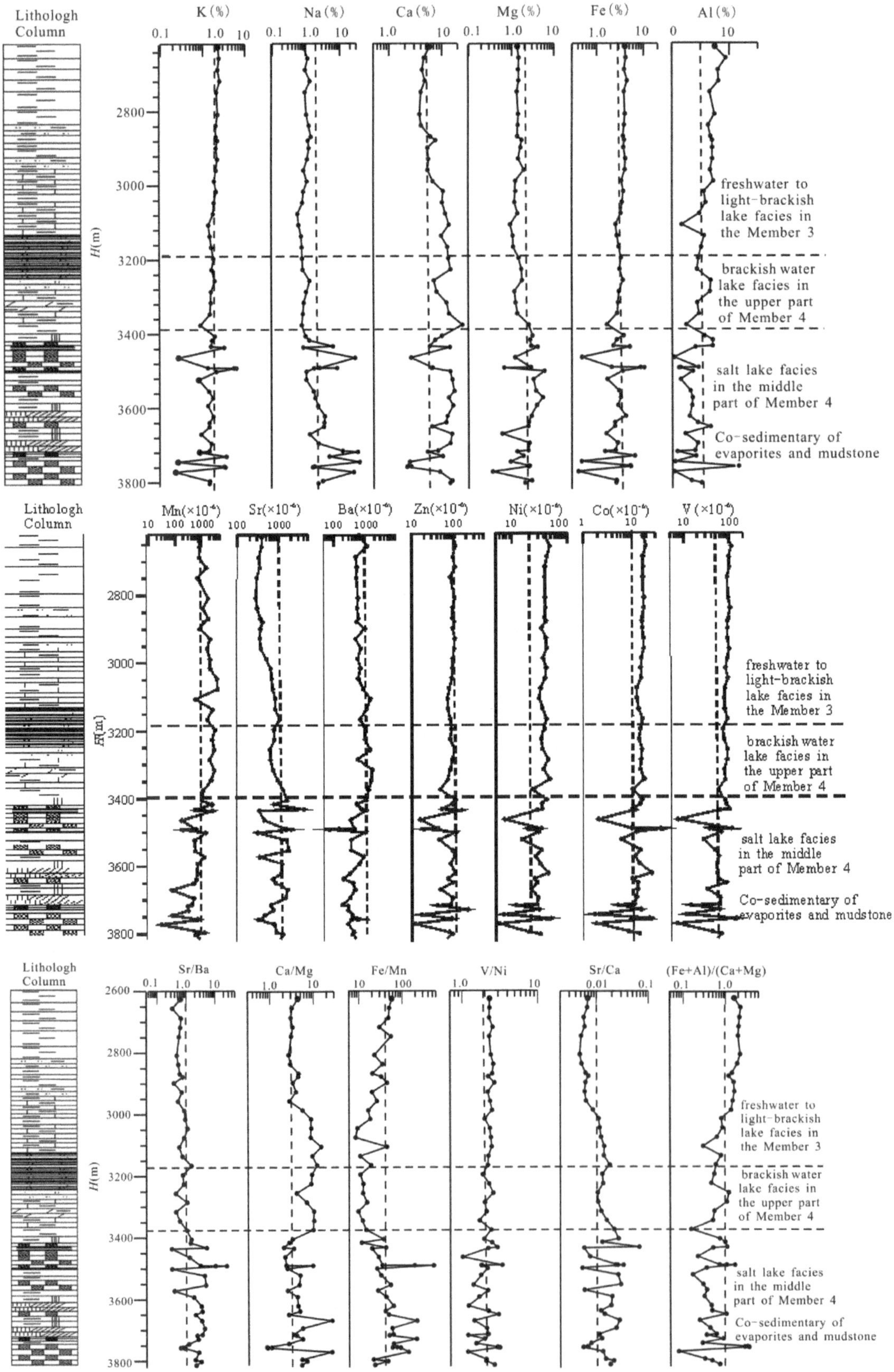

Figure 5: The distribution of the content of mineral elements in the Paleogene system from Hk1 Well in Dongying depression.(a, major minerals; b, trace minerals; c, element ratio).

Depth (m)	Lithology	Na (%)	K (%)	Ca (%)	Mg (%)	Fe (%)	Al (%)	Mn (10^{-6})	Ni (10^{-6})	Zn (10^{-6})	Sr (10^{-6})	Co (10^{-6})	Ba (10^{-6})	V (10^{-6})	Sr/Ba	Ca/Mg	V/Ni	Fe/Mn
3650	Gray mudstone	3.23	1.39	6.38	2.38	2.4	6.67	495.38	20.07	92.75	829.9	12.91	398.2	86.04	2.08	2.68	3.03	48.45
3659	Dark-gray mudstone	2.27	1.56	14.32	2.56	2.75	4.45	470.54	29.32	92.64	1399.7	12.14	324.1	62	4.32	5.59	2.11	58.44
3715	Gypsum-salt	11.52	1.39	8.27	2.4	2.47	3.96	399.93	27.82	81.5	657.7	10.96	184.4	64.59	3.57	3.45	2.32	61.76
3718	Gypsum-salt	32.26	0.81	5.25	1.3	1.47	0.84	293.49	11.85	41.45	652.6	5.05	277.3	14.02	2.35	4.04	1.18	50.09
3730	Gray mudstone	4.65	3.53	10.21	1.83	6.37	4.11	298.16	30.45	204.7	735.3	14.94	281.5	59.88	2.61	5.58	1.97	213.6
3745	Gypsum-salt	36.72	0.25	0.24	0.9	0.46	0.47	73.25	7.14	16.89	458.0	1.69	377.4	10.89	1.21	0.27	1.53	62.80
3772	Gypsum-salt	24.61	0.22	9.02	0.34	0.39	0.33	28.2	5.88	11.76	770.2	1.84	283.9	6.93	2.71	26.5	1.18	138.3
3802	Purple-red mudstone	2.22	1.39	14.11	2.01	2.54	5.08	547.92	32.09	89.59	1247.7	14.1	379.5	66.32	3.29	7.02	2.07	46.36
3856	Purple-red mudstone	3.65	2.01	3.92	2.42	1	6.8	403.09	34.79	117.8	801.8	13.45	816.4	100	0.98	1.62	2.87	24.81
3862	Gypsum-salt	27.08	0.28	3.4	0.58	0.46	0.53	68.91	7.09	16.63	351.5	1.94	254.4	8.91	1.38	5.86	1.26	66.75
3884	Gypsum-salt	34.98	0.2	3.17	0.29	0.33	0.42	28.3	4.67	11.25	368.8	2.45	262.4	7.23	1.41	10.9	1.55	116.6
3896	Purple	5.21	1.79	4.33	1.87	3.09	3.59	422.06	31.69	107.5	1429.4	12.81	583.0	90	2.45	2.32	2.84	73.21
3902	Purple	3.76	1.75	10.82	2.54	3.02	5.35	354.54	28.92	96.73	825.4	13.89	487.8	68.95	1.69	4.26	2.38	85.18
3914	Purple	3.95	1.44	8.67	1.98	2.66	5.65	643.14	25.95	97.61	618.7	12.93	262.6	60.7	2.36	4.38	2.34	41.36

Table 1: The distribution of mineral elements and some ration parameters in the co-sedimentary system of saline and mud in the middle part of Es_4 from the well Hk1 in Dongying depression.

Depth(km)	Lithology	Values of δ¹³C(‰)		
		alkane	aromatic hydrocarbon	non-hydrocarbon
3.963	Dark-gray salted mudstone	-27.6	-25.5	-25.7
3.968	Dark-gray salted mudstone	-27.3	-25.3	-25.6
3.971	Dark-gray salted mudstone	-27.4	-25.3	-25.5
3.973	Dark-gray salted mudstone	-27.4	-25.3	-25.6
4.024	Salted rock	-25.0	-23.9	-24.5
4.295	Laminated mudstone	-24.2	-22.8	-24.3
4.301	Laminated mudstone	-21.0	-21.0	-22.6
4.498	Dark-gray gypsum-salted mudstone	-22.4	-23.6	-23.6
4.499	Dark-gray gypsum-salted mudstone	-25.2	-24.3	-25.3
4.500	Dark-gray gypsum-salted mudstone	-22.1	-23.2	-24.2
4.501	Dark-gray gypsum-salted mudstone	-23.9	-24.3	-25.6

Table 2: The characteristics of carbon isotope of the group composition of the organic matter in the middle part of Member-4 of Shahejie Formation from the Well Fs2 in the Dongying depression.

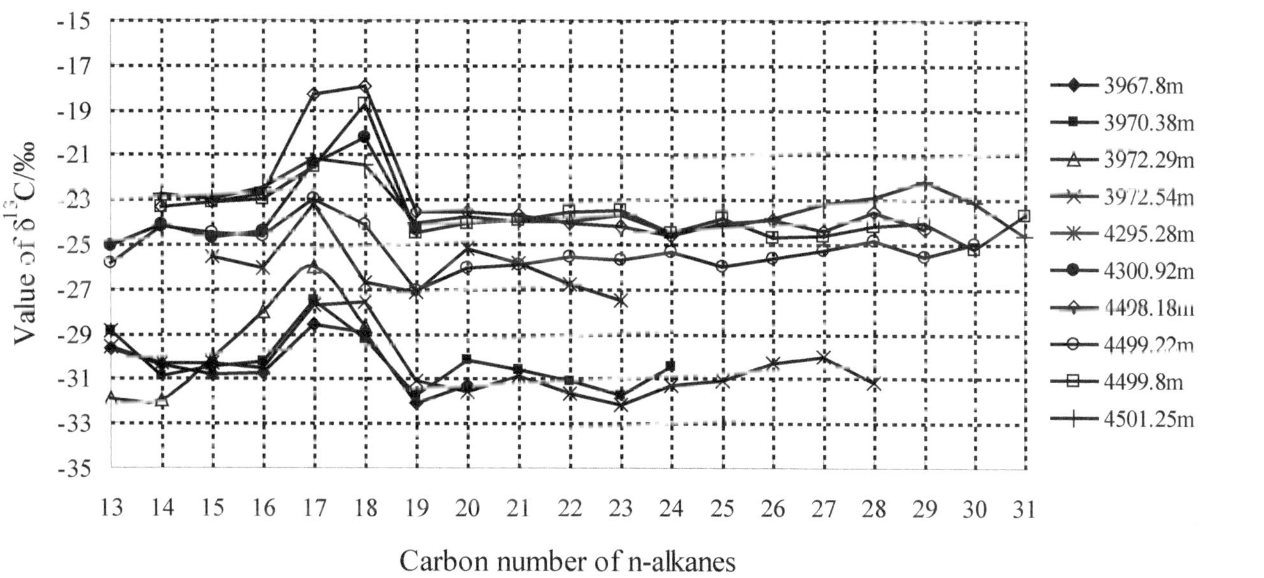

Figure 6: The characteristics of carbon isotope of n-alkanes from the source rocks of the middle part of Member-4 of Shahejie Formation from the Well Fs2 in Dongying depression.

The $\delta^{13}C$ of the n-alkane series only slightly varies, except a remarkably high $\delta^{13}C$ of C_{17} and C_{18}. The $\delta^{13}C$ of n-alkanes increase slightly with increasing carbon number, reflecting a characteristic of the $\delta^{13}C$ from mature alkanes in hypersaline lacustrines.

Comparatively, the values of $\delta^{13}C$ in the interval 3963 m to 3973 m are low, around -27.5‰, -25.5‰, and -25.5‰ for alkane hydrocarbon, aromatic hydrocarbon, and non-hydrocarbon, respectively. By contrast, the values of $\delta^{13}C$ in the interval 4024 m to 4501 m are higher, distributed in the ranges -25‰ to -21‰, -24.3‰ to -21‰, and -25.6‰ to -22.6‰, respectively. These values reflect the effect of thermal evolution on the distribution of $\delta^{13}C$ in similar sedimentary environments. The measured vitrinite reflectance (R_o) is 0.95% at a depth of 3972 m, whereas at 4500 m, R_o reaches 1.2%, indicating much higher thermal maturation.

The stability of various carbon bonds varies: ^{13}C-^{13}C > ^{13}C-^{12}C > ^{12}C-^{12}C. The generated hydrocarbon has lighter isotopic composition than its parent material because the ^{12}C-^{12}C bond is easier to break, resulting in a heavier isotope ($\delta^{13}C$) gathered in residual organic matter. This result explains the heavier carbon isotope distribution in the interval 4024 m to 4501 m.

The distribution of $\delta^{13}C$ was not only controlled by maturity but also resulted from the reaction Thermochemical Sulfate Reduction (TSR). TSR is a function of the interaction between sulfate and hydrocarbons, and in the process, sulfate minerals are deoxidized into acid gases, such as H_2S. In carbonate rocks, the TSR reaction is the widespread occurrence of geological phenomena and has been reported in some China basins [26,27]. A certain thickness of anhydrite, adequate hydrocarbons, and higher paleothermal energy are the prerequisites for TSR reaction. TSR reaction is a very complex chemical reaction that significantly changes carbon isotopes and enriches H_2S [28].

In TSR reaction with the consumption of hydrocarbon, the ^{12}C-^{12}C bond is broken with priority, and more ^{12}C is involved in the reaction. More ^{13}C is retained in the residual hydrocarbons; thus, residual hydrocarbon after the reaction is relatively enriched in ^{13}C, and the gained weight of $\delta^{13}C$ can reach 2‰ to 4‰ in the Sichuan basin [26]. In the interval 4024 m to 4501 m of the Fs2 Well in Dongying Depression, conditions exist for TSR reaction with higher content of sulfate and high temperature. As the differentiation of the $\delta^{13}C$ value in the two intervals (3963 m to 3973 m and 4024 m to 4501 m) reaches more than 5%, the TSR reaction contributes to $\delta^{13}C$ distribution.

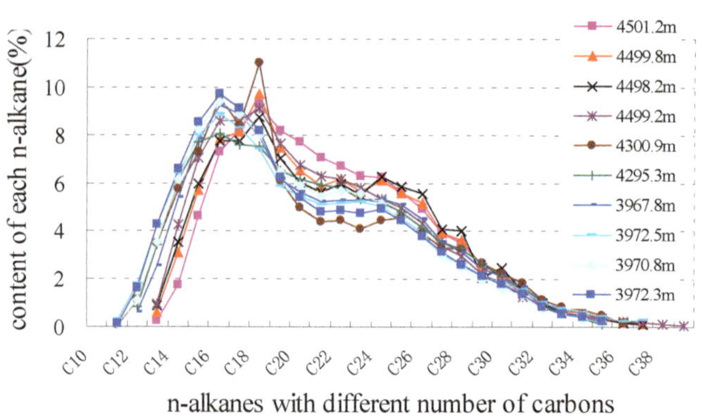

Figure 7: The distribution of n-alkanes from high-matured source rocks deposited in the evaporated system of the Well Fs2 in Dongying depression.

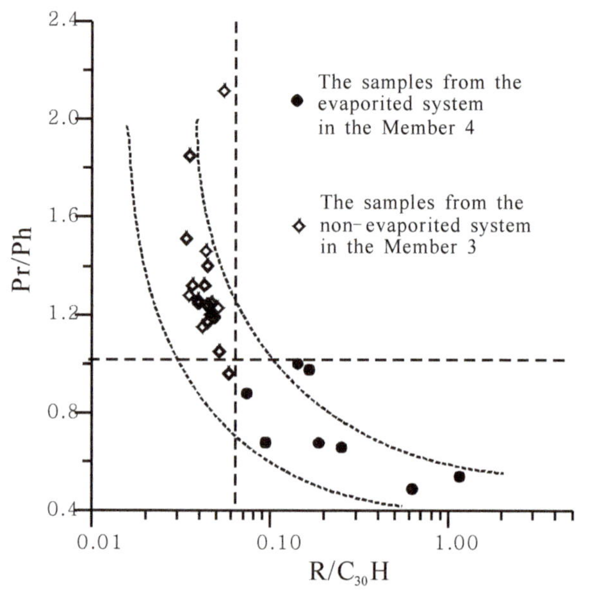

Figure 8: The distribution of Pr/Ph and $\gamma/C_{30}H$ from the source rocks in the evaporated system of E_{s4} and non-evaporated system of E_{s3} in Dongying depression.

Biomarker response in evaporite system

Source-related biomarker parameters: The mature to highly mature hydrocarbon source rocks in the evaporite system tend to have a predominance of even carbon numbers, showing the "single-peak" distribution of n-alkanes (Figure 7). The peak carbon is located between nC_{16} and nC_{20}, and most of the peak carbons are nC_{16} and nC_{18}. The distribution of Pr/Ph is well differentiated by the two class of samples, and the obvious advantage of phytane (Pr/Ph<1.0) is demonstrated in the evaporite samples from Es_4 (Figure 8). Haven et al. [29] found that low Pr/Ph (<1) is typical of hypersaline environments, where the ratio appears to be controlled by the effects of variable salinity on halophilic bacteria [29]. Likewise, Schwark et al. [30] noted a general decrease in Pr/Ph during the deposition of the Permian Kupferschiefer sequence in Germany and interpreted it to indicate increasing paleosalinity [30]. Although the utility of Pr/Ph to accurately describe the redox state of paleoenvironments is limited by several factors, including variable source inputs, different rates of early generation, variations at higher maturity, and analytical uncertainty [31], the supporting relationship between Pr/Ph and the gammacerane index confirms the inferred salinity relationship.

In the series of tricyclic terpanes, gamma content is remarkably high, indicative of marked stratification in a salted water environment with halocline [32,33], because gammacerane indicates a stratified water column in marine and non-marine source-rock depositional environments [6], commonly resulting from hypersalinity at depth (Figure 9). The value 0.065 of $\gamma/C_{30}H$ is a division for source rocks between the evaporite system and non-evaporite system. For the evaporite system, the value of $\gamma/C_{30}H$ is generally higher than 0.065, but not for the non-evaporite system. This result indicates that increased water salinity during the deposition of source rocks increases gammacerane indices and reduces pristine/phytane, corresponding to the result of the investigation of the biomarkers of oils from lacustrine source rocks in Angola [32]. In the distribution of hopane series with a long chain, the characteristics of C_{35}hopane>C_{34}hopane>C_{33}hopane are shown, and the baseline shifts (Figure 9), fully reflecting the characteristics of biomarkers in a depositional environment with high salinity [5, 34-36]. At the same time, the differentiation of biomarker

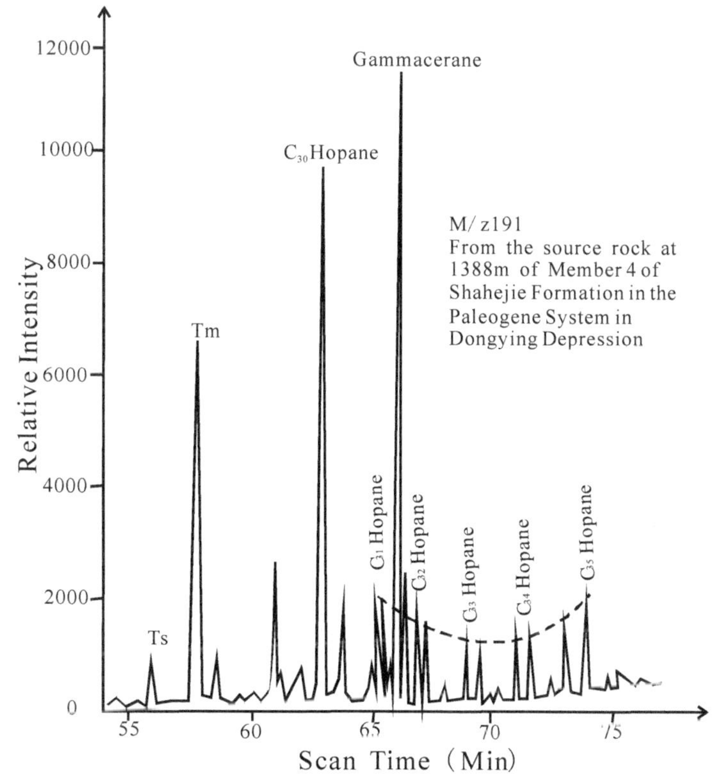

Figure 9: The distribution of a typical response of gammacerane and the hopanes series in the m/z191 in the evaporated system in Dongying depression.

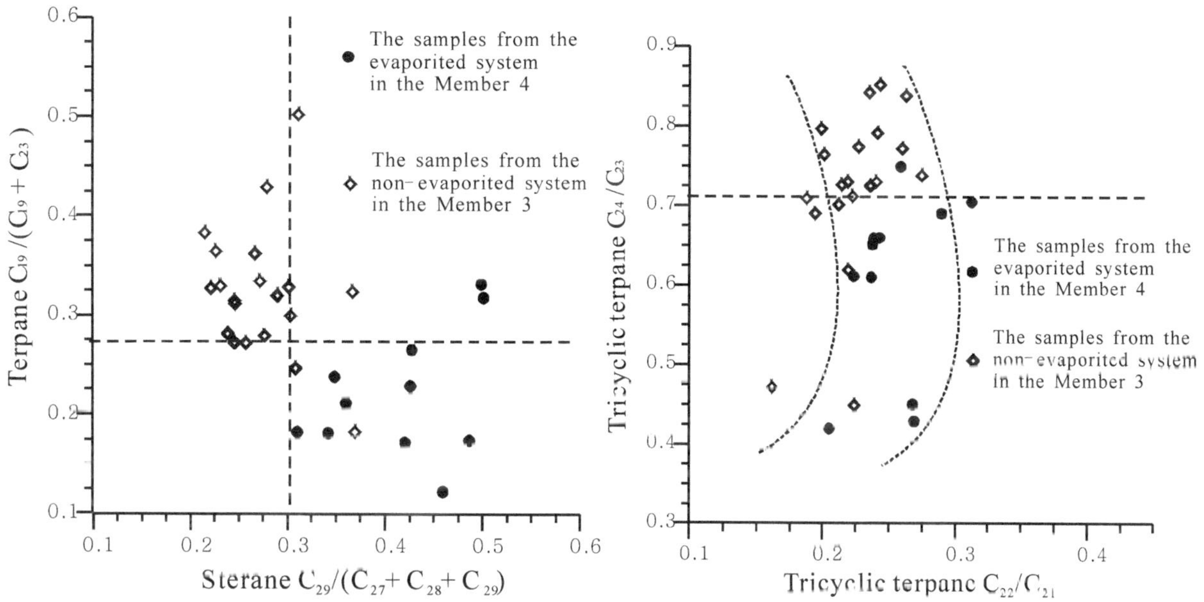

Figure 10: The distribution of $C_{19}/(C_{19}+C_{23})$ from terpanes and $C_{29}/(C_{27}+C_{28}+C_{29})$ from steranes (a) and the distribution of C_{24}/C_{23} and C_{22}/C_{21} from tricyclic terpanes (b) in the different sedimentary systems in Dongying depression.

responses for different source rocks in the evaporite system of Es_4 and the non-evaporite system of Es_3 in the Dongying Depression is exhibited by other indicators, such as $C_{19}/(C_{19}+C_{23})$ from terpanes, $C_{29}/(C_{27}+C_{28}+C_{29})$ from steranes, and C_{24}/C_{23} and C_{22}/C_{21} from tricyclic terpanes (Figure 10). The divided values of indicators C_{24}/C_{23}, $C_{19}/(C_{19}+C_{23})$, and $C_{29}/(C_{27}+C_{28}+C_{29})$ are 0.70, 0.27, and 0.30, respectively. For the source rocks from the evaporite system, the values of C_{24}/C_{23} and $C_{19}/(C_{19}+C_{23})$ are generally lower than 0.70 and 0.27, respectively, whereas the values of $C_{29}/(C_{27}+C_{28}+C_{29})$ are generally higher than 0.30.

Biomarker maturity parameters: An analysis of many samples of the profile of the Paleogene System in the Dongying Depression reveals a distinct response of typical maturity-related biomarker parameters in the evaporite system compared with that in the non-evaporite system. The values of $C_{29}\beta\beta/(\beta\beta+\alpha\alpha)$, C_{29} 20S/(20S+20R), and Ts/(Ts+Tm) began negative reversion at a depth of around 3.5 km to 3.6 km in the trend of their evolution, consistent with the sharp increase of the carbonate content in the interval (Figures 11 and 12).

Regional geological data show that at a shallower burial depth (<3.57 km), the main source rocks are calcareous shale and clayed mudstone and that at a great burial depth (>3.57 km), the gypsum and salt rock content increases significantly and the evaporite and mudstone form a co-sedimentary system. Take the Hk1 Well as an example. Below 3.57 km, gypsum and salt rock developed, and in the symbiotic system of evaporite and mudstone, the content of elements, such as calcium and sodium, increased significantly than that in the upper layer; the ratios of Sr/Ba, Sr/Ca, and Fe/Mn showed a significant high-value response, indicating a typical high-salt environment [36]. In such a high-salt environment, the negative reversion of $C_{29}\beta\beta/(\beta\beta+\alpha\alpha)$, C_{29} 20S/(20S+20R), and Ts/(Ts+Tm) showed good relations with high levels of gypsum-salt, reflecting the effect of inhibition or retardation on thermal evolution by gypsum-salt minerals.

The retarding effect of carbonate minerals in the transforming process from organic matter to oil and gas is an important reason for the delay and backwardness of oil and gas generation in the source rocks of evaporite systems, and such retarding effect is mainly controlled by the absence of the catalysis of clay minerals [37]. The related thermal simulation experiment also showed that carbonate rocks have an anti-catalytic effect on hydrocarbon generation from kerogen and that increasing temperature significantly increases the anti-catalytic function [38]. Clay minerals in shale catalyze the transformation of kerogen, whereas calcites in gypsum-salt rock have a small or no role in the transformation of kerogen. In addition, organic matter in gypsum-salt rock is mainly found within cryptocrystalline or fine-grained carbonate minerals and is not easily released and thus may retard thermal evolution in evaporite systems [39].

Retardation was demonstrated by many immature to low-maturity

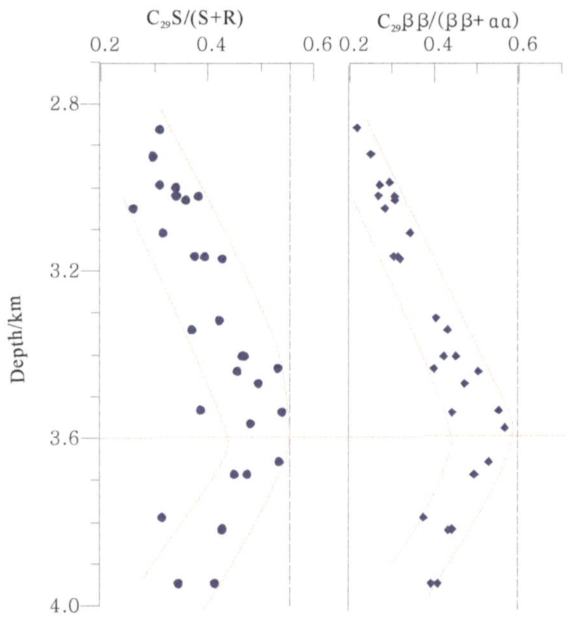

Figure 11: The distribution of the parameters of $C_{29}\beta\beta/(\beta\beta+\alpha\alpha)$ and C_{29}S/(S+R) in profile from E_{s4} in Dongying depression.

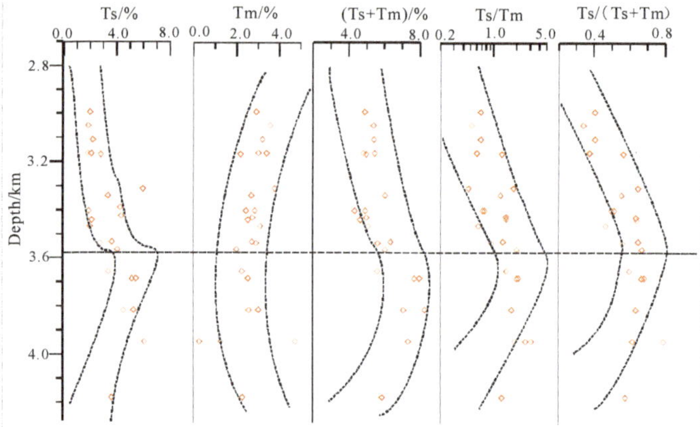

Figure 12: The distribution of the parameters of Ts/Tm, Ts/(Ts+Tm) in profile from E_{s4} in Dongying depression.

oils discovered in several co-sedimentary systems of evaporite and mudstone. Examples include source rocks from the Lower Cretaceous in the South Florida Basin, the Guang-33 Well in the Jianghan Basin, the Yuecan-1 and Kaiqian-67 Wells in the Qaidamu Basin, and the Bamianhe area in the Dongying Depression. For instance, in the Biyang Depression in China, the threshold depth of oil generation from the dolomites was 300 m deeper than that from the mudstone in the Paleogene system, and the maturity indicated by biomarkers was significantly negatively correlated to the carbonate content in rocks.

Thermal retardation in the evaporite system is not only reflected by biomarker parameters but also displayed in the distribution of conventional thermal indicators, including vitrinite reflectance (R_o) and pyrolysis peak temperature (T_{max}). The negative departure of R_o and T_{max} in the overpressured interval (high overpressure, the pressure coefficient reaches 1.8) from 2400 m to 3400 m suggests the inhibition of thermal evolution in response to overpressure [36]. The greater decline of R_o and T_{max} in the hypersaline interval from 3400 m to 4000 m indicates similar retardation because of the presence of carbonate minerals, as with related biomarker maturity ratios [36].

Hydrocarbon Generation and Expulsion from Evaporites

Group composition and organic matter type

The capacity of hydrocarbon generation and expulsion from the co-sedimentary system of evaporites and mudstone (salted lake phase) is of widespread concern. Organic matter and high-quality source rocks are enriched in the depositional sequence of carbonate, sulfate, and chloride salt in salted lakes. In different sequences, the composition of organic matter and thermal evolution show a certain difference.

Evaporites in Es_4 in the Dongying Depression are mainly composed of anhydrite and salt rock, and their paleosalinity is about 12% to 26% [40]. The high-salt environment may limit the growth of parasites and may be conducive to the formation of algal blooms and may thus increase the productivity of organic matter [13,41]. Moreover, the heat-conductive property of gypsum-salt is much higher than that of sand and shale; thus, source rocks in thick layers of gypsum rock can obtain heat earlier [9]. This early heating facilitates hydrocarbon generation.

To investigate the ability to generate and expel hydrocarbons from this co-sedimentary of evaporites and mudstone, a series of indicators was investigated from the Fs2 Well. The basic composition of the organic matter is sapropel (the dominant component) and vitrinite (Table 3).

The sapropel and vitrinite content is 78% to 92% and 8.3% to 22.3%, respectively, and few exinite and inertinite exist. The index of the organic matter type is 58.9% to 85.4%, representing types I and II of kerogen. The group composition of the organic matter is mainly alkane (content: 74.49% to 80.98%) and aromatics and non-hydrocarbons in lower proportion (aromatics: 3.13% to 11.36%; non-hydrocarbons: 4.71% to 15.25%). The maturity parameters of vitrinite reflectance (R_o) and pyrolysis peak temperature (T_{max}) are greater than 0.93% and 400°C, respectively, indicating that the samples have passed the peak period of oil generation. However, the distribution of maturity parameters in the co-sedimentary system shows an irregular linear increase with burial depth. The increased thermal conductivity of the evaporites reduces the maturity parameters in the evaporites compared with those in the mudstone. For instance, R_o and T_{max} pyrolysis in the mudstone at 4300.92 m are 1.17% and 419°C, respectively, whereas in the gypsum-salted mudstone at 4498.8 m to 4499.8 m, the three samples of R_o are distributed from 1.13% to 1.14%, and the T_{max} of two samples are 403 and 417°C, respectively.

Capacity for hydrocarbon generation and expulsion

A calculation model of hydrocarbon generation and expulsion was built based on geochemical information from pyrolysis [10]. A series of indicators was calculated to study the capacity for hydrocarbon generation and expulsion from the co-sedimentary system in the two intervals of the Fs1 Well.

Results show the stronger capacity of hydrocarbon generation and expulsion in the interval 3966 m to 3975 m (Figure 13): residual organic carbon content (C_t) ranged from 0.6% to 2.63% with an average of 1.54%; original organic carbon content (C_o) was distributed in the range 1.8% to 4.1%; the content of residual soluble hydrocarbons (S_1) and pyrolysis hydrocarbons (S_2) is distributed in the ranges 0.85 mg/g to 8.24 mg/g and 0.48 mg/g to 5.69 mg/g, averaging 4.31 and 2.83 mg/g, respectively; the remaining hydrocarbon generation potential (S_1+S_2) is distributed from 1.33 mg/g to 13.54 mg/g with an average of 7.14 mg/g; the soluble hydrocarbon index (*HI*, equivalent to $S_1 \times 100/C_t$) is distributed in the range 142 mg/g to 320 mg/g with an average of 255 mg/g, indicating that the organic matter entered the peak period of hydrocarbon generation. The conversion rate of pyrolysis (K_t, equivalent to $S_1 \times 100/(S_1+S_2)$ is high, generally greater than 55%, whereas the hydrogen index (*IH*, equivalent to $S_2 \times 100/C_t$) is significantly low (less than 200 mg/g). The original hydrocarbon generating potential (S_p) and the amount of hydrocarbon expelled (S_o) are distributed in 10 mg/g to 40 mg/g and 10 mg/g to 20 mg/g, respectively. The efficiency of hydrocarbon expulsion (K_p, equal to $S_o \times 100/S_p$) is higher than 60% generally, indicating that the source rocks in the co-sedimentary system can generate and expel hydrocarbon.

In the 4498 m to 4504 m interval, source rocks enter the high-maturity stage, and the indicators are quite different, showing that the high thermal maturity reduced the residual capacity for hydrocarbon generation and expulsion: *HI* is no higher than 150 mg/g, and the *HI* of most of the samples is lower than 50 mg/g; *IH* is less than 50 mg/g; K_t is over 60% mostly; and K_p is more than 90%, whereas some samples even reach 95%. S_1 and S_2 are lower than 1.0 and 0.5 mg/g, respectively; S_1+S_2 is no greater than 1.5 mg/g. The original potential of hydrocarbon generation and expulsion is also limited, lower than 15 mg/g, and that of most of the samples is below 6 mg/g.

Discussion: Significance of the Evaporite

The formation of evaporites

Regarding the formation of the evaporites here, Chen et al. [42] and Jian-hui et al.[43] suggested that evaporites form in a shallow lake, whereas mudstones form in a deep lake [42,43]. Laiyi et al. [44] considered that co-sediment is formed in alternating sedimentary environments from salified water to fresh water [44]. However, the current dominant view is that the co-sediment is formed in deep water, whereas a salted lake background is characterized by "mountains and deep basin" [45,46]. This point is based on the fact that the Paleogene Dongying Depression is surrounded by mountains, buried hills, and alluvial fans, which form a comparatively closed sedimentary system.

Salt lake in the western Qaidam basin in China is a typical inland salt lake, in which has formed extensive evaporite, and its formation mechanism can be refered. In the co-sedimentation of the evaporites and source rocks in age of Oligocene and Miocene in the western Qaidam basin, the prospective oil source rocks were just distributed stratigraphically and tectonically in the locations where halite and gypsum were deposited. A genetic model was postulated for evaporites and source rocks, which was described as "big shallow water saline lake surrounding small deep water brine lake" [47]. Jin and Zha [9] believed that water depth controlled both on deposit ion of the evaporites and the accumulation of organic matter in the salt lake environment [47].

The significance of evaporites for the organic matter

At the beginning of the evaporite formation, due to the salinity increase or the mixture of different salinity water, the bottom water is at stagnation, thus causing the death of various organisms. But because the river continuously supplied biological and organic substances, lakes contained a large number of various biological residue and organic material. When the lake salinity increased, the various creatures will die and drown in the lake, and in consequence humus mud phase at

Depth (m)	Lithology	Maturity parameters		Composition (%)		Organic matter type		Group (%)		
		T_{max} (°C)	Ro (%)	sapropel	vitrinite	Type index	type	alkanes	aromatics	Non-hydrocarbon
3967.8	Salted mudstone	422	0.93	70.7	21.3	62.7	II_1	80.53	11.32	4.74
3970.7	Salted mudstone	418	1.00	89.7	10.3	81.9	I	79.90	11.08	5.15
3972.3	Salted mudstone	417	0.95	91.7	8.3	85.4	I	80.11	11.36	4.83
3972.5	Salted mudstone	419	0.98	91.3	8.7	83.8	I	75.65	10.21	4.71
4300.9	Laminated mudstone	419	1.17	85.7	14.3	74.9	II_1	—	—	—
4498.2	Gypsum-salted mudstone	403	1.13	90.7	9.3	83.7	I	79.26	3.13	9.09
4499.2	Gypsum-salted mudstone	421	1.14	87.7	12.3	78.4	II_1	80.98	4.35	7.61
4499.8	Gypsum-salted mudstone	417	1.14	78.3	20.3	61.8	II_1	76.53	4.50	12.54
4501.2	Gypsum-salted mudstone	427	1.24	82.7	15.7	69.3	II_1	64.49	6.54	25.23

Table 3: The characteristics of the organic matter in the Es_4 from the Well Fs_2 in the Dongying depression.

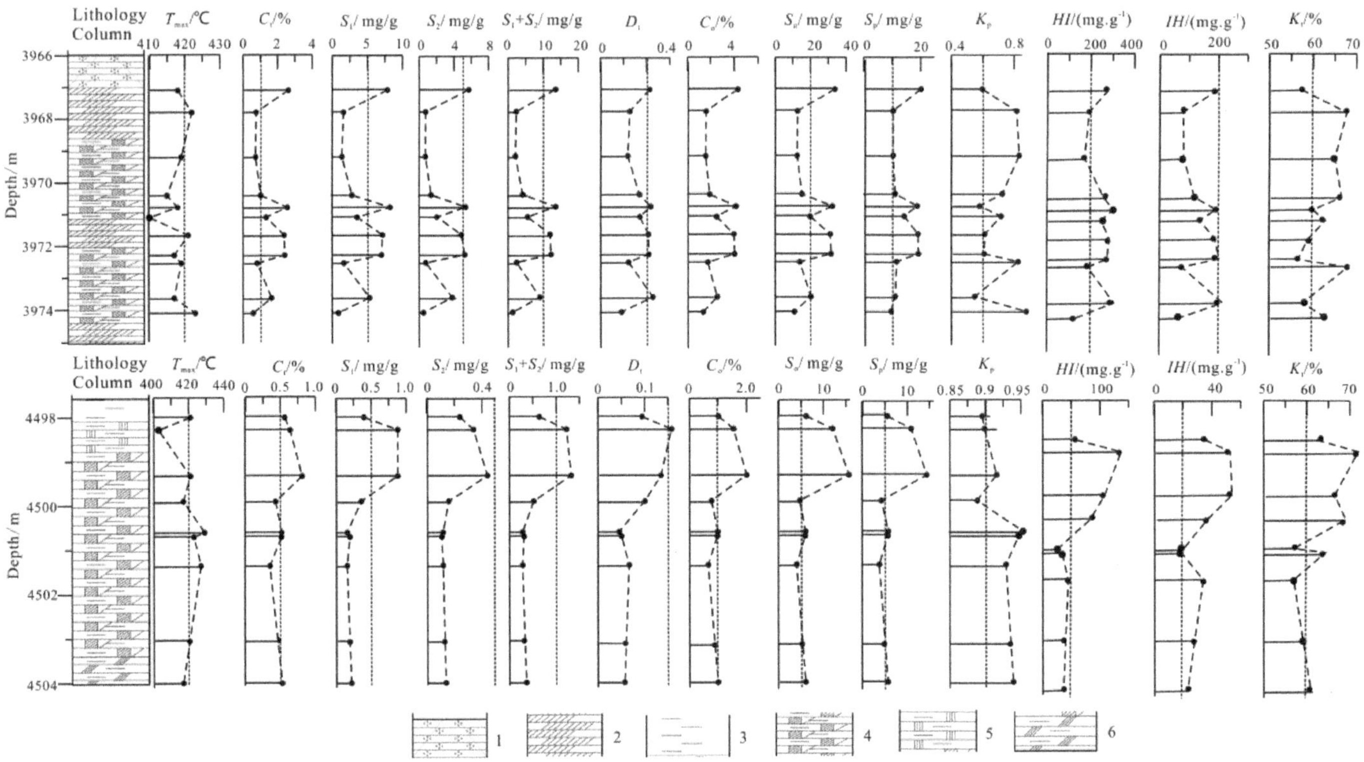

Figure 13: The hydrocarbon generation and expulsion from the source rocks in evaporated system of Es_4 from the Well Fs2 in Dongying depression.
1-halite,gypsum; 2-salt rock; 3-mudstone; 4-salted-mudstone; 5-ypsumed-mudstone; 6-mudstone included salt
(Notes: T_{max}, the supreme temperature in thermolysis; C_t, residual organic carbon content; S_1, residual dissolved hydrocarbon; S_2, pyrolysed hydrocarbon; S_1+S_2, residual potential of hydrocarbon generation; D_t, coefficient of residual hydrocarbon degraded; C_o, original organic carbon content; S_o, original capacity of hydrocarbon generation; S_p, amount of hydrocarbon expulsed; K_p,efficiency of hydrocarbon expulsion; HI, index of dissolved hydrocarbon; IH, hydrogen index; K_t, transformation ratio in thermolysis)

the bottom of the lake basin formed in a weak redox environment. The continuously conveying biological and organic substances from rivers provide material base for the hydrocarbon formation.

Generally, the formation and preservation of effect organic matter in salt lake is in an anaerobic environment (strong reducing environment). The high salinity in salt lake makes a halocline permanent hierarchy easy to form. The anoxic environment forms under bottom water characterized by high salinity, and the organic matter avoids the decomposition from oxidation, resulting in the formation of high quality source rocks [48]. This is reflected in the Chaerhan modern salt lakes in Qaidam basin: 1 m below the salt water is a permanent brine layers, in the depth of 3~4 m is strong reducing environment, where can form black mud.

Numerous continental salt basins in the eastern China show that at the second or third sedimentary cycle is the stage of source rocks formation. Many practice indicated that such a set of evaporite formation commonly was before the inevitable development of source rock. Peters et al. [32] pointed out that in barrier type of sedimentary basins; the content of organic matter was associated with the concentration of salt that deposited at the same time [49]. Tissot [50] observed that the organic matter content of rock salt is second only to clay (or shale) based on analysis on 108 pieces of rock salt [50].

Zhanhua depression is adjacent to the northeast of Dongying depression, thus the significance of evaporates for the organic matter in this depression can be compared. In the Zhanhua depression also develops the evaporate sediment combination up to 500 m on the upper Es_4. Research shows the evaporite exists not only with source rock, but also with high quality one. In salt lake on saline lake, the salinity difference between surface and bottom layers could result in the salt transition layer and the division of upper and lower layers makes a long-rang delamination of salinity. Bottom water circulation is in an absolutely stagnant state and results in anoxia environment. This environment restrains the activity of benthon and makes a maximum preservation of sediment organic matters. Oil-source correlation shows that salt lake saline lake source rock is an important one in Zhanhua Depression and controls the formation and distribution of many pools.

The significance of evaporites for the thermal maturation and oil generation

The thermal conductivity of gypsum and cream salt layer is significantly higher than other rocks, and the heating rate is significantly lower than other rocks. Test of the rock from the Hk1 Well shows the average thermal conductivity of the gypsum halite layer dominated by salt rocks is 5.48 W/m.K, and the average heat production rate of gypsum and salt rock are only 0.07 μW/m³ and 0.02μW/m³ respectively (Table 4). Compared to the evolution of normal mature hydrocarbon source rock (mudstone and shale) the salt layer shows lower vitrinite reflectance (R_o), and correspondly the paleogeothermal is lower than normal source rocks (Figure 14). Therefore, the existence of the halite layer makes the lower strata heat easily lose, lowers the temperature of the deep under the salt, suppresses the diagenesis, causes the diagenetic evolution of salt formation relatively lags behind the diagenetic evolution in the region without salt or gypsum, and correspondly

Samples	depth(m)	stratum	U(ppm)	Th(ppm)	a (μW/m³)	b (W/m.K)	Lithology
S1	3431.07	Es_4	0.13	0.19	0.05	5.22	salt rock
S2	3433.08	Es_4	1.29	3.22	0.572	2.84	gypsum mudstone
S3	3435.6	Es_4	0.065	0.088	0.024	5.74	salt rock
S4	3437.11	Es_4	0.087	0.15	0.034	5.41	gypsum salt rock
S5	3438.3	Es_4	0.029	0.05	0.012	5.76	gypsum salt rock
S6	3440.4	Es_4	0.023	0.041	0.009	5.28	salt rock
S7	3447.12	Es_4	0.023	0.031	0.009	5.45	salt rock
S8	3756.77	Es_4	3.45	14.7	1.939	2.53	mudstone
S9	3770.65	Es_4	0.033	0.062	0.013	5.52	salt rock
S10	3977.04	Es_4	4.75	20.6	2.77	2.16	mudstone

a: radioactive heat generation of rocks; b: thermal conductivity

Table 4: The radioactive heat generation and thermal conductivity of rocks in the Es_4 from the Well Hk1 in the Dongying depression.

prolong the oil generation window of the source rocks under salt layers.

The significance of evaporites for the oil reservoir and preservation

Salt-gypsum layer lead to the formation of overpressure sealing strip, which suppresses the compaction and is conducive to the preservation of primary pores. The inhibitory effect of abnormal high pressure on the thermal evolution of organic matter and hydrocarbon generation, make the release space of organic acids in the abnormal high pressure basin and the influence of organic acids on sandstone diagenesis greatly exceed the normal pressure basin. Organic acid and CO_2 will release in the process of thermal evolution of organic matter, reduce the pH value of pore water, further strengthen the dissolution of minerals, make the cement and clastic particles in sandstone dissolve ,and promote the formation of secondary pore. The pressured seal formed by Salt-gypsum layers is beneficial to the growth of secondary porosity under salt and preserved. For example, in the Paleogene *Ek* (Kongdian Member) and Es_4 there develop two overpressure belts and two interval depths (3398-3754m and 4052-4275.5m) for the development of secondary porosity accordingly, which are associated with the development of two halite layers respectively (Figure 15).

Figure 15 indicated that the development of secondary porosity under salt-gypsum layers was affected by rock types, structure and the development of primary pore at the same time. When primary pore relatively developed, acid formation water can smoothly enter between the particles and dissolve rock components.

The concentration layers of salt-gypsum in the Paleogene Es_4 in Dongying depression forms a good regional cap rocks for the oil and gas below. This is indicated by the occurring of cracking gas and even industrial air flow in the Fs1, Fs2 and Fs3 Wells in the Minfeng sag in the north of Dong depression. The thick salt-gypsum layer becomes a good cover to prevent dissipation of oil and gas (Figure 16). In the salt-gypsum layer in the north central zone of Dongying depression there develops abnormal high pressure, and the local pressure coefficient reaches 2.05. The existence of abnormal high pressure caused by the salt paste interval further enhances the sealing capability of the rocks.

Conclusions

This study showed the geochemical response in a co-sedimentary system by evaporites and mudstone in the Es_4 of the Dongying Depression in Eastern China.

The total salinity was generally high. The type water of $CaCl_2$ dominated. The Al, Mn, and Ba contents and the parameters of V/Ni and (Fe+Al)/(Ca+Mg) showed low value responses. The Na, Ca, and Mg contents and the parameters of Sr/Ba, Fe/Mn, and Sr/Ca showed high values. Sr content was remarkably high compared with that in the non-evaporite system. The distribution of mineral elements had visible characteristics of discretion and fluctuation, and mineral elements, except for Na, had discrepancies in saline deposits and were enriched in adjacent mud.

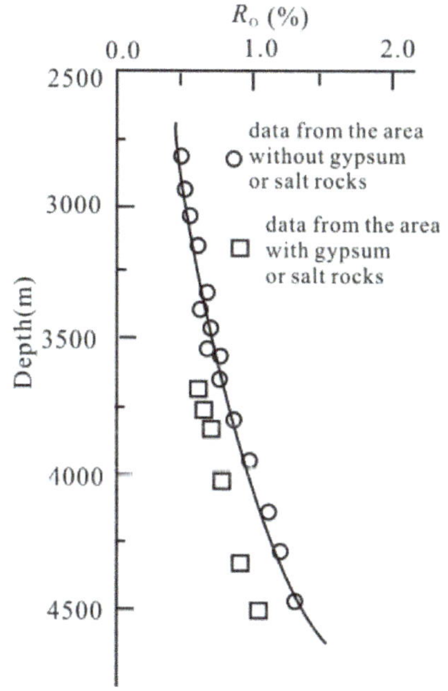

Figure 14: The distribution of Ro with depths in the area with/without evaporates in the Dongying depression.

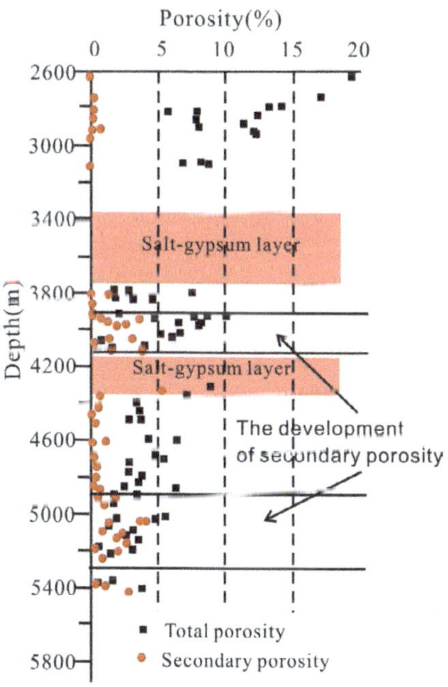

Figure 15: The distribution of total porosity and secondary porosity with depths in the Well Haoke1 in the Dongying depression.

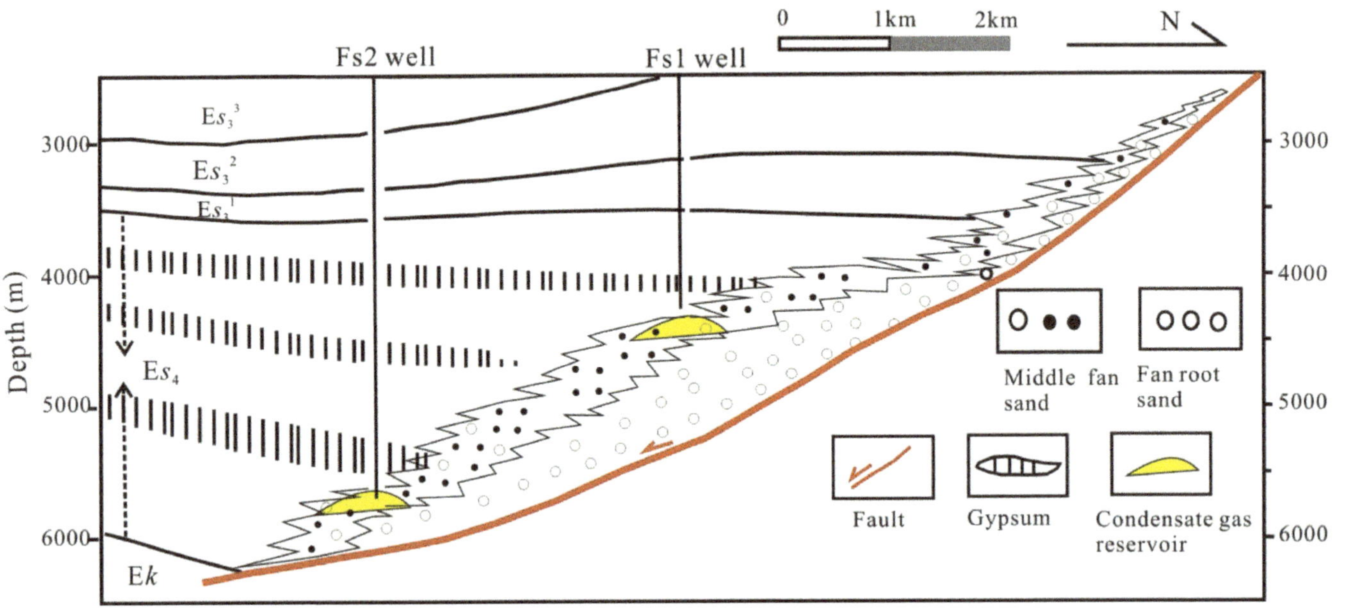

Figure 16: The geological section across the Wells Fs1 and Fs2 showing the relation of gypsum and cracking gas in the Lijin–Minfeng sag of Dongying depression.

A heavier isotope of carbon (high value of $\delta^{13}C$) was displayed, and the value of $\delta^{13}C$ for the high-maturity evaporite source rocks was significantly increased by thermal evolution and TSR reaction. High content of gammacerane and low value of Pr/Ph were exhibited. The indicators $C_{19}/(C_{19}+C_{23})$ from terpanes and C_{24}/C_{23} from tricyclic terpanes were lower, whereas $C_{29}/(C_{27}+C_{28}+C_{29})$ from steranes were higher than those in the non-evaporite system in the Dongying Depression.

A distinct negative reversion of biomarker maturity parameters $C_{29}\beta\beta/(\beta\beta+\alpha\alpha)$, $C_{29}20S/(20S+20R)$, and Ts/(Ts+Tm) was demonstrated, indicating that thermal evolution was restrained and retarded for the high content of gypsum-salt minerals. The co-sedimentary system can generate and expel hydrocarbon, and high maturation can reduce the capacity for these actions.

Acknowledgments

This study was supported by the National Natural Science Foundation of China (Grant No. 41272140 and 40802026) and Shandong Province Natural Science Foundation (Grant No. ZR2011DM004).

References

1. Trippetta F, Collettini C, Vinciguerra S, Meredith PG (2010) Laboratory measurements of the physical properties of Triassic Evaporites from Central Italy and correlation with geophysical data. Tectonophysics 492: 121-132.
2. Trippetta F, Collettini C, Meredith PG, Vinciguerra S (2013) Evolution of the elastic moduli of seismogenic Triassic Evaporites subjected to cyclic stressing. Tectonophysics 592: 67-79.
3. Trippetta F, Collettini C, Barchi MR, Lupattelli A, Mirabella F (2012) A multidisciplinary study of a natural example of a CO2 geological reservoir in central Italy. International Journal of Greenhouse Gas Control 12: 72-83.
4. Carminati E, Corda L, Mariotti G, Scifoni A, Trippetta F (2013) Mesozoic Syn- and Postrifting Evolution of the Central Apennines, Italy: The Role of Triassic Evaporites. The Journal of Geology 121: 327-354.
5. Tissot BP, Welte DH (1978) Petroleum Formation and Occurrence. Springer Berlin Heidelberg Germany.
6. Sinninghe Damste JS, Kenig F, Koopmans MP, Koster J, Schouten S, et al. (1995) Evidence for gammacerane as an indicator of water column stratification. Geochim Cosmochim Acta 59: 1895-1900.
7. Waples DW (1983) Reappraisal of anoxia and organic richness, with emphasis on Cretaceous of North Atlantic. AAPG Bulletin 67: 963-978.
8. Hofmann P, Leythaeuser D, Carpentier B (1993) Palaeoclimate controlled accumulation of organic matter in Oligocene evaporite sediments of the Mulhouse basin. Org Geochem 20: 1125-1138.
9. Qiang J, Guang-you Z (2006) Progress in Research of Deposition of Oil Source Rocks in Saline Lake and Their Hydrocarbon Generation. Acta Metallurgica Sinica 12: 483-492.
10. Zhonghong C, Ming Z, Qiang J (2004) An Investigation on Generation and Expulsion of Hydrocarbon from Source Rocks of the Shahejie Formation in the Well Niu-38, Dongying Depression. Chinese Journal of Geology 39: 356-366.
11. Zhonghong C, Ming Z, Qiang J (2006) teristics of the source rocks in fluctuation from lake facies: an example from the Well Niu-38 in the Dongying Depression, China. Journal of Lake Sciences 1: 29-35.
12. Zhonghong C, Ming Z (2008) Mechanism of overpressured fluid compartment and its controlling on hydrocarbon migration and accumulation in faulted lacustrine basin:A case study from the Dongying Sag,Bohai Bay Basin. Chinese Journal of Geology 43: 50-64.
13. Guang-you Z, Qiang J, Jin-xing D, Shui-chang Z, Lin-ye Z, et al. (2004) Investigation on the Salt Lake Source Rocks for Middle Shasi Column of Dongying Depression. Acta Metallurgica Sinica 10: 257-266.
14. Mango FD (1991) The stability of hydrocarbons under the time–temperature conditions of petroleum genesis. Nature 352: 146-148.
15. Tijani MN (2004) Evolution of saline waters and brines in the Benue-Trough, Nigeria. Appl Geochem 19: 1355-1365.
16. Zhong-hong C, Ming Z (2007) Correlation between inorganic elements and abnormal vitrinite reflectance in lacustrine source rocks. Geochimica 36: 275-278.
17. Hardie LA, Eugster HP (1970) The evolution of closed-basin brines. Mineral Soc Am Spec Pap 3: 273-290.
18. Eugster HP (1980) Geochemistry of Evaporitic Lacustrine Deposits. Annu Rev Earth Pl Sc 8: 35-63.
19. Warren JK (2010) Evaporites through time: Tectonic, climatic and eustatic controls in marine and nonmarine deposits. Earth-Sci Rev 98: 217-268.
20. Yangming Z, Huanxin W, Aiguo S, Digang L, Dehua P (2005) Geochemical characteristics of Tertiary saline lacustrine oils in the Western Qaidam Basin, northwest China. Appl Geochem 20: 1875-1889.
21. Chun-rong L, Kai-yuan C (2004) The Application of C and O Isotope Analyse

in Sediment Environment of Qianjiang Saline Lake. Natural Gas Geoscience 15: 320-322.

22. Zhenheng C, Zhonghong C, Yuti Z, Jun P, Zhenhua J, et al. (2007) An Investigation on the Geochemical Characteristics of Crude Oil from Wenliu Area in the Dongpu Sag,Bohai Bay Basin. Petroleum Geology & Experiment 29: 178-187.

23. Schouten S, Hartgers WA, Lòpez JF, Grimalt JO, Sinninghe Damsté JS (2001) A molecular isotopic study of 13C-enriched organic matter in evaporitic deposits: recognition of CO2-limited ecosystems. Org Geochem 32: 277-286.

24. Murray AP, Summons RE, Boreham CJ, Dowling LM (1994) Biomarker and n-alkane isotope profiles for Tertiary oils: relationship to source rock depositional setting. Org Geochem 22: 521-542.

25. Popp BN, Laws EA, Bidigare RR, Dore JE, Hanson KL, et al. (1998) Effect of phytoplankton cell geometry on carbon isotopic fractionation. Geochim Cosmochim Ac 62: 69-77.

26. Zhu GY, Zhang SC, Liang YB, Dai JX, Li J (2005) Alteration of thermochemical sulfate reduction to hydrocarbons. Acta Petrolei Sinica 26: 48-52.

27. Ma YS, Cai XY, Li GX (2005) Basic Characteristics and concentration of the Puguang gas field in the Sichuan Basin. Acta Geologica Sinica 79: 858-865.

28. Zhi-Yong C, Qi-Ming L, Bin X (2007) Study on the Mechanism and Accumulation Effect of Thermochemical Sulfate Reduction in Northeast Area of Sichuan Absin. Natural Gas Geoscience 18: 743-749.

29. Haven HLT, Leeuw JWD, Rullkötter J, Sinninghe Damsté JS (1987) Restricted utility of the pristane/phytane ratio as a palaeoenvironmental indicator. Nature 330: 641-643.

30. Schwark L, Vliex M, Schaeffer P (1998) Geochemical characterization of Malm Zeta laminated carbonates from the Franconian Alb, SW-Germany (II). Org Geochem 29: 1921-1952.

31. Peters KE, Walters CC, Moldowan JM (2005) The Biomarker Guide: Biomarkers and isotopes in petroleum systems and Earth history, Volume 2. Cambridge University Press UK.

32. Peters KE, Moldowan JM (1993) The Biomarker Guide: Interpreting Molecular Fossils in Petroleum and Ancient Sediments. Prentice Hall Englewood Cliffs NJ USA.

33. Wang TG (2002) A contribution to some sedimentary environmental biomarkers in crude oils and source rocks in China. Geological Review 48: 256-262.

34. Seifert WK, Moldowan JM (1979) The effect of biodegradation on steranes and terpanes in crude oils. Geochim Cosmochim Ac 43: 111-126.

35. Seifert WK, Moldowan JM (1981) Paleoreconstruction by biological markers. Geochim Cosmochim Ac 45: 783-794.

36. Chen Z, Zha M, Jin Q, Ren Y (2011) Distribution of sterane maturity parameters in a lacustrine basin and their control factors: A case study from the Dongying Sag, East China. Petroleum Science 8: 290-301.

37. Tuo JC, Huang XZ, Ma WY (1994) The Lagging phenomenon of the petroleum generation in carbonate rocks. Petroleum Exploration and Development 21: 1-5.

38. Guiyu Z, Shuyuan L, Luofu L (2005) A Study on Characteristics and Kinetics of Catalytic Degradation from Kerogen in Carbonate Rocks. Chinese Journal of Geology 40: 47-54.

39. Xie QL, Zhou ZY, Lu MY (2000) Organic matter enclosed in carbonate minerals; a kind of important hydrocarbon producing matter. Acta Mineralogica Sinica 20: 59-62.

40. Yuan J, Qin K (2001) Characteristics of Evaporite Generated in Deep Water of Sha-4 Member in Dongying Sag. University of Petroleum China 25: 9-11.

41. Chuanlian L, Quanhong Z, Pinxian W (2001) Paleolimnological Significance of Carbon and Oxygen Isotopic Ratios of Ostracoda from Oil Source Rocks in Dongying Depression. Earth Science 26: 441-445.

42. Chen FL, Zhu H, Li HT (2000) Partition of sequence strata and discussion about salt-rock resource in Shahejie formation of Eogene, Dongpu depression. Acta Sedimentologica Sinica 18: 384-388.

43. You-liang J, Jian-hui F, Sheng-lang W, Hong-an Z (2003) High Frequency Lake-Level Change of 3rd Member of Shahejie Formation and Prediction of Lowstand Sand Body in Dongpu Depression. Acta Metallurgica Sinica 9: 99-112.

44. Laiyi R, Guifang L, Zhiqing Z, Xingwu W (2000) Early Tertiary marine transgression in Dongpu depression. Acta Palaeontologica Sinica 39: 553-557.

45. Qiang J, Xinghan H (1985) Studies on the Origin of the Early Tertiary Salt Lake Dongpu Depression——A Postulated Deep Water Model. Journal of the University of Petroleum 9: 1-11.

46. Zhu YM, Su AG, Liang DG (2003) Distribution characterization and origin of n-alkanes in saline lacustrine source rocks of Qaidam Basin. Geochimica 32: 117-123.

47. Qiang J, Ming Z (2000) Co-Sedimentation of Tertiary Evaporites and Oil Source Rocks in the Western Qaidam Basin. Chinese Journal of Geology 35: 465-473.

48. Taberner C, Cendón DI, Pueyo JJ, Ayora C (2000) The use of environmental markers to distinguish marine vs. continental deposition and to quantify the significance of recycling in evaporite basins. Sediment Geol 137: 213-240.

49. Rohrback BG, Peters KE, Kaplan IR (1984) Geochemistry of artificially heated humic and sapropelic sediments; II, Oil and gas generation. AAPG Bulletin 68: 961-970.

50. Tissot BP (1984) Recent Advances in Petroleum Geochemistry Applied to Hydrocarbon Exploration. AAPG Bulletin 68: 545-563.

Molecular Characterization of Bacterial Phylogenetic and Functional Groups at Terrebonne Bay along the Coastline of the Gulf of Mexico

Ola A Olapade*

Department of Biology and the Center for Sustainability and the Environment, Albion College, 611 East Porter Street, Albion, MI 49224, USA

Abstract

The detection and quantification of bacterial phylogenetic and functional groups as well as community diversity at the site of the Deepwater Horizon oil spill in Terrebonne Bay along the Gulf of Mexico were carried out using nucleic acid staining, Fluorescence *in situ* Hybridization (FISH) and 16S rRNA gene cloning and sequencing approaches. Results from the 16S rRNA gene clone library analysis revealed high occurrences of bacterial members belonging to the *Cyanobacteria* (28%), *β- Proteobacteria* (21%), *Bacteroidetes* (17%), *Actinobacteria* (12%) and the *α- Proteobacteria* (10%). Particularly, bacterial members identified within the clone library as belonging to the *β- Proteobacteria* subclass were mostly hydrocarbon degraders, including *Methylibium petroleiphilum, Burkholderia cepacia, Hydrogenophaga taeniospiralis* and *Methylobacillus flagellates.* Simultaneous analyses of both planktonic and benthic bacterial communities by FISH revealed the numerical dominance of members of the type I Methanotrophic Bacteria (MB) over the type II populations. The results from the study clearly reveal a shift in the bacterial community structure and composition in response to the tragic methane and crude oil discharges from the Deepwater Horizon rig along the Gulf of Mexico.

Keywords: 16S rRNA gene; Fluorescence *in situ* hybridization; Pollution; Coastal marine

Introduction

The diversity and global distributions of bacterial populations within indigenous microbial assemblages in marine environments have been well documented because of their significant ecological importance within various milieus [1-8]. For instance, there is currently ample and incontrovertible evidence that bacterial assemblages within coastal marine milieus do not only rapidly respond to oil spills, but also contribute their wide arrays of hydrocarbon degrading capabilities to the effective bioremediation of oil residues in contaminated environments [9-11]. Given the presence of diverse degradative genes needed for *in situ* clean up of complex hydrocarbon pollutants, accurate delineation of *in situ* microbial assemblages is therefore paramount in order to effectively understand the overall dynamics of microbial response and biodegradation process in oil polluted sites. Even more so that it is common knowledge that microbial assemblages are influenced by various controlling factors, including pollutant type and bioavailability, nutrient dynamics as well as continuous fluxes in site-specific hydrodynamic conditions within marine environments [12-15].

This study elucidated *in situ* microbial compositions in response to the tragic crude oil and methane discharges that resulted due to the severance of the Deep water Horizon rig from its well offshore on the Gulf of Mexico (GOM) in 2010. This tragic pollution event ultimately released approximately 1.3 X 10^{10} moles of methane and 205 million gallons (i.e.780,000 m^3) of crude oil into surrounding environments of affected GOM sites [14]. Shortly after the incidence, the widespread dispersal of hydrocarbon plume was linked to the stimulation of indigenous bacterial populations, especially the γ- *Proteobacteria* members known to be closely associated with petroleum degradation [15,16]. Furthermore, these studies also found strong correlations between the occurrences of several hydrocarbon-degrading genes and various components of the hydrocarbon plume at the GOM sites examined.

Therefore, combinations of 16S ribosomal RNA gene sequencing, nucleic acid staining and fluorescence *in situ* hybridization (FISH) analyses were employed to qualitatively and quantitatively examine the phylogenetic composition and community diversity within the bacterioplankton assemblages at one of the most contaminated coastal location along the Gulf of Mexico. Quantitative analysis by FISH was particularly employed to target bacterial phyla with hydrocarbon-utilizing capabilities, including two subclasses (i.e. α- and γ-) of *Proteobacteria* and members of the methanotrophs (i.e. type I and II), since methane was documented as the most abundant hydrocarbon released into the GOM sites during the spill [15]. Generally, the methanotrophs are bacterial groups capable of both aerobic and anaerobic methane oxidation and belong mainly to either the α-*Proteobacteria* (type II methanotrophs) or the γ- *Proteobacteria* (type I methanotrophs) as well as some acidophilic members of the *Verrucomicrobiae* [17,18]. However, several other recent studies have also documented widespread anaerobic methane oxidation among diverse groups and consortia of marine microbial populations [18-22]. In this study, Terrebonne Bay in southern Louisiana was selected as the study site, based on the extent of the Deepwater Horizon oil spill at this particular milieu, combined with several past ecological antecedents and the direct connection of this coastal region to the Mississippi River [23].

***Corresponding author:** Ola A Olapade, Department of Biology and the Center for Sustainability and the Environment, Albion College, 611 East Porter Street, Albion, MI 49224, USA; Tel: 517-629-0296; Fax: 517-629-0264

Materials and Methods

Description of study sites and sample collection

Water and sediment samples were collected in triplicates at three separate locations along the coastline of the Gulf of Mexico on July 19th, 2011, specifically at Terrebonne Bay (29.14134°, -90.56258°) in southern Louisiana (USA) as previously described [7]. Subsamples were later removed from each sample and preserved in 8% (w/v) paraformaldehyde and 1X phosphate-buffered saline solution for nucleic acid (4', 6'-diamidino-2-phenylindole {DAPI} staining and FISH analyses. During sampling, various water characteristics including temperature, pH, conductivity, oxidation-reduction potential were measured using the YSI model 556 MPS multi-probe system (YSI Incorporated, USA).

DNA extraction, PCR and clone sequencing

Approximately 500 mL of water samples were filtered through sterile, 47 mm, 0.2 um pore-size filters before storing at -80°C until processed. Total DNA was later extracted from the preserved filters using FastDNA SPIN Extraction kit (MP Biomedicals, Solon, OH, USA) and eluted in 50 uL of sterile deionized water. PCR amplification was then carried out on the extracted and pooled DNA from the replicate samples by targeting the almost full-length 16S rRNA gene with the universal bacterial primer pair 8F (5' AGA GTT GTA TCC TGG CTC AG 3') and 1492R (5'GGT TAC CTT GTT ACG ACT T3') as previously described [7]. The amplified PCR products were then confirmed on agarose gel and purified with a QIAquick PCR purification kit (QIAGEN, Valencia, CA) before subsequently utilized for cloning and sequencing. Clone libraries were constructed using the TOPO TA cloning kit (Invitrogen, Carslbad, CA). Colony PCR was used in screening transformed cells with vector-specific primers [24] and size of products verified by agarose gel electrophoresis.

Phylogenetic and diversity analysis

A total of 83 clones were successfully sequenced and analyzed using the Sequencher program (version 4.5; Gene Codes Co., Ann Arbor, MI). They were then compared with previously published GenBank sequences using the BLAST system [25] in order to determine their close relatives. Alignment and other manual editing were carried out with ClustalW [26]. The clones were then classified into 66 Operational Taxonomic Units (OTUs) using sequence identity values ranging from ≥ 98% to 100%. The OTUs were analyzed for species richness, Shannon Index, Simpson's (Reciprocal) Index of diversity, species evenness and Chao-1 richness indicator [27-29]. Rarefaction analysis was also performed to determine the diversity of the clone libraries using the freeware program Analystical Rarefaction version 1.3 (S. M. Holland: www.uga.edu/strata/software/Software.html).

Bacterial enumeration

Direct Counts (DAPI Staining): Total bacterial numbers in the preserved samples were determined by concentrating onto 0.2 μm pore-size black polycarbonate filters (Poretics, Livermore, CA) and staining with DAPI solution for 5 minutes. Filters were rinsed with sterile water and then mounted onto glass slides with Type FF immersion oil [30]. Bacterial cells in 10 separate fields were then counted using an epifluorescence microscope.

Fluorescence In Situ Hybridization: FISH analyses were used to determine the abundance of different bacterial phylogenetic groups as described in Lomans et al. [31] and Mills et al. [32]. Briefly, bacterial cells in the preserved water samples were concentrated onto 0.2 μm pore-size anodisc or polycarbonate filters (Whatman, Maidstone, UK), and then rinsed with deionized water, treated with 1 mL 0.1% Nonidet P-40 (Sigma Aldrich, St. Louis, MO). 40 μL of Texas red-labeled probe (Sigma Genosys, The Woodlands, TX; [5 ng/μL final concentration]) dissolved in hybridization buffer (6X standard saline citrate {SSC}, 0.02 M TRIZMA base at pH7.0, 0.1% sodium dodecyl sulfate {SDS}, 0.01% polyadenylic acid, and 30% formamide) were then added to the filters before incubating for 4 h at the appropriate temperature (Table 1). After the incubation, filters were washed twice with 400 μL of wash buffer (0.9 M NaCl, 0.02 M Tris-pH7, 0.1% SDS) and incubated with 80 μL of wash buffer for 10 minutes at the hybridization temperature. The filters were then rinsed twice with 400 μL sterile deionized water before they were mounted on glass slides with immersion oil. Cells that hybridized to each probe were enumerated using the epifluorescence microscopy by counting at least 300 fields on triplicate slides.

Nucleotide sequence accession numbers

Nucleotide gene sequences obtained were already submitted to DDBJ/GenBank/EMBL under accession numbers AB691143 to AB691225.

Results

Environmental variables at study site

Mean values for the various water characteristics that were measured in triplicates at the study sites are: temperature (30.74°C), pH (7.85), dissolved oxygen (109.8%), conductivity (14.27mS/cm) and oxidation-reduction potential (10.87 mV).

Clone library composition and community diversity assessment

Analysis of the 83 16S ribosomal RNA gene sequences revealed eight distinct phylogenetic groups, including bacterial members belonging to the *Cyanobacteria*, four subclasses (i.e. α-, β-, γ- δ-) of the *Proteobacteria, Bacteroidetes, Actinobacteria,* and the *Firmicutes* (Figure 1). Overall, members of the *Cyanobacteria,* accounted for the highest clone representation with 28%, followed by the β-*Proteobacteria* (20.5%), *Bacteroidetes* (17%) and *Actinobacteria* (12%). The percent representations by the different OTUs detected within the clone library constructed after comparing to their closest relatives in the NCBI database are presented in Table 2.

Probe	Taxa	Sequence (5'-3')	Hybridizing Temp (°C)	Reference
EUB338	Domain *Bacteria*	GCTGCCTCCCGTAGGAGT	48	Amann et al. [46]
ALF1b	*α-proteobacteria*	CGTTCG (C/T)TCTGAGCCAG	54	Amann et al. [47]
GAM42a	*γ-proteobacteria*	GCCTTCCCACATCGTTT	57	Manz et al. [48]
SRB385	Sulfate-Reducing-Bacteria	CGGCGTCGCTGCGTCAGG	53	Amann et al. [46]
β -AO233	Ammonia-oxidizing-Bacteria	AGCTAATCAGRCATCGG	44	Stephen et al. [49]
M-450	Type I Methanotrophs	ATCCAGGTACCGTCATTATC	46	Eller et al. [4]
M-84	Type II Methanotrophs	CCACTCGTCAGCGCCCGA	46	Eller et al. [4]

Table 1: Oligonucleotide sequences, target and hybridization conditions for probes used in this study.

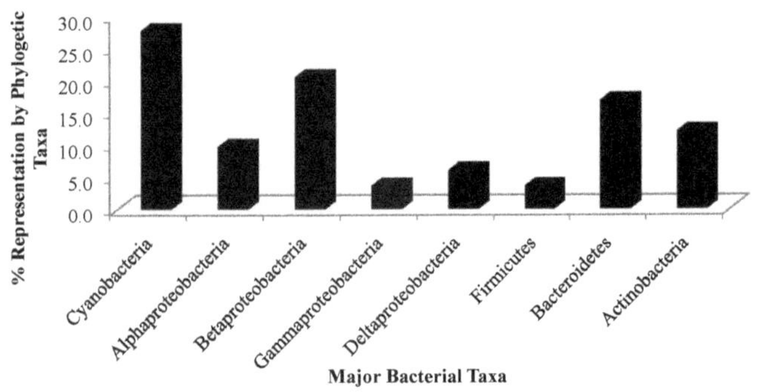

Figure 1: Percent distribution of major bacterial phylogenetic groups based on 16S rRNA gene sequences from the coastal bacterioplankton assemblages in Terrebonne Bay along the Gulf of Mexico.

Among the *Proteobacteria,* members of the *β-Proteobacteria* subclass dominated especially including several hydrocarbon-utilizing bacterial species as *Methylibium petroleiphilum, Burkholderia cepacia* and *Methylophilus methylotrophus.* Conversely, bacterial members of the *α-Proteobacterial* subclass accounted for only 9.6% of the total clone library composition including species *of Oceanicola pacificus* and *Shinella zoogloeoides* belonging to the pyrene and pyridine-degrading consortia. Clones belonging to the γ- and δ- *Proteobacteria* accounted for only 3.6% and 6.0% of total populations, respectively, including species such as *Hydrocarboniphaga effuse* and *Desulfofaba fastidiosa.*

Results from the various diversity measures analyzed revealed high bacterial diversity within the bacterioplankton assemblage at the bay site examined. Specifically, the Simpson's (Reciprocal) index of diversity was 46.2, while the species evenness was 0.969. Equally high were the numbers calculated to be 217.3 and 4.06 for the Chao 1 estimate of species richness and the Shannon Weiner Index, respectively. Rarefaction analysis revealed that the amounts of clones sequenced and screened are probably not sufficient for the estimation of the bacterial diversity within the clone libraries (Figure 2).

Abundances of bacterial phylogenetic and functional groups

Numbers of total bacteria within the bacterioplankton and benthic assemblages averaged about 3.5 x 10^7/mL and 1.5 x 10^8/g, respectively. While, Domain Bacteria occurrence accounted for between 11% and 20% of total bacterial counts in the water and sediment at average abundance of 4.0 x 10^6/mL and 3.0 x 10^7/g respectively (Figure 3a and 3b). When two subclasses (*i.e.* α-, and γ-) of the *Proteobacteria* were enumerated, their numbers were comparable between both the bacterioplankton and benthic assemblages, although members of α-*Proteobacteria* were found to be numerically more dominant within both habitats at the bay site (Figure 4a and 4b).

The occurrences of both type 1 and II methanotrophic functional bacterial populations followed the same trend and on average were at least one order of magnitude higher within the sediment than in the bacterioplankton communities (Figure 4a and 4b). Comparatively, the type 1 group were more numerically dominant than the type II, in both water and sediment samples examined. In contrast, the abundances of the other two functional groups examined i.e., the sulfate-reducing and the ammonia-oxidizing bacterial populations differed in their pattern of occurrences within both the bacterioplankton and benthic assemblages. Specifically, numbers of AO233-hybridized cells were found to be higher than those detected with the SRB385 probe in the sediment; conversely, the entire opposite in occurrence was the case for both populations enumerated in the water samples (Figure 4a and 4b).

Discussion

In this study, by applying combinations of several culture-independent (i.e. nucleic acid staining, fluorescence *in situ* hybridization

	Closest Phylogenetic Taxa from NCBI	GenBank Acc. Number	Abundance (n/%)a
A	**Cyanobacteria**		
1	*Uncultured Cyanobacterium*	HQ242211	1/1.20
2	*Uncultured Cyanobacterium*	JF966676	1/1.20
3	*Uncultured Cyanobacterium*	EU930687	1/1.20
4	*Uncultured Cyanobacterium*	AM690936	6/7.23
5	*Uncultured Cyanobacterium*	AB491631	2/2.41
6	*Uncultured Cyanobacterium*	FJ352328	1/1.20
7	*Uncultured Cyanobacterium*	GQ349130	1/1.20
8	*Uncultured Cyanobacterium*	FM995186	1/1.20
9	*Uncultured Cyanobacterium*	KC545747	1/1.20
10	*Uncultured Cyanobacterium*	FJ763779	1/1.20
11	*Uncultured Cyanobacterium*	EU780238	1/1.20
12	*Uncultured Cyanobacterium*	JF966674	1/1.20
13	*Uncultured Cyanobacterium*	EU800916	1/1.20
14	*Uncultured Cyanobacterium*	HM057705	1/1.20
15	*Uncultured Cyanobacterium*	AM259752	1/1.20
16	*Uncultured Cyanobacterium*	HQ914635	1/1.20
17	*Uncultured Cyanobacterium*	GU074287	1/1.20
B	**Alphaproteobacteria**		
18	*Rhodobacter veldkampii*	NR043405	1/1.20
19	*Shinella zoogloeoides*	NR041341	1/1.20
20	*Shinella zoogloeoides*	NR041342	1/1.20
21	*Rhodoplanes serenus*	NR040936	1/1.20
22	*Anderseniella baltica*	NR042626	2/2.41
23	*Skermanella aerolata*	NR043929	1/1.20
24	*Oceanicola pacificus*	NR043915	1/1.20
C	**Betaproteobacteria**		
25	*Methylibium petroleiphilum*	NR041768	1/1.20
26	*Burkholderia ginsengisoli*	NR041288	1/1.20
27	*Massilia lutea*	NR043310	2/2.41
28	*Burkholderia cepacia*	NR041719	1/1.20
29	*Methylophilus methylotrophus*	NR041257	1/1.20
30	*Denitratisoma oestradiolicum*	NR043249	2/2.41
31	*Burkholderia endofungorum*	NR042584	1/1.20
32	*Massilia dura*	NR043307	1/1.20
33	*Methylobacillus flagellatus*	NR043691	1/1.20
34	*Hydrogenophaga pseudoflava*	NR028717	1/1.20
35	*Azoarcus buckelii*	NR027190	1/1.20
36	*Methylobacillus flagellatus*	NR043691	1/1.20
37	*Hydrogenophaga taeniospiralis*	NR028716	1/1.20
38	*Burkholderia cepacia*	NR041719	1/1.20
39	*Methylobacillus flagellatus*	NR043691	1/1.20
D	**Gammaproteobacteria**		
40	*Thioalkalivibrio denitrificans*	NR028745	1/1.20
41	*Singularimonas varicoloris*	NR042175	1/1.20
42	*Hydrocarboniphaga effusa*	NR029102	1/1.20
E	**Deltaproteobacteria**		
43	*Desulfofaba fastidiosa*	NR025746	1/1.20
44	*Desulfuromusa bakii*	NR026175	1/1.20
45	*Geobacter thiogenes*	NR028775	1/1.20
46	*Desulfuromonas alkaliphilus*	NR043709	1/1.20
47	*Desulfuromusa bakii*	NR026175	1/1.20

F	**Firmicutes**		
48	*Anaeroarcus burkinensis*	NR025298	1/1.20
49	*Parasporobacterium paucivorans*	NR025390	1/1.20
50	*Thermincola carboxydiphila*	NR043010	1/1.20
G	**Bacteroidetes**		
51	*Flavisolibacter ginsengisoli*	NR041500	2/2.41
52	*Owenweeksia hongkongensis*	NR040990	5/6.02
53	*Flavobacterium sp*	NR040990	1/1.20
54	*Perexilibacter aurantiacus*	NR041534	2/2.41
55	*Robiginitalea myxolifaciens*	NR041514	1/1.20
56	*Lishizhenia caseinilytica*	NR041043	1/1.20
57	*Fluviicola taffensis*	NR041911	1/1.20
58	*Haliscomenobacter hydrossis*	NR042316	1/1.20
H	**Actinobacteria**		
59	*Streptomyces hebeiensis*	NR029091	1/1.20
60	*Ferrimicrobium acidiphilum*	NR041768	1/1.20
61	*Ilumatobacter fluminis*	NR041633	2/2.41
62	*Ferrimicrobium acidiphilum*	NR041798	2/2.41
63	*Streptomyces hebeiensis*	NR029091	1/1.20
64	*Ilumatobacter fluminis*	NR041633	1/1.20
65	*Kitasatospora saccharophila*	NR041538	1/1.20
66	*Patulibacter minatonensis*	NR041254	1/1.20

a = total of 83 clones

Table 2: Percent occurrences of bacterial phylogenetic groups in the bacterioplankton assemblage at Terrebonne Bay along the Gulf of Mexico.

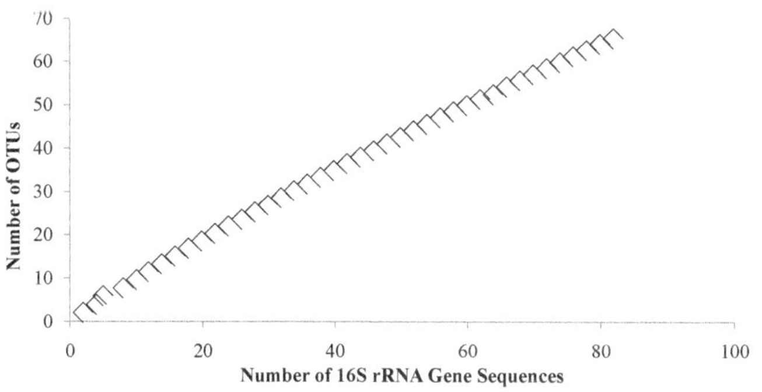

Figure 2: Percent abundance of different OTUs presented by genus from Terrebone Bay along the Gulf of Mexico.

(FISH), gene cloning and sequencing approaches, the impact of the Deepwater Horizon oil spill on indigenous microbial assemblages was assessed qualitatively and quantitatively at the study site in Terrebonne Bay. Sequence diversity analyses revealed the presence of eight distinct bacterial phyla comprising mostly of various hydrocarbon-utilizing phylotypes that strongly suggests that such a shift was stimulated by the oil disaster at the GOM. This result corroborates earlier documentation that also showed strong stimulation of indigenous bacterial populations, especially the γ- *Proteobacteria* members closely associated with petroleum degradation at the same polluted GOM sites [33,15]. Interestingly, results documented in some earlier studies that were conducted along the same coastlines of the GOM, prior to the oil spillage [1,32,7] appeared to show different compositions of the microbiota compared to those reported in this current study, therefore indicating that a shift in the dominant microbial groups must have occurred. Particularly, Olapade [7] observed majority of sequences (*i.e.* between 30% and 60% of cloned libraries) within the bacterioplankton communities in three GOM sites to be closely related to gene sequences belonging to unknown bacteria. This shift in the major microbial phylogenetic groups probably further validate the suggestion that the phyla with hydrocarbon-degrading abilities were relatively more stimulated and supported by the oil plume at these GOM locations [14,33,15].

Generally, variations of sequences belonging to globally distributed bacterial taxa are typically found within microbial communities in coastal marine environments [34,2,5]. However, in this study the relatively high occurrence of β-*Proteobacteria* found is quite suggestive of the oil plume influence, especially given the relatively rare occurrence of this particular taxa in marine milieus as compared to freshwater environments [28,7,8]. Moreover, the bacterial members identified as belonging to the *β- Proteobacteria* subclass within the clone library were mostly hydrocarbon (e.g., methanol) degraders, including *Methylibium petroleiphilum, Burkholderia cepacia, Hydrogenophaga taeniospiralis* and *Methylobacillus flagellates* species [35-37].

Furthermore, bacterial members of the δ- *Proteobacteria* and the *Firmicutes* that represented about 6% and 3% of total clone populations comprised of several species such as *Desulfofaba fastidiosa, Desulfuromusa bakii* and *Parasporobacterium paucivorans* which are capable of utilizing various crude-oil derived compounds including methyloxylated aromatics, propionate and sulfur [38-40]. Typically, majority of bacterial species belonging to these two taxa are reportedly associated with soil and sediment in coastal marine environments especially during tidal events [1,7]. Therefore, the presence of high numbers of hydrocarbon-utilizing bacterial phyllotypes in this study within the bacterioplankton communities strongly suggest a possible change in the physiological and metabolic profiles of some of the taxa in response to the available hydrocarbon substrates.

The relatively high representations by the α-*Proteobacteria* (10%) and the *Bacteroidetes* (17%) were not at all surprising and further

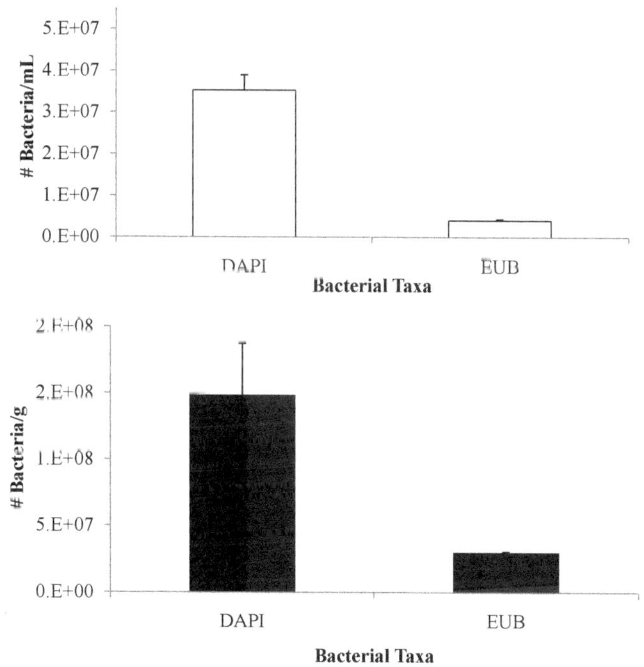

Figure 3: Numbers of total bacteria (as determined by DAPI staining) and domain bacteria (determined by FISH) in the coastal bacterioplankton [A] and benthos [B] assemblages in Terrebonne Bay along the Gulf of Mexico. Values represent mean ± 1 SE (n = 3).

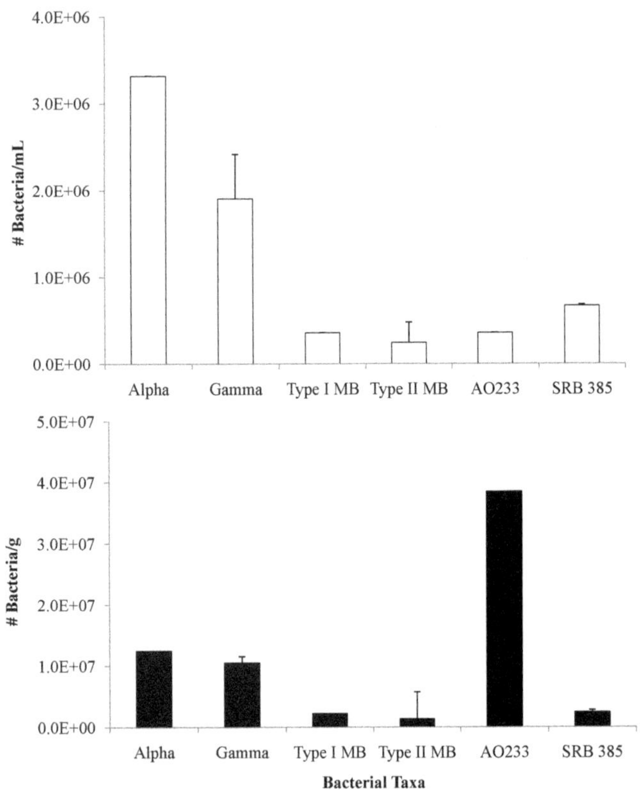

Figure 4: Numbers of α- and γ- *Proteobacteria*; Type I and Type II methanotrophs and Ammonia-oxidizing and sulfate-reducing bacterial populations (determined by FISH) in the coastal bacterioplankton [A] and benthos [B] assemblages in Terrebonne Bay along the Gulf of Mexico. Values represent mean ± 1 SE (n = 3).

corroborate earlier documentations of the dominance of the two bacterial taxa in coastal marine systems [5]. Species such as *Shinella zoogloeoides* that are capable of pyridine degradation [41] dominated α-*Proteobacteria* clones, while members of the *Bacteroidetes* that typically play significant roles in the decomposition of high-molecular-weight organic compounds in marine environments [42], were mostly represented by *Owenweeksia hongkongensis,* within the bacterioplankton communities.

The numerical dominance of both types of the methanotrophs examined within the sediment matrixes as compared to populations in the bacterioplankton is further indicative of their typical rapid response towards the source of methane release [43], which in this case was close to the ocean sub-seafloor [28]. The close proximity to methane source may also partly explain the higher presence of bacterial members belonging to the ammonia-oxidizing and sulfate-reducing bacterial populations, despite the frequent methane oxidization in both aerobic and anaerobic environments [22,44]. The disparity observed here regarding the occurrence of the AOB populations strongly agrees with an earlier similar study from the Northwestern Mediterranean Sea which indicated that members of this functional group might probably occupy distinct environmental niches in marine systems, because of their lower representations in planktonic libraries as compared to those that were particle-associated [45]. Similar contrasting patterns have also been observed especially in the diversity of aerobic methanotrophs between planktonic and sediment- associated assemblages in marine environments [17]. In this study conducted in two seep systems, they reported that while the sediment associated methanotrophs were less diverged and dominated by single taxa, in contrast the diversity within the planktonic community were found to be limited to a small number of moderately diverged clades within the same *Methylococcaceae* genus.

In conclusion, the results from both FISH and 16S rRNA gene clone sequences clearly reflects the subsequent shift in the bacterial community structure and composition at Terrebonne Bay in southern Louisiana in response to the tragic methane and crude oil discharges from the Deepwater Horizon rig along the Gulf of Mexico. This obvious shift in bacterial community diversity to mostly hydrocarbon-degrading phylotypes at the GOM site examined, further highlights both the ecological importance as well as various degradative potentials of autochthonous bacterial assemblages within contaminated coastal marine milieus.

Acknowledgements

The study was supported mostly by the Albion College Hewlett-Mellon Faculty Development Funds. Sincere thanks to Lori Duff, Freyja Davis, Dave Carey and Kurt Hellman for various support during the study period.

References

1. Chauhan A, Cherrier J, Williams HN (2009) Impact of sideways and bottom-up control factors on bacterial community succession over a tidal cycle. Proc Natl Acad Sci USA 106: 4301-4306.
2. Cottrell MT, Kirchman DL (2000) Community composition of marine bacterioplankton determined by 16S rRNA gene clone libraries and fluorescence in situ hybridization. Appl Environ Microbiol 66: 5116-5122.
3. Giovannoni SJ, Rappé M (2000) Evolution, diversity and molecular ecology of marine prokaryotes. In Microbial ecology of the oceans DL Kirchman Wiley-Liss, New York USA.
4. Eller G, Stubner S, Frenzel P (2001) Group-specific 16S rRNA targeted probes for the detection of type I and type II methanotrophs by fluorescence in situ hybridisation. FEMS Microbiol Lett 198: 91-97.
5. Giovannoni SJ, Stingl U (2005) Molecular diversity and ecology of microbial plankton. Nature 437: 343-348.
6. Hazen TC, Dubinsky EA, DeSantis TZ, Andersen GL, Piceno YM, et al. (2010) Deep-sea oil plume enriches indigenous oil-degrading bacteria. Science 330: 204-208.
7. Olapade OA (2010) Molecular analyses of the diversity in marine bacterioplankton assemblages along the coastline of the northeastern Gulf of Mexico. Can J Microbiol 56: 853-863.
8. Rappé MS, Vergin K, Giovannoni SJ (2000) Phylogenetic comparisons of a coastal bacterioplankton community with its counterparts in open ocean and freshwater systems. FEMS Microbiol Ecol 33: 219-232.
9. James R Bragg, Roger C Prince, James Harner E, Ronald M Atlas (1994) Effectiveness of bioremediation for the Exxon Valdez oil spill. Nature 368: 413-418.
10. Röling WF, Milner MG, Jones DM, Fratepietro F, Swannell RP, et al. (2004) Bacterial community dynamics and hydrocarbon degradation during a field-scale evaluation of bioremediation on a mudflat beach contaminated with buried oil. Appl Environ Microbiol 70: 2603-2613.
11. Swannell RPJ, Mitchell D, Lethbridge G, Jones D, HeathD, et al. (1999) A Field Demonstration of the Efficacy of Bioremediation to Treat Oiled Shorelines Following the Sea Empress Incident. Environ Technol 20: 863-873.
12. Atlas RM, Bartha R (1972) Degradation and mineralization of petroleum in sea water: limitation by nitrogen and phosphorous. Biotechnol Bioeng 14: 309-318.
13. Olapade OA (2012) Diel fluctuations in the abundance and community diversity of coastal bacterioplankton assemblages over a tidal cycle. Microb Ecol 63: 96-102.
14. Valentine DL, Kessler JD, Redmond MC, Mendes SD, Heintz MB, et al. (2010) Propane respiration jump-starts microbial response to a deep oil spill. Science 330: 208-211.
15. Kostka JE, Prakash O, Overholt WA, Green SJ, Freyer G, et al. (2011)

Hydrocarbon-degrading bacteria and the bacterial community response in gulf of Mexico beach sands impacted by the deepwater horizon oil spill. Appl Environ Microbiol 77: 7962-7974.

16. Holmes AJ, Roslev P, McDonald IR, Iversen N, Henriksen K, et al. (1999) Characterization of methanotrophic bacterial populations in soils showing atmospheric methane uptake. Appl Environ Microbiol 65: 3312-3318.

17. Tavormina PL, Ussler W 3rd, Orphan VJ (2008) Planktonic and sediment-associated aerobic methanotrophs in two seep systems along the North American margin. Appl Environ Microbiol 74: 3985-3995.

18. Boetius A, Ravenschlag K, Schubert CJ, Rickert D, Widdel F, et al. (2000) A marine microbial consortium apparently mediating anaerobic oxidation of methane. Nature 407: 623-626.

19. Dunfield PF, Yuryev A, Senin P, Smirnova AV, Stott MB, et al. (2007) Methane oxidation by an extremely acidophilic bacterium of the phylum Verrucomicrobia. Nature 450: 879-882.

20. Ettwig KF, Butler MK, Le Paslier D, Pelletier E, Mangenot S, et al. (2010) Nitrite-driven anaerobic methane oxidation by oxygenic bacteria. Nature 464: 543-548.

21. Hanson RS, Hanson TE (1996) Methanotrophic bacteria. Microbiol Rev 60: 439-471.

22. Milucka J, Ferdelman TG, Polerecky L, Franzke D, Wegener G, et al. (2012) Zero-valent sulphur is a key intermediate in marine methane oxidation. Nature 491: 541-546.

23. Penland S, Sutter JR, and McBride RA (1987) Delta plain development and sea level history in Terrebone coastal region, Lousiana. In: Kraus NC Coastal Sediments 87. American Society of Civil Engineers, NewYork 1689-1704.

24. Messing J (1983) New M13 vectors for cloning. Methods Enzymol 101: 20-78.

25. Altschul SF, Madden TL, Schaffer AA, Zhang J, Zhang Z, et al. (1997) Gapped BLAST and PSI-BLAST: a new generation of protein database search programs. Nucleic Acids Res 25: 3389-3402.

26. Perrière G, Gouy M (1996) WWW-query: an on-line retrieval system for biological sequence banks. Biochimie 78: 364-369.

27. Chao A (1984) Nonparametric Estimation of the Number of Classess in a Population. Scand J Statist 11: 265-270.

28. Chao A (1987) Estimating the population size for capture-recapture data with unequal catchability. Biometrics 43: 783-791.

29. Schloss PD, Handelsman J (2006) Introducing SONS, a tool for operational taxonomic unit-based comparisons of microbial community memberships and structures. Appl Environ Microbiol 72: 6773-6779.

30. Porter KG, Feig YS (1980) The Use of DAPI for Identifying and Counting Aquatic Microflora. Limnol Oceanogr 25: 943-948.

31. Lomans BP, Leijdekkers P, Wesselink JJ, Bakkes P, Pol A (2001) Obligate sulfide-dependent degradation of methoxylated aromatic compounds and formation of methanethiol and dimethyl sulfide by a freshwater sediment isolate, Parasporobacterium paucivorans gen. nov., sp. nov. Appl Environ Microbiol 67: 4017-4023.

32. Mills HJ, Hunter E, Humphrys M, Kerkhof L, McGuinness L, et al. (2008) Characterization of nitrifying, denitrifying, and overall bacterial communities in permeable marine sediments of the northeastern Gulf of Mexico. Appl Environ Microbiol 74: 4440-4453.

33. Kessler JD, Valentine DL, Redmond MC, Du M, Chan EW, et al. (2011) A persistent oxygen anomaly reveals the fate of spilled methane in the deep Gulf of Mexico. Science 331: 312-315.

34. Britschgi TB, Giovannoni SJ (1991) Phylogenetic analysis of a natural marine bacterioplankton population by rRNA gene cloning and sequencing. Appl Environ Microbiol 57: 1707-1713.

35. Chistoserdova L, Lapidus A, Han C, Goodwin L, Saunders L, et al. (2007) Genome of Methylobacillus flagellatus, molecular basis for obligate methylotrophy, and polyphyletic origin of methylotrophy. J Bacteriol 189: 4020-4027.

36. Lambo AJ, Patel TR (2006) Isolation and characterization of a biphenyl-utilizing psychrotrophic bacterium, Hydrogenophaga taeniospiralis IA3-A, that cometabolize dichlorobiphenyls and polychlorinated biphenyl congeners in Aroclor 1221. J Basic Microbiol 46: 94-107.

37. Nakatsu CH, Hristova K, Hanada S, Meng XY, Hanson JR, et al. (2006) Methylibium petroleiphilum gen. nov., sp. nov., a novel methyl tert-butyl ether-degrading methylotroph of the Betaproteobacteria. Int J Syst Evol Microbiol 56: 983-989.

38. Abildgaard L, Ramsing NB, Finster K (2004) Characterization of the marine propionate-degrading, sulfate-reducing bacterium Desulfofaba fastidiosa sp. nov. and reclassification of Desulfomusa hansenii as Desulfofaba hansenii comb. nov. Int J Syst Evol Microbiol 54: 393-399.

39. Werner Liesack, Kai Finster (1994) Phylogenetic Analysis of Five Strains of Gram-Negative, Obligately Anaerobic, Sulfur-Reducing Bacteria and Description of Desulfirornusa gen. nov., Including Desulfuromusa kysingii sp. nov., Desulfurornusa bakii sp. nov., and Desulfuromusa succinoxidans sp. nov. Int J Syst Bacteriol 44: 753-758.

40. González JM, Moran MA (1997) Numerical dominance of a group of marine bacteria in the alpha-subclass of the class Proteobacteria in coastal seawater. Appl Environ Microbiol 63: 4237-4242.

41. Bai Y, Sun Q, Zhao C, Wen D, Tang X (2009) Aerobic degradation of pyridine by a new bacterial strain, Shinella zoogloeoides BC026. J Ind Microbiol Biotechnol 30: 1391-1400.

42. Cottrell MT, Kirchman DL (2000) Natural assemblages of marine proteobacteria and members of the Cytophaga-Flavobacter cluster consuming low- and high-molecular-weight dissolved organic matter. Appl Environ Microbiol 66: 1692-1697.

43. Ronald S Oremland, Charles W Culbertson (1992) Importance of methane-oxidizing bacteria in the methane budget as revealed by the use of a specific inhibitor. Nature 356 421-423.

44. McDonald IR, Bodrossy L, Chen Y, Murrell JC (2008) Molecular ecology techniques for the study of aerobic methanotrophs. Appl Environ Microbiol 74: 1305-1315.

45. Phillips CJ, Smith Z, Embley TM, Prosser JI (1999) Phylogenetic differences between particle-associated and planktonic ammonia-oxidizing bacteria of the beta subdivision of the class Proteobacteria in the Northwestern Mediterranean Sea. Appl Environ Microbiol 65: 779-786.

46. Amann RI, Krumholz L, Stahl DA (1990) Fluorescent-oligonucleotide probing of whole cells for determinative, phylogenetic, and environmental studies in microbiology. J Bacteriol 172: 762-770.

47. Amann RI, Ludwig W, Schleifer KH (1995) Phylogenetic identification and in situ detection of individual microbial cells without cultivation. Microbiol Rev 59: 143-169.

48. Manz W, Amann R, Ludwig W, Wagner M, Schleifer KH (1992) Phylogenetic Oligodeoxynucleotide Probes for the Major Subclasses of Proteobacteria: Problems and Solutions. Syst Appl Microbiol 15: 593-600.

49. Stephen JR, Kowalchuk GA, Bruns MAV, McCaig AE, Phillips CJ, et al. (1998) Analysis of ubroup Proteobacterial Ammonia Oxidizer populations in soil by denaturing gradient gel electrophoresis analysis and hierarchical phylogenetic probing. Appl Environ Microb 64: 2958 2065.

In Silico Approach for the Bioremediation of Toxic Pollutants

Fazlurrahman Khan, Mohammad Sajid and Swaranjit Singh Cameotra*

Environmental Biotechnology and Microbial Biochemistry Laboratory, Institute of Microbial Technology, Sector 39-A, Chandigarh-160036, India

Abstract

Microbial degradation is considered as one of the environment friendly and cost-effective method for restoration of ecological niches contaminated with chemical pollutants. Thus before the application of microbial system for the degradation of any newly released pollutant in the environment, there is need to in silico study for predicting the possible degradation pathways by using various computational tools. There are large number of databases and computer programs available to perform the computational analysis for assisting the development and implementation of microbial bioremediation. The present review provides a comprehensive account of these databases, software, their respective work methodologies and potential application for the bioremediation. The information collected for the above study from different in silico resources for assisting the environmental degradation studies is discussed.

Keywords: Bioremediation; Database; Pollutant; Metabolic network; Bioinformatics

Introduction

Studies pertaining to re-establishment of polluted ecological niches have led to a generalized acceptance of microbial potential as environment friendly and cost effective measure by their decontamination of pollutant [1-5]. The enrichment of cultures have been proposed as one of the most potent approaches to restore the sites contaminated with pollutants, however, it has been considered as one of the demanding challenge for a long time uses [6,7]. Even though, isolation of culturable microbial strains is also one of the most important bottle-neck for the successful biodegradation [5]. Subsequently, other limitations (chemical complexity of the target pollutant, transformation of pollutant substrate into more toxic intermediates) are known to inhibit the speedy progress of the microbial bioremediation projects [8]. In the recent past, the microbial degradation studies have been pursued with a system biology approach, wherein the degradation studies are devised on the past experiences and information obtained with earlier studies [9-12]. This has resulted in a very strong need for the maintenance and easy access to this information. Consequently, there is a need for computational programs that can utilize all the available information regarding bioremediation. Thus use of various computational programmes, software, tools and database, a lot of things can be subjected to prediction approaches with the eventual objective of developing an applicable bioremediation technology. Conventionally, bioremediation studies have depended upon isolation of degradative microbial strain from the contaminated habitat [5,6]. Further the application of isolated strain for bioremediation in the field principally depends upon the environmental limiting factor [13]. Hence, it is necessary to implement bioremediation process in natural environmental conditions where microorganism faces the different challenges impose by various abiotic and biotic factors [14-16]. However, the efficiency of the degradation process under different mechanism affected by these factors [13]. Thus, to understand the mechanism of degradation several studies has been designed to investigate the effect of all environmental factor(s) [17,18]. To study the biodegradation from lab to field, there should be ecological sustainability of the degradative strain in the field study. Thus, before going to ex situ and in situ bioremediation of any pollutants there is need to develop in silico method for the study of degradation. The present review article describes the collection of all databases, tools and softwares which helps in the in silico analysis/prediction of toxicity of any chemical along with elucidation of feasible microbial degradation pathways.

In silico toxicity of the compounds

For environmental cleanup of the toxic compounds by microorganisms many technologies have been developed, but without the knowledge of the toxicity level of the compounds it cannot be fully successful as toxicity affects the survival of the degradative strains [19]. Thus before making many efforts in development of in silico bioremediation technology of any chemicals there is also need to predict its toxicity levels by in silico approaches. Predicting toxicity of a compound by in silico toxicological methods is a developing field with great potential. More than 70 million chemicals were identified till 29 May 2013 (CAS) [20]. Exposures of these toxic chemicals e.g., pesticides, products of chemical industries like cosmetics and drugs etc. leads to various health effects [21]. Several in silico procedures have been developed and rooted by pharmaceutical industry to understand the pharmacodynamic, pharmacokinetic and toxicological profile of a compound [22]. Tables 1 and 2 summarises various databases, methods, homology models, pharmacophores and several other molecular modelling approaches for determining the toxicity of any chemical. Although there are many data bases developed about toxicity of the compounds [23], toxicity value hierarchy from Environmental Protection Agency's (EPA), and toxicity assay method [24-26]. However, various industries are synthesizing new chemicals more rapidly than academic research and the norms lay down by the regulatory agencies. The quantitative structure-activity relationships (QSARs) are the one such in silico method for determining the quantitative structure-activity relationship model, which is based on

***Corresponding author:** Swaranjit Singh Cameotra, Environmental Biotechnology and Microbial Biochemistry Laboratory, Institute of Microbial Technology, Sector 39-A, Chandigarh-160036, India, Tel: +91-6665223/224; Fax: +91-172-2690632

S.No	Name of Programme	URLCode	Properties/Reference
1.	Derek (Lhasa Ltd)	http://www.lhasalimited.org/	Lhasa Limited specialises in the development of *in silico* prediction and database systems for use in metabolism, toxicology and related sciences.
2.	HazardExpert (CompuDrug)	http://www.compudrug.com/	CompuDrug is software tool to estimate toxicity of an organic compound in higher animals.
3.	ACD/Tox Suite (ToxBoxes)	http://www.acdlabs.com/products/pcadmet/tox/tox/	ACD/Tox Suite is a collection of software modules that predict probabilities for basic toxicity endpoints.
4.	ADMET Predictor	http://www.simulations-plus.com/Products.aspx?pID=13	ADMET acronym indicates all the parameters associated with absorption, metabolism, distribution, elimination, and toxicity of chemical in the human tissue.
5.	OncoLogic (USEPA)	http://www.epa.gov/oppt/sf/pubs/oncologic.htm	OncoLogic™ is a computer program that evaluates the likelihood that a chemical may cause cancer.
6.	Toxtree (JRC)	http://ihcp.jrc.ec.europa.eu/our_labs/predictive_toxicology/qsar_tools/toxtree	Toxtree open-source application that classifies toxic chemicals into various categories and predicts their toxic effect by using decision tree approaches.
7.	MolCode Toolbox	http://www.molcode.com	Molcode Toolboxe applies to simulate various experimentally unknown properties of compounds including physical, chemical, biological, ADME-Tox, ecological pathways/ eco-toxicity and adverse drug effects.
8.	TerraQSAR™	http://www.terrabase-inc.com/	TerraQSAR™ computation programs are designed for the quick and reliable estimation of biological effects and physico-chemical properties of organic compounds.
9.	Toxicity Estimation software tool (T.E.S.T)	http://www.epa.gov/nrmrl/std/qsar/qsar.html	TEST will enable users to easily estimate acute toxicity using several different QSAR methodologies including a hierarchical method, FDA method, Single-model method, Group contribution method, Nearest neighbour method, **Random forest method, multiple linear regressions, and Consensus method.**
10.	ORCHESTRA	http://www.orchestra-qsar.eu	A project funded by EC to disseminate recent research on computer-based methods to evaluate the toxicity of hazardous chemicals [64].
11.	VirtualToxLa	http://www.biograf.ch	It is an *in silico* tool for predicting the toxicity of drugs, chemicals, and natural products. It simulates and quantifies their relations toward a series of proteins known to trigger adverse effects using automated, flexible docking combined with multidimensional QSAR [65].

Table 1: Software tools for toxicity prediction.

S.No.	Name of Database	URL Code	Properties/Reference
1.	Acutoxbase	https://acubase.amwaw.edu.pl	Acutebase has been developed to manage all information relevant to the EU integrated project 'ACuteTox'. It provides *in vitro* testing approach for predicting human toxicity of a compound [66].
2.	ChemIDplus	http://chem.sis.nlm.nih.gov/chemidplus/	ChemIDplus is web-based search application which allows opening structure and nomenclature authority files of chemical substances cited in National Library of Medicine (NLM) databases including the TOXNET® system.
3.	Chemical Effects in Biological Systems (CEBS)	http://www.niehs.nih.gov/research/resources/databases/cebs/index.cfm	CEBS is applying to view data in the form of biology and study design and permit data integration across studies.
4.	Terra-Base	http://www.terrabase-inc.com/	TerraTox™ databases provide for the quick search of compounds with structure/fragment-specific biological effects and properties.
5.	GENE-TOX	http://toxnet.nlm.nih.gov/cgi-bin/sis/htmlgen?GENETOX	Databank developed by the Environmental Protection Agency (EPA) through genetic toxicology.
6.	Hazardous Substances Data Bank	http://toxnet.nlm.nih.gov/cgi-bin/sis/htmlgen?HSDB	Hazardous substance data Bank is toxicology database on National library of Medicine. It has information regarding exposure routes, industrial hygiene, emergency handling procedures, environmental fate and regulatory details.
7.	SuperToxic	http://bioinformatics.charite.de/supertoxic/index.php?site=home	Recently, this database has information of approximately 60,000 compounds which are classified according to their toxicological profiles.
8.	Aggregated Computational Toxicology Resource)	http://actor.epa.gov/actor/faces/ACToRHome.jsp	ACToR is openly access chemical toxicity database and can be used to discover potential chemical hazards to human health and the environment.
9.	Comparative Toxicogenomics Database	http://toxnet.nlm.nih.gov/cgi-bin/sis/htmlgen?CTD	CTD™-elucidates human toxicity of a compound and its molecular mechanisms by which environmental chemicals acts.
10	Carcinogenic Potency Database	http://toxnet.nlm.nih.gov/cgi-bin/sis/htmlgen?CPDB.htm	CPD provides standardized analyses of chronic, long-term animal cancer tests and reported in the general published literature or by the National Cancer Institute and the National Toxicology Program.

Table 2: Databases containing toxicity information.

experimental or calculated data [27] (Figure 1). The toxicity of allele-chemical like pesticides can be also predicted by applying combination of methods such as 3D-QSAR, docking, Local Binding Energy (LBE) and GRID [28]. Further, OSIRIS property explorer is another in silico method for assessing the toxicity of any chemical compound. Discovery of newer drug may also bring health hazards. The information regarding toxicological profile of known toxins will make a basis to assign the toxicity of unknown compound. The Structural Bioinformatics Group, Institute of Molecular Biology and Bioinformatics, Charite (Cbf), Berlin, Germany created a SuperToxic database containing information of various toxic compounds (~60,000).

Prediction of environmental destiny of toxic compounds

Biodegradation of toxic chemicals is most important parameter influencing whether compound will be biodegraded and, if so, will the biodegradation proceed slowly or quickly. To study the biodegradability in silico there are several softwares and databases available (Table 3). Among softwares most important one is BESS (Biodegradability Evaluation and Simulation System) [29]. The prediction of biodegradation of a compound can be made possible by using BESS software based on structural description of the compound and the existing environmental condition. BESS uses promising enzymatic

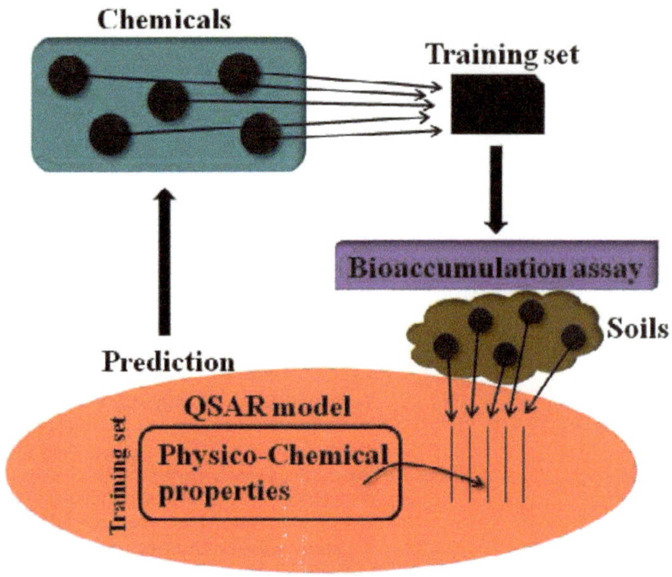

Figure 1: QSAR model for bioremediation.

S.No.	URL Code	Properties	References
1.	http://www.labmed.umn.edu/umbbd/ index.html	Prediction of biodegradation pathway	[67]
2.	http://www.labmed.umn.edu/umbbd/predictbt/	Predicting Biotransformation build on existing bio degradation information contained in the UM-BBD.	[68]
3.	http://umbbd.ahc.umn.edu/index.html	UM-BBD: University of Minnesota Biocatalysis/Biodegradation Database.	[69]
4.	http://umbbd.msi.umn.edu/predict/	UM-PPS: predicts microbial catabolic reactions using substructure searching, a rule-base and atom-to-atom mapping.	[70]
5.	http://www.genome.ad.jp/kegg/kegg2.html	KEGG: Kyoto Encyclopaedia of Genes and Genomes.	[71]
6.	http://www.expasy.org/cgi-bin/search-biochem-index	Boehringer Mannheim Biochemical Pathways on the ExPASy server, Switzerland.	[72]
7.	http://emp.mcs.anl.gov/	Enzyme and Metabolic Pathway (EMP) Database at Argonne National Laboratories.	[73]
8.	http://www.issx.org/	International society for study of Xenobiotics.	
9.	http://biocyc.org/	Biocyc: Knowledge Library of Pathway/ Genome Databases.	[74]
10.	http://www.ncgr.org/pathdb/	Path DB: Metabolic Pathways Database at NCGR.	[75]
11.	http://www.tcd.ie/Biochemistry/IUBMB-Nicholson/	Metabolic Pathway Minimaps at Trinity College, Dublin, Ireland.	[76]
12.	http://www.daylight.com/smiles/f_smiles.html	SMILES is a system for coding chemical compounds as linear strings of ASCII characters.	[77]
13.	http://biorad.igib.res.in	BioRad Base is a database for bioremediation of radioactive waste.	[78]
14.	http://bsd.cme.msu.edu	Biodegradative Strain Database	[47]
15.	http://www.epa.gov/opptintr/exposure/pubs/episuite.htm	A Windows®-based, EPI (Estimation Programs Interface) Suite™ provides physical/ chemical properties and environmental fate.	[79]

Table 3: Programme, databases and web resources containing biodegradability information.

reactions that are hierarchically organised according to knowledge of microbial physiology and ecology. Such type of prediction reduces the potential large number of enzymatic conversion which is most likely to provide anabolic intermediates or energy to micro-organisms (Figure 2a). Biochemical Network Integrated Computational explorer (BNICE) is also known for the prediction of possible microbial degradation pathways (14). It involves rules of enzyme classification system and predicts thermodynamically favoured reactions rather than using microbial physiological and ecological conditions (Figure 2b). The multi computer automated structure evaluation/ (MultiCASE/META) system combines a group-contribution model and an expert system to simulate aerobic biodegradation pathways [30]. MultiCASE approach has been also used to model anaerobic aquatic biodegradation rates [31]. Mineralization of organic compound is a significant factor when considering their fate in the environment. A model was developed and integrated into an expert software system named CATABOL which is a knowledge based expert system [32]. It can simulate the likelihood of biodegradation of organic compounds directly from their structure. However, this probabilistic model can also be useful in determining the probabilities for overall Biochemical Oxygen Demand (BOD) and extent of CO_2 production in bioremediation process [32,33]. Fate of a specific chemical spilled in a given site and even show interventions aimed at accelerating the process can be described by MetaRouter system [34]. The MetaRouter allows visualization through a web interface of all probable pathways that a large number of intractable compounds can take through known steps of all the reactions taken from the UM-BBD. The system searches in the database for all possible combination of enzymes (and wherever available, their cognate

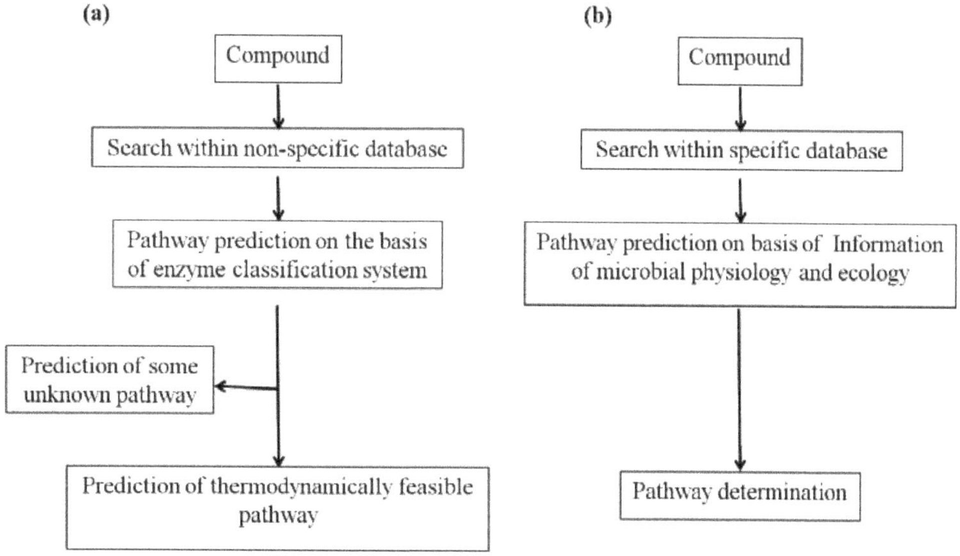

Figure 2: Comparison between pathway prediction rules between BNICE (2a) and BESS (2b).

genes) required to convert a certain substrate into their metabolic intermediates or into any other products. Through MetaRouter system a virtual pathways having meshwork of genes/enzymes can be predicted which come from different bacteria, sometimes having very different lifestyles (for instance, aerobic and anaerobic). Nevertheless, such combinations may not exist or may have not been exposed yet in nature, these cross pathways reflect probable processes that can occur at different stages and locations by divergent microorganisms. The MetaRouter system does not give information regarding kinetics or thermodynamics of the proposed pathways, although it can certainly guide metabolic engineering attempts. Although MetaRouter describes only biodegradation information for a compound and gives a picture of how given chemicals could be degraded if passed through the complex metabolism of a complex community rather than how they could be metabolized by one specialist strain [34]. Metabolic Knowledgebase (MKB) uses ontology to formalize metabolic data and apply large flexible and scalable metabolic knowledgebase to capture several levels of chemical or biological information. It develops inference tools to support complex metabolic queries to semantically integrate data from chemical structures to complete pathways; incorporate data from public domain sources. Biocatalysis Classification Scheme (BCS) uses MKB to identify subset of biocatalytic function based on pathway of interest and to find relevant compounds and infer their sub structural features. Biocatalysis Assignment Tool (BAT) uses MKB to identify relevant sets of proteins and their biocatalytic functions. It also identifies conserved features among proteins and establishes correlation between function and conserved features. Metabolic Pathway Synthesis (MPS) provides information regarding predicting enzymatic activity from the cellular environment and helps to classify pathways with respect to cellular parameters and to obtain information about metabolic pathway and its regulation. It uses BCS/MKB to find out the possible transformation from native compound to intermediates. In case of database KEGG (Kyoto Encyclopedia of Genes and Genomes), commonly used resource provides information on genes and metabolic pathways in a wide range of species [35]. KEGG consists of three databases: PATHWAY provides network of interacting molecules, GENES contains catalogs of all full and partial genome sequence and LIGAND for the collection of chemical compounds in the cell, enzyme molecules and enzymatic reactions [36].

Influence of environmental factors on biodegradation

Earlier studies of bioremediation trials were not performed under natural environmental conditions. Therefore, the impact of environmental factors on the bioremediation process was never expected. However, after the investigation of in situ bioremediation approaches now it is feasible to understand the bioremediation process is influenced significantly by environmental factors such as the physiological and chemical ambience of the contaminated environment, bioavailability of nutrients, concentration and properties of cocontaminants, level of contamination, community organization of the indigenous microbial communities [13,37-40]. Various abiotic and biotic factors play important role in bioremediation. Their dynamic interactions occur in concrete abiotic conditions which are defined by physico-chemical conditions like O_2 supply, electron transport, water, temperature, pH, salt concentration, many of which [41-43]. The above environmental factors determine the dynamic of endogenous microbial community structures along with the availability of given chemical and energy source [41].

Ecological consequences on the biodegradative strains

Knowledge of indigenous microflora is required which may or may not affect the degradation of toxic compounds by test microorganism [13]. In situ bioremediation is directly or indirectly affected by the indigenous micro flora, thus it is most important to evaluate the ecological consequences on the biodegradative strains [13,41]. There are several non related phenomenons to study these ecological consequences. However, for the development of bioremediation technology, it is required to analyze the effect on indigenous microbial community structure by the bioremediation process in a particular ecological conditions [41,42,44-46]. Biodegradative Strain Database (BSD) provides information of degradative bacteria and hazardous chemicals degraded by these bacteria [47]. It also includes corresponding literature citation, relevant patents and link to additional web based biological and chemical data. The BSD is being developed within the phylogenetic framework of the Ribosomal Database Project

II to provide a biological complement to the chemical and degradative pathway [47]. Ecological Structure Activity Relationships (ECOSAR) is a foretelling system providing information for aquatic toxicity of a compound [48]. The program estimates a chemicals acute toxicity and chronic toxicity to aquatic organisms by using computerized Structure Activity Relationships (SARs).

Discussion

As a result of worldwide extensive application of pesticides, it may get released into different environmental compartment (e.g. soil, sediment and water bodies) through waste streams and causing threat to the various life forms [3]. Thus public concern has been prompted to develop robust technology for the eradication of pesticides and restoration of environmental health. There are two ways for the decontamination of pesticides from the environment; one is nonbiological (e.g. incineration, land filling, hydrolysis, photolysis, chemical lysis and thermal decomposition) and the second is biological means [2,3,49,50]. Mostly the microbial transformation/degradation of the pesticides/ or chemical is considered as one of the most effective, ecofriendly and technologically challenging approaches for the bioremediation of toxic pollutants from the different environmental compartment [3,4,51,52]. The bioremediation database provides by University of Minnesota provides most comprehensive platform for nearly all the bioremediation pathway and helps in developing various prediction programmes using these databases [53]. Pathway prediction system (PPS) predicts possible bioremediation pathway for xenobiotics using biotransformation parameters provided by UM-BBD database as well as scientific literature [54]. Other programmes similar to PPS are METEOR, MetabolExpert, DEREK, StAR, CATABOL system, MetaRouter and MultiCASE/META [32,34,55-58]. To establish strain designation in prediction of bioremediation within phylogenetic perspective, Biodegradative Strain Database (BSD) was developed by integrating metabolic data of UM-BBD and phylogenetic data of Ribosome Database project (RPD-II) [47]. Till now dogma of the biodegradation is one strain degrades one pollutant, but information available in literature of bioremediation lack essential aspect of natural scenarios, like interchange of gene between bacteria or their metabolic network co-operation [51,59,60]. It has also been known a single microbe exhibit metabolically versatile in nature due to the presence of large size of genome and plasmid containing large number of metabolic genes, resulting in expansion of the multiple metabolic pathways [43]. However, insufficient biological information regarding the regulation of growth and metabolism in various microbial communities restricts development in the site-specific mineralization process. Similarly, the bioremediation of xenobiotic compounds either ex situ or in situ by pure isolates does not represent an actual behaviour of the microorganisms; however, it depends on cooperative metabolic activities of mixed microbial population [61-63]. Under these conditions biodegradative potential of all microbes can be crossed with the all known compounds and in silico bioremediation helps to predict the destiny of a compound whether partially or fully degraded to non toxic compounds. Bioinformatics based search will facilitate and speed up the analysis of microbial degradation of hazardous compounds. In situ bioremediation can be applied in a number of action modes including aerobic, anaerobic, anoxic (nitrate respiration) and co-metabolic. Bioinformatics have huge amount of data collected from various resources such as chemical structure and reactivity properties of compounds; protein sequence, structure and function; comparative genome analysis; phylogenetic analysis; environmental biotechnology. These informations will collectively provide a bigger picture regarding degradation of a compound in the environment (Figure 3). The above summarized computational database, software and tools and their collective integration will help to determine the environmental fate of any compounds more precisely and accurately. After predicting the more accurate the environmental fate of any compound by computational approaches, future work will be carried out to validate

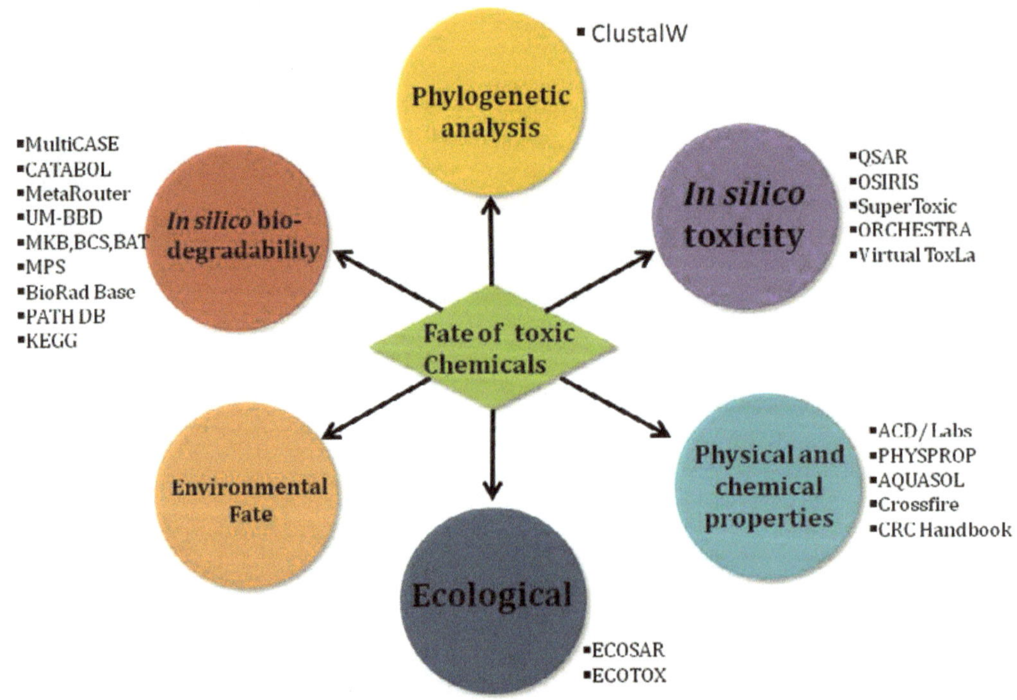

Figure 3: Schematic representations of prediction softwares and databases involve in destiny of a toxic compounds.

the complete fate of the compound by single isolate as well as by mixed microbial consortium. The pathway predicting software will also help in minimizing the number of possible combination for the development of microbial consortia.

Acknowledgements

FK thanks DBT for the Research Associate ship and MS acknowledge his research fellowship from UGC, India. This is the IMTECH communication no 0123/2013.

References

1. Lovley DR (2003) Cleaning up with genomics: applying molecular biology to bioremediation. Nat Rev Microbiol 1: 35-44.
2. Paul D, Pandey G, Pandey J, Jain RK (2005) Accessing microbial diversity for bioremediation and environmental restoration. Trends Biotechnol 23: 135-142.
3. Khan F, Pal D, Vikram S, Cameotra SS (2013) Metabolism of 2-chloro-4-nitroaniline via novel aerobic degradation pathway by Rhodococcus sp. strain MB-P1. PLoS One 8: e62178.
4. Khan F, Vyas B, Pal D, Cameotra SS (2013) Aerobic Degradation of N-Methyl-4-Nitroaniline (MNA) by Pseudomonas sp. Strain FK357 Isolated from Soil. PLoS One 8: e75046.
5. Fazlurrahman, Batra M, Pandey J, Suri CR, Jain RK (2009) Isolation and characterization of an atrazine-degrading Rhodococcus sp. strain MB-P1 from contaminated soil. Lett Appl Microbiol 49: 721-729.
6. Khan F, Pandey J, Vikram S, Pal D, Cameotra SS (2013) Aerobic degradation of 4-nitroaniline (4-NA) via novel degradation intermediates by Rhodococcus sp. strain FK48. J Hazard Mater 254-255: 72-8.
7. Mohn WW, Tiedje JM (1992) Microbial reductive dehalogenation. Microbiol Rev 56: 482-507.
8. Boopathy R (2000) Factors limiting bioremediation technologies. Bioresource Technol 74: 63-67.
9. Finley SD, Broadbelt LJ, Hatzimanikatis V (2010) In silico feasibility of novel biodegradation pathways for 1,2,4-trichlorobenzene. BMC Syst Biol 4: 7.
10. Hou BK, Wackett LP, Ellis LB (2003) Microbial pathway prediction: a functional group approach. J Chem Inf Comput Sci 43: 1051-1057.
11. Ellis LB, Roe D, Wackett LP (2006) The University of Minnesota Biocatalysis/Biodegradation Database: the first decade. Nucleic Acids Res 34: D517-521.
12. Wackett LP, Ellis LB (1999) Predicting biodegradation. Environ Microbiol 1: 119-124.
13. Pandey J, Chauhan A, Jain RK (2009) Integrative approaches for assessing the ecological sustainability of in situ bioremediation. FEMS Microbiol Rev 33: 324-375.
14. Ward DM, Brock TD (1976) Environmental factors influencing the rate of hydrocarbon oxidation in temperate lakes. Appl Environ Microbiol 31: 764-772.
15. Dinkla IJ, Gabor EM, Janssen DB (2001) Effects of iron limitation on the degradation of toluene by Pseudomonas strains carrying the tol (pWWO) plasmid. Appl Environ Microbiol 67: 3406-3412.
16. Kim HJ, Graham DW (2003) Effects of oxygen and nitrogen conditions on the transformation kinetics of 1,2-dichloroethenes by Methylosinus trichosporium OB3b and its sMMOC mutant. Biodegradation 14: 407-414.
17. Hoyle BL, Scow KM, Fogg GE, Darby JL (1995) Effect of carbon:nitrogen ratio on kinetics of phenol biodegradation by Acinetobacter johnsonii in saturated sand. Biodegradation 6: 283-293.
18. Lovanh N, Hunt CS, Alvarez PJ (2002) Effect of ethanol on BTEX biodegradation kinetics: aerobic continuous culture experiments. Water Res 36: 3739-3746.
19. Copley SD (2009) Evolution of efficient pathways for degradation of anthropogenic chemicals. Nat Chem Biol 5: 559-566.
20. 2013 CAS Spring Meeting Vancouver, Canada.
21. Rider CV, Carlin DJ, Devito MJ, Thompson CL, Walker NJ (2012) Mixtures research at NIEHS: An evolving program. Toxicology .
22. Ekins S, Mestres J, Testa B (2007) In silico pharmacology for drug discovery: applications to targets and beyond. Br J Pharmacol 152: 21-37.
23. Schmidt U, Struck S, Gruening B, Hossbach J, Jaeger IS, et al. (2009) SuperToxic: a comprehensive database of toxic compounds. Nucleic Acids Res 37: D295-299.
24. Ohga T, Masduki A, Kato J, Ohtake H (1993) Chemotaxis away from thiocyanic and isothiocyanic esters in Pseudomonas aeruginosa. FEMS Microbiol Lett 113: 63-66.
25. Sager BM, Sekelsky JJ, Matsumura P, Adler J (1988) Use of a computer to assay motility in bacteria. Anal Biochem 173: 271-277.
26. Nikata T, Sumida K, Kato J, Ohtake H (1992) Rapid method for analyzing bacterial behavioral responses to chemical stimuli. Appl Environ Microbiol 58: 2250-2254.
27. Schultz TW, Cronin MTD, Walker JD, Aptula AO (2003) Quantitative structure–activity relationships (QSARs) in toxicology: a historical perspective. J Mol Struc-THEOCHEM 622: 1-22.
28. Fratev F, Lo Piparo E, Benfenati E, Mihaylova E (2007) Toxicity study of allelochemical-like pesticides by a combination of 3D-QSAR, docking, Local Binding Energy (LBE) and GRID approaches. SAR QSAR Environ Res 18: 675-692.
29. Williams PF (2000) BESS, A System for Predicting the Biodegradability of New Compounds. Environmental Protection Agency USA.
30. Klopman G, Tu M, Talafous J (1997) META. 3. A genetic algorithm for metabolic transform priorities optimization. J Chem Inf Comput Sci 37: 329-334.
31. Rorije E, Peijnenburg WJGM, Klopman G (1998) Structural requirements for anaerobic biodegradation of organic chemicals: A fragment model analysis. Environ Toxicol Chem 17: 1943-1950.
32. Jaworska J, Dimitrov S, Nikolova N, Mekenyan O (2002) Probabilistic assessment of biodegradability based on metabolic pathways: catabol system. SAR QSAR Environ Res 13: 307-323.
33. Jaworska J, Van Genderen-Takken H, Hanstveit A, van de Plassche E, Feijtel T (2002) Environmental risk assessment of phosphonates, used in domestic laundry and cleaning agents in The Netherlands. Chemosphere 47: 655-665.
34. Pazos F, Guijas D, Valencia A, De Lorenzo V (2005) MetaRouter: bioinformatics for bioremediation. Nucleic Acids Res 33: D588-592.
35. Kanehisa M, Goto S (2000) KEGG: kyoto encyclopedia of genes and genomes. Nucleic Acids Res 28: 27-30.
36. Goto S, Nishioka T, Kanehisa M (2000) LIGAND: chemical database of enzyme reactions. Nucleic Acids Res 28: 380-382.
37. Leahy JG, Colwell RR (1990) Microbial degradation of hydrocarbons in the environment. Microbiol Rev 54: 305-315.
38. Mills DK, Fitzgerald K, Litchfield CD, Gillevet PM (2003) A comparison of DNA profiling techniques for monitoring nutrient impact on microbial community composition during bioremediation of petroleum-contaminated soils. J Microbiol Methods 54: 57-74.
39. Moreels D, Bastiaens L, Ollevier F, Merckx R, Diels L, et al. (2004) Effect of in situ parameters on the enrichment process of MTBE degrading organisms. Commun Agric Appl Biol Sci 69: 3-6.
40. Coulon F, McKew BA, Osborn AM, McGenity TJ, Timmis KN (2007) Effects of temperature and biostimulation on oil-degrading microbial communities in temperate estuarine waters. Environ Microbiol 9: 177-186.
41. Katsivela E, Moore ER, Maroukli D, Strömpl C, Pieper D, et al. (2005) Bacterial community dynamics during in-situ bioremediation of petroleum waste sludge in landfarming sites. Biodegradation 16: 169-180.
42. Wenderoth DF, Rosenbrock P, Abraham WR, Pieper DH, Höfle MG (2003) Bacterial community dynamics during biostimulation and bioaugmentation experiments aiming at chlorobenzene degradation in groundwater. Microb Ecol 46: 161-176.
43. Eyers L, Smoot JC, Smoot LM, Bugli C, Urakawa H, et al. (2006) Discrimination of shifts in a soil microbial community associated with TNT-contamination using a functional ANOVA of 16S rRNA hybridized to oligonucleotide microarrays. Environ Sci Technol 40: 5867-5873.
44. Röling WF, Milner MG, Jones DM, Fratepietro F, Swannell RP, et al. (2004) Bacterial community dynamics and hydrocarbon degradation during a field-scale evaluation of bioremediation on a mudflat beach contaminated with buried oil. Appl Environ Microbiol 70: 2603-2613.

45. Viñas M, Sabaté J, Espuny MJ, Solanas AM (2005) Bacterial community dynamics and polycyclic aromatic hydrocarbon degradation during bioremediation of heavily creosote-contaminated soil. Appl Environ Microbiol 71: 7008-7018.

46. Paul D, Pandey G, Meier C, van der Meer JR, Jain RK (2006) Bacterial community structure of a pesticide-contaminated site and assessment of changes induced in community structure during bioremediation. FEMS Microbiol Ecol 57: 116-127.

47. Urbance JW, Cole J, Saxman P, Tiedje JM (2003) BSD: the Biodegradative Strain Database. Nucleic Acids Res 31: 152-155.

48. Levet A, Bordes C, Clément Y, Mignon P, Chermette H, et al. (2013) Quantitative structure-activity relationship to predict acute fish toxicity of organic solvents. Chemosphere 93: 1094-1103.

49. Dua M, Singh A, Sethunathan N, Johri AK (2002) Biotechnology and bioremediation: successes and limitations. Appl Microbiol Biotechnol 59: 143-152.

50. Khan F, Prakash D, Jain R (2011) Development of an HPLC method for determination of pentachloronitrobenzene, hexachlorobenzene and their possible metabolites. BMC Chem Biol 11: 2.

51. Khan F, Pal D, Ghosh A, Cameotra SS (2013) Degradation of 2,4-dinitroanisole (DNAN) by metabolic cooperative activity of Pseudomonas sp. strain FK357and Rhodococcus imtechensis strain RKJ300. Chemosphere 93: 2883-2888.

52. Khan F, Kumari M, Cameotra SS (2013) Biodegradation of the Allelopathic Chemical m-Tyrosine by Bacillus aquimaris SSC5 Involves the Homogentisate Central Pathway. PLoS One 8: e75928.

53. Hou BK, Ellis LB, Wackett LP (2004) Encoding microbial metabolic logic: predicting biodegradation. J Ind Microbiol Biotechnol 31: 261-272.

54. Klopman G, Zhang Z, Balthasar DM, Rosenkranz HS (1995) Computer-Automated predictions of aerobic biodegradation of chemicals. Environ Toxicol Chem 14: 395-403.

55. Klopman G, Dimayuga M, Talafous J (1994) META. 1. A program for the evaluation of metabolic transformation of chemicals. J Chem Inf Comput Sci 34: 1320-1325.

56. Greene N, Judson PN, Langowski JJ, Marchant CA (1999) Knowledge-based expert systems for toxicity and metabolism prediction: DEREK, StAR and METEOR. SAR QSAR Environ Res 10: 299-314.

57. Oh M, Yamada T, Hattori M, Goto S, Kanehisa M (2007) Systematic analysis of enzyme-catalyzed reaction patterns and prediction of microbial biodegradation pathways. J Chem Inf Model 47: 1702-1712.

58. Darvas F (1987) Metabolexpert: An Expert System for Predicting Metabolism of Substances. QSAR in Environmental Toxicology - II 71-81.

59. [No authors listed] (1991) Clinical results using two new antibiotics--clarithromycin and temafloxacin. Oncology (Williston Park) 5: 142, 145-146.

60. de Souza ML, Newcombe D, Alvey S, Crowley DE, Hay A, et al. (1998) Molecular basis of a bacterial consortium: interspecies catabolism of atrazine. Appl Environ Microbiol 64: 178-184.

61. Boonchan S, Britz ML, Stanley GA (2000) Degradation and mineralization of high-molecular-weight polycyclic aromatic hydrocarbons by defined fungal-bacterial cocultures. Appl Environ Microbiol 66: 1007-1019.

62. Wang Z, Zhang J, Zhang Y, Hesham AL, Yang M (2006) Molecular characterization of a bacterial consortium enriched from an oilfield that degrades phenanthrene. Biotechnol Lett 28: 617-621.

63. Barreiros L, Nogales B, Manaia CM, Ferreira AC, Pieper DH, et al. (2003) A novel pathway for mineralization of the thiocarbamate herbicide molinate by a defined bacterial mixed culture. Environ Microbiol 5: 944-953.

64. Mays C, Benfenati E, Pardoe S (2012) Use and perceived benefits and barriers of QSAR models for REACH: findings from a questionnaire to stakeholders. Chem Cent J 6: 159.

65. Vedani A, Smiesko M, Spreafico M, Peristera O, Dobler M (2009) VirtualToxLab - in silico prediction of the toxic (endocrine-disrupting) potential of drugs, chemicals and natural products. Two years and 2,000 compounds of experience: a progress report. ALTEX 26: 167-176.

66. Kinsner-Ovaskainen A, Rzepka R, Rudowski R, Coecke S, Cole T, et al. (2009) Acutoxbase, an innovative database for in vitro acute toxicity studies. Toxicol In Vitro 23: 476-485.

67. Ellis LB, Hershberger CD, Wackett LP (2000) The University of Minnesota Biocatalysis/Biodegradation database: microorganisms, genomics and prediction. Nucleic Acids Res 28: 377-379.

68. Ellis LB, Hershberger CD, Wackett LP (1999) The University of Minnesota Biocatalysis/Biodegradation Database: specialized metabolism for functional genomics. Nucleic Acids Res 27: 373-376.

69. Ellis LB, Hershberger CD, Bryan EM, Wackett LP (2001) The University of Minnesota Biocatalysis/Biodegradation Database: emphasizing enzymes. Nucleic Acids Res 29: 340-343.

70. Jaimini D, Shabnam AA, Sarkar C (2012) In-Silico Feasibility of Novel Biodegradation Pathways For1-Naphthyl Methylcarbamate. American-Eurasian Journal of Toxicological Sciences 4: 89-93.

71. Kanehisa M, Goto S, Kawashima S, Nakaya A (2002) The KEGG databases at GenomeNet. Nucleic Acids Res 30: 42-46.

72. HarshaRani GV, Vayttaden SJ, Bhalla US (2005) Electronic data sources for kinetic models of cell signaling. J Biochem 137: 653-657.

73. Selkov E Jr, Grechkin Y, Mikhailova N, Selkov E (1998) MPW: the Metabolic Pathways Database. Nucleic Acids Res 26: 43-45.

74. Caspi R, Altman T, Dale JM, Dreher K, Fulcher CA, et al. (2010) The MetaCyc database of metabolic pathways and enzymes and the BioCyc collection of pathway/genome databases. Nucleic Acids Res 38: D473-D479.

75. http://www.icgeb.res.in/whotdr/cd1/PreCourseReading/NAR_DB2003/www3.oup.co.uk/nar/database/summary/115.html

76. Azzi A (2005) IUBMB Publications: An Overview. IUBMB Life 57: 249-250.

77. O'Boyle NM (2012) Towards a Universal SMILES representation - A standard method to generate canonical SMILES based on the InChI. J Cheminform 4: 22.

78. Reena M, Majhi C, Arya AK, Kumar R, Kumar A (2012) BioRadBase- A Database for Bioremediation of Radioactive waste. 11: 8718-8721.

79. Estimation Program Interface (EPI) Suite (2012) Environmental Protection Agency USA.

Polyhydroxy Alkanoates - A Sustainable Alternative to Petro-Based Plastics

Divya Gupta,[1]* Archana Tiwari [1] and Ramirez Alejandro Manzano[2]

[1]*School of Biotechnology, Rajiv Gandhi Proudyogiki Vishwavidyalaya, Bhopal, M.P., India*
[2]*Department of Material Sciences CINVESTAV-IPN, Unidad Querétaro Libramiento Norponiente, Querétaro, México*

Abstract

Industrial growth, urbanization and wrong agricultural practices are responsible for pollution and loss of environmental quality. Petro plastics, produced from mineral resources, are one of the most important materials, but the production process destructs the environment. Bio-based and biodegradable plastics can form the basis for environmentally preferable and sustainable alternative to current materials based exclusively on petroleum feed stocks. Bioplastic helps us to overcome the problem of pollution caused by synthetic plastics as they originate from renewable resources and are a new generation of plastics able to significantly reduce the environmental impact in terms of energy consumption and greenhouse effect. Polyhydroxyalkanoate (PHA) emerges as a potential candidate in a way to be used as a biopolymer material which not only possesses the characteristic similar to the traditional plastic, but it is also biodegradable in nature without any toxic production .A way to reduce the production cost is the use of alternative substrates, such as the agro industrial wastes. This review gives an overview of economical strategies to reduce production costs of PHA as well as its applications in various fields.

Keywords: Polyhydroxyalkanoates; Bioplastics; Environment

Introduction

Plastic can be regarded as one of the greatest inventions and an indispensable commodity of human's life ever since it has been developed into a major industry [1]. It contributes to sustainable development and brings quality life to citizens. As example plastics makes many goods in our daily life more affordable and reduce the wastage of many valuable resources.

Plastics have eased the everyday life; its usage is increasing and annual production has increased substantially over the last 60 years, it has hiked by 3.8 % to around 280 million tons in 2011 over the world. The statistics of European market clearly reveal the annual turnover of 300 billion Euros [2]. The main source for the production of plastics are petrochemicals which are non renewable in nature. A recent estimation done on the earth's mineral resources showed an alarming rate of depletion of these valuable natural assets. This has created a renewed impetus to search for various other sustainable alternatives. The increasing cost and the awareness of consumers of the negative environmental impact of the fluid mineral fuels and related products like recalcitrance to biodegradation [3], toxicity after incineration and massive waste accumulation into the landfills as well as growing water and land pollution problems have led to concern about plastics.

Biopolymer

With the excessive use of plastics, rising pressure is getting placed to meet the ever increasing demand of petrochemicals coupled with the search for a safe plastic waste disposal process. This awareness of the waste problem and its impact on the environment has awakened new interest in the area of economic and efficient biodegradable polymers sources for production of plastic or popularly known as the ***"bioplastic"***. Table 1 shows the comparitive analysis between biopolymer and peto based plastics.

Numerous types of biodegradable polymers are under development that popularly includes Polylactides, Polyglycolic acids, Polyhydroxyalkanoates (PHAs), aliphatic polyesters, polysaccharides. On the other hand, natural renewable polymers include porous sponges (from cellulose wood fibres), fibres (made from natural fibres), hydrogels, starch, cellulose, chitin, chitosan, lignin and proteins. Among these numerous aforementioned biodegradable polymers, PHAs is being considered as the most potential renewable substitute to petrochemical plastics because of its resemblance to commercially available plastic in context to physical and chemical properties [4,5].

Polyhydroxyalkanoates (PHAs) are polymers synthesized entirely by a biological process that involves conversion of carbon sources directly into PHAs through microbial fermentation [6,7]. In contrary, most of the other biopolymers like polybutylene succinate (PBS), Polytrimethylene Terephthalate (PTT) and Polylactic Acid (PLA) are chemically synthesized using fermentation-derived monomers. For example, PLA is prepared by Ring Opening Polymerization (ROP) of lactide, a cyclic dimer of dehydrated lactic acid that is produced by fermentation [8-11].

Classification of PHAs

PHAs are lipoidic material [12] accumulated intracellularly as insoluble inclusions by a wide variety of microorganisms In the presence of an abundant carbon source. The assimilated carbon sources are biochemically processed into hydroxyalkanoate units under stressed conditions, polymerised and stored in the form of water insoluble inclusions in the cell cytoplasm. The ability of cells to carry out the polymerization process rests on the availability of a key enzyme known as PHA synthase. The product of this enzyme is high Molecular Weight (MW) optically active crystalline polyester [13]. The molecular weight of PHAs depends upon the type of growth

***Corresponding author:** Divya Gupta, School of Biotechnology, Rajiv Gandhi Proudyogiki Vishwavidyalaya, Bhopal, India, Tel: 91-9584491252

Properties	Biopolymer	Petro-based plastics
Renewability	Yes or partially	No
Sustainability	Yes	No
Degradability	Biodegradable and/or compostable	Usually non-degradable (Some are degradable through polymer oxidation)
Polymer range	Biopolymers (lipids, proteins, Starch etc.), Bacterial polymers (Polyhydroxyalkanoates, polyhydroxybutyrate)	Extensive
Green house gas emissions	Low emission	High emission
Utilization of fossil fuels	Limited utilization	High utilization
Agriculture land utilization	Expected to increase	No utilization

Table 1: Comparison between Biopolymer and Petro-based plastics.

conditions and microorganism, which range between 2×10^5 to 3×10^6 daltons [6, 14, 15].

Among these Polyhydroxybutyrate, poly(3-hydroxybutyrate) [P(3HB)] was the first and most common type of PHA to be identified by the French microbiologist Maurice Lemoigne [16]. P(3HB) is an optically active biological linear polyester which is insoluble in water and exhibit a high degree of polymerization that ranges from 10^5 to approximately 10^7. The biosynthesized P(3HB) is thus perfectly isotactic and upon extraction from the microorganisms has a crystallinity of about 55-80% with a melting point at around 180°C [17-20].

Structurally, these polymers are classified on the basis of the number of carbon atoms that ranges from 3-5, 6-14, mostly and some are more than 14 [7,20,21] and the type of monomeric units, producing homopolymers or heteropolymers. PHAs with 3–5 carbon atoms are considered as short chain length PHA's (scl-PHAs). Examples of this class include poly(3-hydroxybutyrate) [P(3HB)] and poly(4-hydroxybutyrate) [P(4HB)]. Medium chain length PHAs (mcl-PHAs) contains 6-14 carbon atoms which includes homopolymers poly(3-hydroxyhexanoate) [P(3HHx)], poly(3-hydroxyoctanoate) [P(3HO)] and heteropolymers such as P(3HHx-co-3HO) [22]. PHAs with more than 14 carbon atoms are considered as long chain length PHAs (lcl-PHAs) and they are very uncommon and least studied. For example P(3HB-3HV-3HHD-3HOD) copolymer in *Pseudomonas aeruginosa* [21].

Characteristics of Polyhydroxyalkanoates

Physical characteristics

Although aliphatic polyesters have been studied extensively but their remarkable properties such as sufficiently high molecular mass coupled with polymerization characteristics were realized only recently which can be exploited in order to replace conventional plastics such as polypropylene. Naturally occurring PHA's are optically active linear polyesters with each repeating unit in the stereochemical R-configuration. Monomeric compositions, chemical structure as well as the molecular weight are the key factors influencing the physical properties of the polymer. The molecular mass of PHAs varies per PHA producer but is generally in the order of 50,000 to 1,000,000 Dalton.

Scl-PHAs such as P(3HB) are crystalline polymers which are quite brittle and rigid, with high melting points and low glass transition temperatures [23]. These unique characteristics of this biologically synthesized P(3HB) arises due to its exceptional purity. The brittleness is due to the formation of large crystalline domains in the form of spherulites. On the other hand, P(4HB), P(3HB-co-4HB) P(HB-co-HV), scl-PHAs are strong and malleable thermoplastic polyesters [24],especially P(HB-co-HV). It is created by incorporating PV into PHB which is less stiff and brittle than PHB, as a result it can be used to prepare films with excellent water and gas barrier properties similar to polypropylene. It can also be processed at lower temperature along with preserving most of mechanical properties of PHB [25]. Moreover, Mcl-PHAs are the thermoplastic elastomers with low crystallinity and tensile strength but high elasticity. They also have a lower melting point and glass transition temperatures in contrast to scl-PHAs and polypropylene.

However, for the replacement of petroleum-based plastics, PHA copolymers consisting of both SCL and MCL monomers are considered better choices than PHB due to the presence of certain enhanced properties, such as flexibility and ease of processing [26]. Table 2 gives an outline of comparison of physical properties of different PHAs with conventional polymers [6,27]

Biological characteristics

The diversity of numerous hydroxy acid monomers that can be incorporated into PHAs, in conjunction with a biological polymerization system capable of generating high-molecular weight materials, has resulted in a manifold rise in the availability of wide range of new polymers. Important biological attributes of PHA includes:

Biodegradability: Besides the typical polymeric properties described above, biodegradability is conceivably the most versatile property of PHAs. These biological polyesters are inert, insoluble in water, not affected by moisture and indefinitely stable in air [28].

These polymers are biologically processed by microbial enzymes into inorganic molecules such as carbon dioxide and water under aerobic conditions or methane and water under anaerobic conditions [29]. Such a processing is a part of the natural carbon recycling cycle that exists on the earth shown in (Figure 1). The ability to degrade PHA is broadly carried out by bacteria and fungi, which depends on secretion of specific extracellular PHA depolymerases into the environment or by the intracellular mobilization of PHA in the accumulating strain itself. Both Intracellular mobilization of PHA and extracellular degradation differ from each other due to difference in the biophysical conformations of extracellular (denatured) PHA from those of intracellular (native) PHA. Intracellular PHA depolymerases hydrolyze an endogenous carbon reservoir which are in amorphous state, the native PHA granules, consisting of the polymer with a surface layer of proteins and phospholipids which is sensitive to physical or chemical stress. Extracellular PHA depolymerases degrade denatured extracellular granules which are partially crystalline and are lacking a surface layer [30,31].

Intracellular depolymerases degrades poly[D-(-)-3-hydroxybutyrate] (PHB) within the granules into 3-hydroxybutyric acid. It acts on the latter and oxidises it to acetylacetate and a β-ketothiolase acts on acetylacetate to break it down to acetyl-CoA. Under aerobic

Polymer	Melting temperature (°C)	Young's modulus (GPa)	Tensile strength (MPa)	Elasticity (%)
P(3HB)	175-180	3.5–4	40	3–8
P(4HB)	53	149	104	N.A.
P(3HB-co-3HV) (3 mol% HV)	170	2.9	38	N.A.
P(3HB-co-3HV) (20 mol% HV)	145	1.2	32	50–100
P(3HB-co-4HV) (3 mol% HV)	166	N.A.	28	45
P(3HB-co-4HV) (10 mol% HV)	159	N.A.	24	242
P(3HO)	61	N.A.	6–10	300–450
Polystyrene (PS)	80–110	3.0–3.1	50	3–4
Polypropylene (iPP)	176	1.7	34.5	400
Nylon-6,6	265	2.8	83	60

Table 2: A comparison between the physical properties of different PHA's and commonly used conventional polymers.

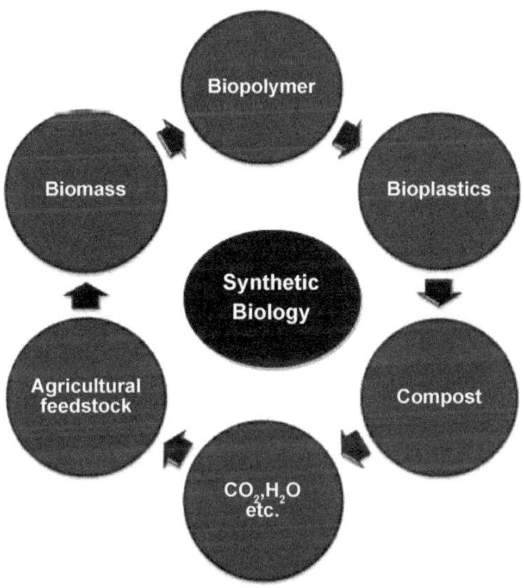

Figure 1: Synthetic biology, Bioplastics and Environment: Bioplastics derived from synthetic biology approaches using renewable feed stock, are good candidates for sustainable development and eco-friendly environment, due to their biodegradability and compostability.

conditions, the acetyl-CoA enters the citric acid cycle and is oxidized to CO_2 [32]. Very little is known about the intracellular depolymerases since they are always found to be intimately connected to the P(3HB) granules and the overall process is very complex [31,33]. Extracellular depolymerases degrade polyhydroxyalkanoates present in the environment [31]. The environmental microbes attack the polymers on the surface [7] and secrete extracellular enzymes that solubilise the polymer and these soluble products are then absorbed through their cell walls and utilized. The PHA depolymerase enzymes act on the polymer mainly by hydrophobic interactions. Degradation by these depolymerases initially produces oligomers. Some microbes produce an additional dimer hydrolase, which further breaks down the oligomers into the corresponding monomer [34]. These extracellular enzymes are quite well understood. The activities of these enzymes may vary depending on the composition of the polymer, its physical form (amorphous or crystalline), dimensions and most importantly, the environmental conditions.

Biocompatibility: According to Williams DF the biocompatibility may be defined as the ability of a material to perform with an appropriate host response in a specific application [35] without any toxic or injurious after effects on biological functions. PHA is highly biocompatible in nature as no toxic compound is generated during polymer degradation. The biocompatibility of PHA particularly P(3HB), is no surprise, as 3HB is a normal blood constituent and is found in the cell envelope of eukaryotes [36]. The ease of crystallization of PHB during precipitation makes entrapment of the drug difficult [37]. Hence, copolymers with HV have been used for drug formulation more than the pure PHB. Copolymers of 3-hydroxybutyrate and 3-hydroxyvalerate (PHBV), P(4HB), copolymers of 3-hydroxybutyrate and 3-hydroxyhexanoate (PHBHHx) have been tested in animals. These tests revealed that all these polymers were biocompatible in various host systems [38]. PHAs have great contribution in medicine ,it has been used in different medical applications, such as in wound management (sutures, skin substitutes), vascular system devices (heart valves, vascular grafts), orthopedic (scaffolds for cartilage engineering, screws, bone graft substitutes), controlled drug delivery (through macro and nanocarriers), urological stents etc [39]. Microcapsules from PHB and P(HB-co-HV) has been prepared by various techniques and investigated for the release of bovine serum albumin [40] P(3HB 3HV) capsules can be used as a suitable biodegradable matrix for drug delivery in veterinary medicine that can remain in the rumen of cattle [41,42].PHB and P(HB-co-HV) are both crystalline materials [42] but the actual degree of crystallinity varies considerably depending on crystallization conditions and HV content [43]. This will, in turn, affect the drug release characteristics. Drug release rates have been reported to increase with increasing HV content [43-45]. For instance, microspheres from P(HB-co-HV) containing tetracycline for the treatment of periodontal diseases showed decreasing encapsulation efficiencies and increasing delivery rates with a higher HV content. Sulbactam-cefoperazone antibiotic was integrated into rods made of poly(3-hydroxybutyrate-co-22 mol%-3-hydroxyvalerate) and studies were carried out to use these rods as antibiotic-loaded carriers to treat implant related and chronic osteomyelitis. The rods were implanted into a rabbit tibia that was artificially infected by *Staphylococcus aureus*. After 15 days, the infection subsided and within 30 days there was complete healing. The high melting points of P(3HB) and P(HB- co-HV) is sometimes a disadvantage and the use of medium chain PHAs with lower melting points could provide a considerable advancement in drug delivery [46]. P(4HB) as biocompatible material has been used as a cardiovascular patching material with great success. It has also been used to make heart valves and in vascular grafting. Sutures, ligaments, surgical meshes and pericardial substitutes made with P(4HB) are now being used [47]. PHA matrices have also been tested for hemocompatibility by inspecting the response of mammalian blood when incubated with polymer films. It was shown that PHB or P(HB-co-HV), when in contact with blood, did not affect platelet responses, nor did the polymer activate complement system.

Renewable nature: The production of PHA is solely based on the usage of renewable resources. Fermentative production of PHAs is based on carbon and energy sources obtained from agricultural products such as sugars and fatty acids. These agricultural feed stocks are derived from CO_2 and water, and after their conversion to biodegradable PHA, the breakdown products are again CO_2 and water. Thus, apart from being biodegradable and biocompatible, PHAs receive general attention because they are based on renewable compounds instead of diminishing fossil fuel stockpiles [48].

It is the consequence of PHAs inherent physical and biological properties along with the striking similarity of physical and chemical profile with conventional polymers such as polypropylene, it has attracted commercial interest and has been extensively studied in the last two decades [49,50].

Sustainable PHAs: Using Low Cost Production Technology

The global interest of many companies in large scale production of PHA polymers, has been established [51]. The biopolymer production cost is the major stumbling block in the commercialization of PHA. The high production cost of biopolymer is mainly depends upon the methods adopted for its synthesis and the composition of media which limit the widespread use of PHA [52].

By utilizing using inexpensive renewable carbon source and bacterial strains through a fermentation process, it is plausible to produce all sorts of related biopolymers having properties ranging from stiff and brittle polymers to rubbery polymers [22].

Microorganisms

Application of genetic engineering is essentially paving the way for the development of super PHA production strains that are able to grow to a high cell density within a short period of time on low cost substrates under less demanding fermentation conditions, such as micro-aerobic conditions [6,53]. This approach has significantly reduced the production costs. Engineering approaches including incorporation of engineered PHA synthases and engineered precursor supplying enzymes into the pathways have been applied to achieve enhanced production.

PHA-producing bacteria can be divided into two groups according to the culture conditions required for PHA synthesis. First group is constituted by the wild type forms such as *Ralstonia eutropha* (>80%), *Alcaligenes latus* (>75%), and *Pseudomonas oleovorans* (>60%) that requires limited essential nutrient(s) for the production of PHAs and quantity of PHAs (ranging from 50% to 80% of the dry cell mass) [51]. The second group synthesizes PHAs alongside with the growth in the cultivation medium. Bacteria's in this group includes *Alcaligenes latus* and recombinant *E. coli* containing the PHA biosynthetic genes.

Genetic engineering strategies have proven miraculous in the field of microbiology, as it has been used to reduce the generation time of the recombinant microbes and simplify PHA recovery. Most remarkable example is construction of several refactored microbes producing PHA is among these Recombinant *E. coli* has been of the most favored host developed by harboring the PHAs biosynthesis gene. It has the ability to growth fast facilitate short generation time, best understanding in genetics and biochemistry, abroad substrate utilization ability. Example molasses and whey, and easy purification of the polymer from *E. coli* contribute to its popularity [54,55]. Recombinant *E.coli* harbouring PHB biosynthetic genes from *C.necator* were able to accumulate P(3HB) with a yield of 80–90% dcw in fed-batch cultivation while a P(3HB) content of 76% dcw was obtained in a pH-stat fed-batch culture [56]. Moreover, possibilities for the formation of 'polymeric blends' such as blending of P(3HB) with mcl-PHA has also been investigated in recombinant *Pseudomonas oleovorans* [57]. Three key enzymes for biosynthesis of PHAs are b-ketothiolase, NADPH-dependent acetoacetyl-CoA reductase, and PHA synthase encoded by genes phaA, phaB, and phaC, respectively [7]. Factors regulating PHAs synthesis are phasins protein, it bind directly to PHA and possibly to the PHA-cycling enzymes, PHA synthase and/or PHA depolymerase [58,59], The second type of regulator is PhaR implicates in negative regulation of a phasin protein [60,61]. It was found that P(3HB) acts as an inducer for *phaP* expression in a PhaR-mediated regulatory system. By direct binding PhaR interacts with PHA. PhaR-mediated regulatory mechanism may be responsible for PHA accumulation in cells is common in SCL PHA-producing bacteria [62] it was also found that mutation in PHB synthesis enzyme acetyl coenzyme A (CoA), ketothiolase (PhaA) and NADPH-linked acetoacetyl-CoA reductase (PhaB), revealed not only a PHB deficiency but also a deficiency in growth on C1and C2 compounds [63].

Waste materials as inexpensive substrate

The major cost of the production of PHA is decided by the type of medium that is used [64]. Thus, the selection of media is a vital factor as it not only provides optimal conditions for production of a variety of PHAs by different strains of bacteria but it also deals with high volumetric productivity to provide an end product that is economically feasible. The PHAs can be produced from a wide variety of substrates such as low cost renewable resources (for example, sucrose, starch, cellulose, triacylglycerols), fossil resources (methane, mineral oil, lignite, hard coal), by products (molasses, whey, glycerol), chemicals (propionic acid, 4-hydroxybutyric acid) and carbon dioxide [65].

Innovative processes are getting investigated for producing PHAs from a variety of low cost substrate such as waste water, municipal wastes, agricultural and industrial residues and their byproducts. Waste water would help to improve environment conditions as organic matter present in wastewater (in the form of organic pollutant) can be converted to useful substances or polyhydroxyalkanoates (PHAs) during various stages of its treatment to facilitate the removal of these pollutants [66-68].

Plant oils: Plant oils are derived from oil-bearing crops, such as soy bean, rapeseed, palm, sunflower, and corn. Lately, interest abounded to utilize these plant oils in the bio-based product industries, more specifically, for its efficient conversion into PHA that would make PHA based product's prices competitive with their petroleum counterparts. Plant oils have been investigated and were found to be most desirable carbon source for large scale PHA production. Recent studies have suggested that plant oils give higher yield for both, cell biomass and PHA production (0.6 to 0.8 g of PHA per g of oil) in comparison with other tested substrates such as sugars because of their complex mix of triglycerides [69]. Various plant oils have been used such as P(3HB) homopolymer and P(3HB3co35 mol% 3HHx) copolymer has been synthesize by *Cupriavidus necator* H16 and its recombinant strain (harboring the PHA synthase gene from *A. caviae*) using soybean oil as the sole carbon source [70,71]. Similarly, Lee and colleagues observed the production of 80% dcw P (3HB-co-3HV) from *C. necator* H16 when grown on plant oils such as palm kernel oil [72]. *C. testosteroni* has been studied for its ability to synthesize mcl-PHA from vegetable oils such as castor seed oil, coconut oil, mustard oil, cotton seed oil, groundnut oil, olive oil and sesame oil [73]. *C. necator* organism is also reported to produce PHAs from the waste frying oil [74]. Although these are some very good producers of PHAs, on average the PHA accumulations are still low. Thus, more explorations are needed regarding the development of superior vegetable oil utilizing strains and to further improve fermentation technology for the utilization of waste vegetable oils and other plant oils to further enhance growth and PHA accumulation using these substrates.

Rice bran: Rice is the most important cereal product of Asian countries and second largest produced cereal in the world. India and China are the largest producers of rice, as they account for more than one-third of global population supply. Rice bran is a byproduct of the rice milling process and it accounts for 8 -12 % of the rice kernel [75]. It is associated with environmental impact due to presence of impurities like arsenic and silica, as well as the presence of free fatty acids. Study conducted by Ting-Yen Huang et al. showed procurement of PHA successfully from extruded rice bran and corn starch from *Haloferax mediterranei* under a hyper-saline condition [76]. Using rice bran (a low cost agro-industry residue), 3.63 g/L of biomass, 1.75 g/L of intra-cellular polymer Polyhydroxyalkanoate, (PHA) and 1.2 g/L of extra-cellular polymer (exopolysaccharide, EPS) was produced by *Sinorhizobium meliloti* [77].

Molasses: Molasses is the low value, unpalatable and final residual syrup generated in sugar-refining mills after repeated sugar extraction by carrying out crystallization of sugarcane or sugar beet juice. It is mainly used as a cattle feed supplement, in specialized yeast propagation or as a flavoring agent in some foods [78]. Depending on the grades and sources, sugar molasses cannot be further used in foods or feeds and is, thus, ideal for consideration as an inexpensive carbon source for fermentative processes. According to Tate & Lyle Molasses Germany GMBH [79] the quantities of cane molasses and beet molasses generated worldwide in 2004 were 39 and 12 million tons, respectively. Molasses has negative environmental impact due to its accumulation during sugar processing, thus cane molasses and beet molasses are ideal for consideration as an inexpensive carbon source for PHA production which could alleviate its disposal problems. Soy molasses have also been tested as alternative substrate, which is a by-product of soybean processing industry. It contains various soluble carbohydrates especially sucrose, raffinose and stachyose [80].

A great deal of work has been carried out into the utilization of molasses for PHA production. Both Gram negative and Gram positive bacteria such as recombinant *E. coli* and *Klebsiella* strains [81], *R. eutropha* [82], *Bacillus* strains [83] has used molasses for PHA production.

Cane molasses was found to be an excellent substrate for the growth of *Bacillus megaterium* for subsequent PHA production by [83] a total biomass and PHA content of 72.7 g · L-1 and 42% (wt/wt) [84] demonstrated the production of 61% dcw PHA using molasses as a complex feedstock in 2-stage CSTR [85] reported the production of P(3HB) at 3 g polymer L −1 and 50% dcw P (3HB) content by *Bacillus megaterium* grown on beet molasses [86] have tested *Pseudomonas corrugata* on inexpensive soy molasses carbon substrate for the production of medium-chain length PHA and achieved a cell dry mass concentration of 3 .4 g L -1, with a PHB content of 5 -17% dcw.

Dairy whey: During cheese and casein production at the food processing industries, about 80-90% of the processed milk volume is converted to whey as a byproduct [87]. Whey is rich in lactose, proteins, lipids, and lactic acid [88]. It is the soluble part of milk that is separated from the casein during cheese manufacture and is stabilized by disulfide bonds which makes it insoluble in water. For smaller dairies, disposal of whey is cost demanding as only about half of the produced whey is converted into useful products such as human and animal feed while the rest is disposed of as waste causing environmental problems due to its high oxygen demand.

Many isolates of bacteria were evaluated for the production of different polyester members of the PHA family from lactose or cheese whey as carbon source [89]. Recombinant *Escherichia coli* strains have been used for the production of PHA [90,91]. The production of a heteropolymer from whey by *Thermus thermophilus* HB8 has also been described [92]. Role of *Cupriavidus necator* has been extensively established as a platform for PHA production and it is one of the best industrially useful strain which is able to use cheap carbon sources. However it is unable to hydrolyse the disaccharide lactose and galactose which is the main sugar contained in whey [93] hydrolyzed milk whey in order to make glucose available for *C. necator*. Scientist have proposed an alternative process consisting of two steps for PHB production from milk whey: lactose was first converted to lactic acid by lactobacilli, and the resulting lactic acid was then used as carbon source by *R. eutropha* for PHA production [94]. Further scientist have developed recombinant strains of *C.necator* which are able to produce the polymer directly from lactose and or whey permeate in one step [95]. Hence, milk whey can be successfully used as carbon source for PHAs production which would significant reduce the production cost of the PHAs.

Downstream processing

Recovery of PHA is another process that significantly affects the overall process economics. The development of a cheaper safer downstream process for the recovery of PHA will have considerable impact on industrial production of this versatile biopolymer. The extraction and the purification of bacterial polyhydroxyalkanoates are the key step of the process profitability in the fermentation system. The ideal method should lead to a high purity and recovery level at a low production cost.

Traditional PHA downstream processing methods requiring the use of solvents, enzymatic digestion, or mechanical disruption are expensive and impractical for industrial-scale recovery. A common practice to recover PHA is through the organic solvent extraction method. This process has many advantages over other cell disruption processes, For example solvent extraction do not degrade the polymer and can be useful for several medical applications by the elimination of Endotoxin which can be found in Gram negative bacteria [96]. However, the use of solvents creates hazards for the operators and for the environment which limits its success in pilot-plant and large scale processing [97]. To alleviate this short coming, an extraction method using non-halogenated solvent such as sodium hypochlorite has been used for the aqueous process [98]. Although the use of sodium hypochlorite significantly increased PHA degradation, polymer purity greater than 95% is achieved [96].

Mechanical cell disruption is widely used to release intracellular protein [99]. It is mainly favored due to economic advantageous such as it causes mild damage to the products and does not involve any chemicals making it environmental friendly. But, it also has drawbacks like it decrease the molecular weight of the biopolymer, long processing time and difficulty in scaling up.

Enzymatic digestion method is a gentle and selective alternative separation method, which has attracted the interest of many researchers [100]. It eliminates the need of hazardous solvents but the enzymes used are expensive and the process is also more complex and economically unattractive.

Due to the problem associated with solvent extraction process, much attention has been given to centrifugal fractionation as a simple and economical process for separating specific resin components from the recovered PHA. Continuous centrifugal fractionation gives 85%

of PHA with purity more than 95% [101,102]. Pretreatment of cells, mechanical disruption, solubilisation of non-PHA cell material by chemical or enzymatic treatment in the system, spontaneous liberation of poly(3-hydroxybutyrate), dissolved air flotation or air classification and extraction using supercritical CO_2 have been recently reviewed [103].

Over the years, a variety of PHA recovery methods have been developed, although none of the downstream procedures posses all the necessary requirements for a significant efficient and economical large-scale process [104]. The major drawbacks are cost, safety and scalability. Still, further researches have been made for the development of simple fermentation strategies as well as modification of recovery techniques in order to scale-up the PHA production.

Conclusions and Outlook

As plastic materials are of such importance in our society and knowing that materials are a fundamental determinant of sustainability; currently PHAs have rapidly developed as potential substitute material to petro plastics due to its versatile property like biodegradable and biocompatible. Their structural versatility and characteristics of PHAs have been investigated and new areas of exploitation are being discovered due to its various applications in industries and medical field .The major drawback for extensive use of these polymers and its slow growth is their high production cost and inferior thermo-mechanical, physical and processing properties of biodegradable polymers compared with petrochemical plastics.

Consequently, scientists have shown immense progress in searching for new bacterial strains, creating new types of recombinant strains and tailoring various kinds of PHA to reduce the cost of production. Use of low-cost renewable carbon sources in fermentation process has the potential to reduce PHA production costs, which will certainly improve PHAs competitiveness and its production substantially, leading to an enhancement in its widespread usage as they confer no long term effects on environment and could be adorable at social and economic levels.

Acknowledgement

I would like to thank my Parents. Also I heartily acknowledge Mr. Gagan Flora and Ms. Batul Diwan for correcting and editing the manuscript and helping me throughout. Last but not the least I am thankful to my Friends and Almighty.

References

1. Sudesh K, Iwata T (2008) Sustainability of Biobased and Biodegradable Plastics. CLEAN – Soil, Air, Water 36: 433-442.
2. Plastics – the Facts 2012 An analysis of European plastics production, demand and waste data for 2011.
3. Reddy CS, Ghai R, Rashmi, Kalia VC (2003) Polyhydroxyalkanoates: an overview. Bioresour Technol 87: 137-146.
4. Steinbüchel A, Füchtenbusch B (1998) Bacterial and other biological systems for polyester production. Trends Biotechnol 16: 419-427.
5. Sudesh K, Abe H, Doi Y (2000) Synthesis, structure and properties of polyhydroxyalkanoates: biological polyesters. Prog Polym Sci 25: 1503-1555.
6. Lee SY (1996) Bacterial polyhydroxyalkanoates. Biotechnol Bioeng 49: 1-14.
7. Madison LL, Huisman GW (1999) Metabolic engineering of poly(3-hydroxyalkanoates): from DNA to plastic. Microbiol Mol Biol Rev 63: 21-53.
8. Drumright RE, Gruber PR, Henton DE (2000) Polylactic Acid Technology. Adv Mater 12: 1841-1846.
9. Mehta R, Kumar V, Bhunia H, Upadhyay SN (2005) Synthesis of Poly(Lactic Acid): A Review. Journal of Macromolecular Science, Part C: Polymer Reviews 45: 325-349.
10. Sodergard A, Stolt M (2002) Properties of lactic acid based polymers and their correlation with composition. Prog Polym Sci 27: 1123-1163.
11. Vink ETH, Rabago KR, Glassner DA, Gruber PR (2003) Applications of life cycle assessment to NatureWorks&unknown; polylactide (PLA) production. Polym Degrad Stabil 80: 403-419.
12. Burdon KL (1946) Fatty Material in Bacteria and Fungi Revealed by Staining Dried, Fixed Slide Preparations. J Bacteriol 52: 665-678.
13. Barnard GN, Sanders JK (1989) The poly-beta-hydroxybutyrate granule in vivo. A new insight based on NMR spectroscopy of whole cells. J Biol Chem 264: 3286-3291.
14. Zinn M (2003) Tailor-Made Synthesis of Polyhydroxyalkanoate. European Cells and Materials 5: 38-39.
15. Doi Y (1995) Plastics from microbes. Microbial synthesis of polymers and polymer precursors. Edited by D. P. Mobley, Hanser, Munich 1994, hardcover, 288 pp. DM 128.00, ISBN 3-446-17367-6. Adv Mater 7.
16. Lemoigne M (1926) Products of dehydration and polymerization of b–hydroxybutyric acid. Bull Soc Chim Biol 8: 770-782.
17. Cornibert J, Marchessault RH (1972) Physical properties of poly--hydroxybutyrate. IV. Conformational analysis and crystalline structure. J Mol Biol 71: 735-756.
18. Cornibert J, Marchessault RH, Benoit H, Weill G (1970) Physical Properties of Poly(β-hydroxy butyrate). III. Folding of Helical Segments in 2,2,2-Trifluoroethanol. Macromolecules 3: 741-746.
19. Sudesh K, Fukui T, Taguchi K, Iwata T, Doi Y (1999) Improved production of poly(4-hydroxybutyrate) by Comamonas acidovorans and its freeze-fracture morphology. Int J Biol Macromol 25: 79-85.
20. Taguchi S, Doi Y (2004) Evolution of polyhydroxyalkanoate (PHA) production system by "enzyme evolution": successful case studies of directed evolution. Macromol Biosci 4: 146-156.
21. Singh AK, Mallick N (2008) Enhanced production of SCL-LCL-PHA co-polymer by sludge-isolated Pseudomonas aeruginosa MTCC 7925. Lett Appl Microbiol 46: 350-357.
22. Anderson AJ, Dawes EA (1990) Occurrence, metabolism, metabolic role, and industrial uses of bacterial polyhydroxyalkanoates. Microbiol Rev 54: 450-472.
23. Billingham NC, Henman TJ, Holmes PA (1987) Degradation and Stabilisation of Polyesters of Biological and Synthetic Origin. Developments in Polymer Degradation 81-121.
24. Ryan C, Kai G, Martin D, Said R (2012) Medical Devices Containing Dry Spun Non-Wovens of Poly-4-Hydroxybutyrate and Copolymers. US Patent 20120150285.
25. Kulkarni SO, Kanekar PP, Jog JP, Patil PA, Nilegaonkar SS, et al. (2011) Characterisation of copolymer, poly (hydroxybutyrate-co-hydroxyvalerate) (PHB-co-PHV) produced by Halomonas campisalis (MCM B-1027), its biodegradability and potential application. Bioresour Technol 102: 6625-6628.
26. Noda I, Green PR, Satkowski MM, Schechtman LA (2005) Preparation and properties of a novel class of polyhydroxyalkanoate copolymers. Biomacromolecules 6: 580-586.
27. van der Walle GA, de Koning GJ, Weusthuis RA, Eggink G (2001) Properties, modifications and applications of biopolyesters. Adv Biochem Eng Biotechnol 71: 263-291.
28. Ojumu TV, Yu J, Solomon BO (2004) Production of polyhydroxyalkanoates, a bacterial biodegradable polymer. African Journal of Biotechnology 3: 18-24.
29. Novamont (2011) Biodegradability.
30. Jendrossek D, Handrick R (2002) Microbial degradation of polyhydroxyalkanoates. Annu Rev Microbiol 56: 403-432.
31. Tokiwa Y, Calabia BP (2004) Degradation of microbial polyesters. Biotechnol Lett 26: 1181-1189.
32. Raberg M, Peplinski K, Heiss S, Ehrenreich A, Voigt B, et al. (2011) Proteomic and transcriptomic elucidation of the mutant ralstonia eutropha G+1 with regard to glucose utilization. Appl Environ Microbiol 77: 2058-2070.
33. MERRICK JM, DOUDOROFF M (1964) DEPOLYMERIZATION OF POLY-BETA-HYDROXYBUTYRATE BY INTRACELLULAR ENZYME SYSTEM. J Bacteriol 88: 60-71.

34. Tanaka Y, Saito T, Fukui T, Tanio T, Tomita K (1981) Purification and properties of D(-)-3-hydroxybutyrate-dimer hydrolase from Zoogloea ramigera I-16-M. Eur J Biochem 118: 177-182.

35. Guy Donaruma L (1988) Definitions in biomaterials, D. F. Williams, Ed., Elsevier, Amsterdam, 1987, 72 pp. J Polym Sci Pol Lett 26: 414.

36. Reusch RN (2000) Transmembrane ion transport by polyphosphate/poly-(R)-3-hydroxybutyrate complexes. Biochemistry (Mosc) 65: 280-295.

37. Edlund U, Albertsson AC (2002) Degradable Polymer Microspheres for Controlled Drug Delivery. Adv Polym Sci 157: 67-112.

38. Valappil SP, Misra SK, Boccaccini AR, Roy I (2006) Biomedical applications of polyhydroxyalkanoates: an overview of animal testing and in vivo responses. Expert Rev Med Devices 3: 853-868.

39. Chen Q, Liang S, Thouas GA (2013) Elastomeric biomaterials for tissue engineering. Prog Polym Sci 38: 584-671.

40. Atkins TW, Peacock SJ (1996) The incorporation and release of bovine serum albumin from poly(beta-hydroxybutyrate-hydroxyvalerate) microspheres. J Biomater Sci Polym Ed 7: 1065-1073.

41. Kwan L, Steber W (1992) Intermittent release dosage form. US5110598.

42. Engelberg I, Kohn J (1991) Physico-mechanical properties of degradable polymers used in medical applications: a comparative study. Biomaterials 12: 292-304.

43. Sharma R, Ray AR (1995) Polyhydroxybutyrate, Its Copolymers and Blends. Journal of Macromolecular Science, Part C: Polymer Reviews 35: 327-359.

44. Gangrade N, Price JC (1991) Poly(hydroxybutyrate-hydroxyvalerate) microspheres containing progesterone: preparation, morphology and release properties. J Microencapsul 8: 185-202.

45. Sendil D, Gürsel I, Wise DL, Hasirci V (1999) Antibiotic release from biodegradable PHBV microparticles. J Control Release 59: 207-217.

46. Pouton CW, Akhtar S (1996) Biosynthetic polyhydroxyalkanoates and their potential in drug delivery. Adv Drug Deliver Rev 18: 133-162.

47. Martin DP, Williams SF (2003) Medical applications of poly-4-hydroxybutyrate: a strong flexible absorbable biomaterial. Biochem Eng J 16: 97-105.

48. Williams SF, Peoples OP (1996) Biodegradable plastics from plants. Chemtech 26: 38-44.

49. Steinbuchel A, Valentin HE (1995) Diversity of bacterial polyhydroxyalkanoic acids. FEMS Microbiol Lett 128: 219-228.

50. Braunegg G, Lefebvre G, Genser KF (1998) Polyhydroxyalkanoates, biopolyesters from renewable resources: physiological and engineering aspects. J Biotechnol 65: 127-161.

51. Chen GQ (2009) A microbial polyhydroxyalkanoates (PHA) based bio- and materials industry. Chem Soc Rev 38: 2434-2446.

52. Bengtsson S, Pisco AR, Reis MA, Lemos PC (2010) Production of polyhydroxyalkanoates from fermented sugar cane molasses by a mixed culture enriched in glycogen accumulating organisms. J Biotechnol 145: 253-263.

53. Choi J, Lee SY (1997) Process analysis and economic evaluation for Poly(3-hydroxybutyrate) production by fermentation. Bioprocess Eng 17: 335-342

54. Hahn SK, Chang YK, Lee SY (1995) Recovery and characterization of poly(3-hydroxybutyric acid) synthesized in Alcaligenes eutrophus and recombinant Escherichia coli. Appl Environ Microbiol 61: 34-39.

55. Fidler S, Dennis D (1992) Polyhydroxyalkanoate production in recombinant Escherichia coli. FEMS Microbiol Rev 9: 231-235.

56. Kim BS, Lee SY, Chang HN (1992) Production of poly-β-hydroxybutyrate by fed-batch culture of recombinantEscherichia coli. Biotechnol Lett 14: 811-816.

57. Preusting H, Kingma J, Huisman G, Steinbuchel A, Witholt B (1993) Formation of polyester blends by a recombinant strain of Pseudomonas oleovorans: Different poly(3-hydroxyalkanoates) are stored in separate granules. J Environ Polym Degr 1: 11-21.

58. Hanley SZ, Pappin DJ, Rahman D, White AJ, Elborough KM, et al. (1999) Re-evaluation of the primary structure of Ralstonia eutropha phasin and implications for polyhydroxyalkanoic acid granule binding. FEBS Lett 447: 99-105.

59. Jossek R, Reichelt R, Steinbüchel A (1998) In vitro biosynthesis of poly(3-hydroxybutyric acid) by using purified poly(hydroxyalkanoic acid) synthase of Chromatium vinosum. Appl Microbiol Biotechnol 49: 258-266.

60. Maehara A, Ueda S, Nakano H, Yamane T (1999) Analyses of a polyhydroxyalkanoic acid granule-associated 16-kilodalton protein and its putative regulator in the pha locus of Paracoccus denitrificans. J Bacteriol 181: 2914-2921.

61. Maehara A, Taguchi S, Nishiyama T, Yamane T, Doi Y (2002) A repressor protein, PhaR, regulates polyhydroxyalkanoate (PHA) synthesis via its direct interaction with PHA. J Bacteriol 184: 3992-4002.

62. Yamada M, Takahashi S, Okahata Y, Doi Y, Numata K (2013) Monitoring and kinetic analysis of the molecular interactions by which a repressor protein, PhaR, binds to target DNAs and poly[(R)-3-hydroxybutyrate]. AMB Express 3: 6.

63. Korotkova N, Chistoserdova L, Lidstrom ME (2002) Poly-beta-hydroxybutyrate biosynthesis in the facultative methylotroph methylobacterium extorquens AM1: identification and mutation of gap11, gap20, and phaR. J Bacteriol 184: 6174-6181.

64. Salehizadeh H, Van Loosdrecht MC (2004) Production of polyhydroxyalkanoates by mixed culture: recent trends and biotechnological importance. Biotechnol Adv 22: 261-279.

65. Kim BS (2000) Production of poly(3-hydroxybutyrate) from inexpensive substrates. Enzyme Microb Technol 27: 774-777.

66. Rahman NA, Yee PL, Hassan MA, Omar FN, Hafid HS et al. (2011) Utilization of kitchen waste for the production of green thermoplastic polyhydroxybutyrate (PHB) by Cupriavidus necator CCGUG 52238. African Journal of Microbiology Research 5: 2873-2879.

67. Ertesvåg H, Doseth B, Larsen B, Skjåk-Braek G, Valla S (1994) Cloning and expression of an Azotobacter vinelandii mannuronan C-5-epimerase gene. J Bacteriol 176: 2846-2853.

68. Yu J, Si Y, Keung W, Wong R (2002) Kinetics modeling of inhibition and utilization of mixed volatile fatty acids in the formation of polyhydroxyalkanoates by Ralstonia eutropha. Process Biochem 37: 731-738.

69. Akiyama M, Tsuge T, Doi Y (2003) Environmental life cycle comparison of polyhydroxyalkanoates produced from renewable carbon resources by bacterial fermentation. Polym Degrad Stabil 80: 183-194.

70. Fukui T, Doi Y (1998) Efficient production of polyhydroxyalkanoates from plant oils by Alcaligenes eutrophus and its recombinant strain. Appl Microbiol Biotechnol 49: 333-336.

71. Kahar P, Tsuge T, Taguchi K, Doi Y (2004) High yield production of polyhydroxyalkanoates from soybean oil by Ralstonia eutropha and its recombinant strain. Polym Degrad Stabil 83: 79-86.

72. Loo CY, Lee WH, Tsuge T, Doi Y, Sudesh K (2005) Biosynthesis and characterization of poly(3-hydroxybutyrate-co-3- hydroxyhexanoate) from palm oil products in a Wautersia eutropha mutant. Biotechnol Lett 27: 1405-1410.

73. Thakor N, Trivedi U, Patel KC (2005) Biosynthesis of medium chain length poly(3-hydroxyalkanoates) (mcl-PHAs) by Comamonas testosteroni during cultivation on vegetable oils. Bioresour Technol 96: 1843-1850.

74. Verlinden RA, Hill DJ, Kenward MA, Williams CD, Piotrowska-Seget Z, et al. (2011) Production of polyhydroxyalkanoates from waste frying oil by Cupriavidus necator. AMB Express 1: 11.

75. Wells JH (1993) Utilization of rice bran and oil in human diets. La Agr 36: 4-8.

76. Huang TY, Duan KJ, Huang SY, Chen CW (2006) Production of polyhydroxyalkanoates from inexpensive extruded rice bran and starch by Haloferax mediterranei. J Ind Microbiol Biotechnol 33: 701-706.

77. Devi ES, Vijayendra SVN, Shamala TR (2012) Exploration of rice bran, an agro-industry residue, for the production of intra- and extra-cellular polymers by Sinorhizobium meliloti MTCC 100. Biocatalysis and Agricultural Biotechnology 1: 80-84.

78. Wang HY, Cooney CL, Wang DIC (1979) Computer control of bakers' yeast production. Biotechnol Bioeng 21: 975-995.

79. The Origins of Molasses.

80. Pohl J (1988) H. E. Snyder und T. W. Kwon: Soybean Utilization. 346 Seiten, zahlr. Abb und Tab. An AVI Book, published by Van Nostrand Reinhold

Company, New York 1987. Preis: 39,95 £. Food / Nahrung 32: 408.

81. Zhang H, Obias V, Gonyer K, Dennis D (1994) Production of polyhydroxyalkanoates in sucrose-utilizing recombinant Escherichia coli and Klebsiella strains. Appl Environ Microbiol 60: 1198-1205.

82. Oliveira FC, Freire DM, Castilho LR (2004) Production of poly(3-hydroxybutyrate) by solid-state fermentation with Ralstonia eutropha. Biotechnol Lett 26: 1851-1855.

83. Kulpreecha S, Boonruangthavorn A, Meksiriporn B, Thongchul N (2009) Inexpensive fed-batch cultivation for high poly(3-hydroxybutyrate) production by a new isolate of Bacillus megaterium. J Biosci Bioeng 107: 240-245.

84. Albuquerque MG, Concas S, Bengtsson S, Reis MA (2010) Mixed culture polyhydroxyalkanoates production from sugar molasses: the use of a 2-stage CSTR system for culture selection. Bioresour Technol 101: 7123-7133.

85. Omar S, Rayes A, Eqaab A, Vob I, Steinbuchel A (2001) Optimization of cell growth and poly(3-hydroxybutyrate) accumulation on date syrup by a Bacillus megaterium strain. Biotechnol Lett 23: 1119-1123.

86. Solaiman DK, Ashby RD, Foglia TA, Marmer WN (2006) Conversion of agricultural feedstock and coproducts into poly(hydroxyalkanoates). Appl Microbiol Biotechnol 71: 783-789.

87. Wong HH, Lee SY (1998) Poly-(3-hydroxybutyrate) production from whey by high-density cultivation of recombinant Escherichia coli. Appl Microbiol Biotechnol 50: 30-33.

88. Yang ST, Zhu H, Li Y, Hong G (1994) Continuous propionate production from whey permeate using a novel fibrous bed bioreactor. Biotechnol Bioeng 43: 1124-1130.

89. Nath A, Dixit M, Bandiya A, Chavda S, Desai AJ (2008) Enhanced PHB production and scale up studies using cheese whey in fed batch culture of Methylobacterium sp. ZP24. Bioresour Technol 99: 5749-5755.

90. Nikel PI, Pettinari MJ, Méndez BS, Galvagno MA (2005) Statistical optimization of a culture medium for biomass and poly(3-hydroxybutyrate) production by a recombinant Escherichia coli strain using agroindustrial byproducts. Int Microbiol 8: 243-250.

91. Park SJ, Park JP, Lee SY (2002) Metabolic engineering of Escherichia coli for the production of medium-chain-length polyhydroxyalkanoates rich in specific monomers. FEMS Microbiol Lett 214: 217-222.

92. Pantazaki AA, Papaneophytou CP, Pritsa AG, Liakopoulou-Kyriakides M, Kyriakidis DA (2009) Production of polyhydroxyalkanoates from whey by Thermus thermophilus HB8. Process Biochem 44: 847-853.

93. Marangoni C, Furigo Jr A, de Aragao GMF (2002) Production of poly(3-hydroxybutyrate-co-3-hydroxyvalerate) by Ralstonia eutropha in whey and inverted sugar with propionic acid feeding. Process Biochem 38: 137-141.

94. Koller M, Hesse P, Bona R, Kutschera C, AtliÄ‡ A, et al. (2007) Potential of various archae- and eubacterial strains as industrial polyhydroxyalkanoate producers from whey. Macromol Biosci 7: 218-226.

95. Povolo S, Toffano P, Basaglia M, Casella S (2010) Polyhydroxyalkanoates production by engineered Cupriavidus necator from waste material containing lactose. Bioresour Technol 101: 7902-7907.

96. Lee SY, Choi Ji, Han K, Song JY (1999) Removal of endotoxin during purification of poly(3-hydroxybutyrate) from gram-negative bacteria. Appl Environ Microbiol 65: 2762-2764.

97. Gorenflo V, Schmack G, Vogel R, Steinbüchel A (2001) Development of a process for the biotechnological large-scale production of 4-hydroxyvalerate-containing polyesters and characterization of their physical and mechanical properties. Biomacromolecules 2: 45-57.

98. Asrar J, Kurdikar DL, Paster MD, Solodar AJ, Strauser FE (2000) Methods of PHA extraction and recovery using non-halogenated solvents. US 6043063 A.

99. Tamer IM, Moo-Young M, Chisti Y (1998) Optimization of poly(b-hydroxybutyric acid) recovery from Alcaligenes latus: combined mechanical and chemical treatments. Bioprocess Eng 19: 459-468.

100. Ramsay JA, Berger E, Voyer R, Chavarie C, Ramsay BA (1994) Extraction of poly-3-hydroxybutyrate using chlorinated solvents. Biotechnol Tech 8: 589-594.

101. Kapritchkoff FM, Viotti AP, Alli RC, Zuccolo M, Pradella JG, et al. (2006) Enzymatic recovery and purification of polyhydroxybutyrate produced by Ralstonia eutropha. J Biotechnol 122: 453-462.

102. Choi J, Lee SY (1999) Efficient and economical recovery of poly(3-hydroxybutyrate) from recombinant Escherichia coli by simple digestion with chemicals. Biotechnol Bioeng 62: 546-553.

103. Jacquel N, Lo CW, Wei YH, Wu HS, Wang SS (2008) Isolation and purification of bacterial poly(3-hydroxyalkanoates). Biochem Eng J 39: 15-27.

104. Yu J, Chen LX (2006) Cost-effective recovery and purification of polyhydroxyalkanoates by selective dissolution of cell mass. Biotechnol Prog 22: 547-553.

Process Parameters for Influencing Polyhydroxyalkanoate Producing Bacterial Factories: An Overview

Amandeep Girdhar[1]*, Mayuri Bhatia[1], Sunil Nagpal[2], Amol Kanampalliwar[1] and Archana Tiwari[1]

[1]*School of Biotechnology, Rajiv Gandhi Proudyogiki Vishwavidhyalaya, State Technological University, Airport Bypass Road, Bhopal 426033, Madhya Pradesh, India*
[2]*Department of Biochemical Engineering, Indian Institute of Technology, Safdarganj Development Area, Hauz Khas, IIT Campus, New Delhi, Delhi 110016, India*

Abstract

The ever increasing potentialities of petroleum plastics with respect to lack of degradation, inability to recycle and the toxic effects of incineration, has urged to design biodegradable polymers, often called Green Plastics. These biodegradable plastics are promiscuous due to their analogous properties and environmental friendliness. Bacterial factories and Plants being their natural sources for production made them a promiscuous solution. Fermentation is the procedural technology used with certain fillers that are known to enhance the chemo-mechanical properties. The process at the industrial level is not well accepted due to the certain lacunas. The review mainly focuses to assimilate a few researches that implicate the best known process parameters for Batch, Fed-batch, Continuous and Two stage modes of fermentation without compromising the downstream processing at commercial level.

Keywords: Polyhydroxyalkanoate; Bacteria; Carbon sources; Batch culture; Economy

Introduction

In the world of advancements today, almost every product is constituted of some kind of polymer. There is no doubt that the need for polymers and thus the products they constitute is ever increasing. Till date, these needs are being fulfilled by synthetic polymers (often called plastics), which are produced from petrochemicals [1] which makes them eco- 'unfriendly'. The inherent nature of petroleum derived products, calls for a serving approach, in the form of Biopolymers-polymers derived from living organisms/renewable resources [2]. To satisfy the consumers and get acceptability substitute needs to exhibit similar (if not identical) characteristics to the product being replaced and so it is for biopolymers while replacing the synthetic polymers properties, ranging from molecular weight, density, melting point, crystallinity, glass transition temperature to O_2-permeability, UV-resistance, resistance to solvents, tensile strength and elongation to break [3]. Fortunately, a special class of biopolymers called PHAs shows some of the extraordinary similarities to the well known synthetic polymers like polypropylene, polyethylene etc. [4], moreover, their biodegradability has made them renowned as biopolymers of today.

History and chemical nature

In 1920, a French microbiologist Maurice Lemoigne discovered a gram positive bacterium *Bacillus megaterium* [5] that accumulated intracellular granules of polyester called poly(3-hydroxybutyrate) [6]. Table 1 enlists different types of PHA.

PHAs are polymeric compounds biosynthesized by a variety of gram positive and gram negative bacteria [7], as carbon and energy reserves (often called carbonosomes) [8]. Structurally, R-hydroxyalkonic acids act as the monomeric form of PHAs.

Natural sources of production

Polyhydroxyalkanoates are produced in microbes. Although efforts have been made from plant cells through transgenics, but has not achieved much success because low yields of less than 10 % (w/w) of dry cell weight can be sustained whereas, high yield limits growth and development of plants [9,10]. On the other hand, PHAs can be accumulated upto 90% (w/w) in bacterial cells [10] and are thus a priority because of the ease in culturing and economical, in contrast to the complex plant system [10].

Bacterial Polymer Production

PHAs accumulation is an inherent response to the stress conditions faced by bacterial cells [11,12], these are generated *in vitro* by exposing bacteria to nutrient limitations, due to which they switch their metabolic pathways and cause PHA production as their carbon and energy reserves [13], to name a few of these substrates are bagasse [14], molasses [15,16], corn cob [14] and other agricultural wastes [4,17]. In fact significant PHA production has been reported among various bacterial strains when growing on kitchen waste [18], industrial wastes [7,14,19], crude and edible oils as carbon sources [20-23]. The costing of these substrates is low or null, making the process cost effective at the upstream level. The Figures 1 and 2 depicts the biopolymer synthesis from bacteria.

PHA production has been reported in wide variety of bacterial

PHAs	Number of Carbon Atoms in Monomers of PHA	Examples
Short chain length PHAs (scl-PHAs)	3 – 5	P(3HB) P(4HB)
Medium chain length PHAs (mcl-PHAs)	6 – 14	P(3HHx) P(3HO) P(3HHx-co-3HO)

Table 1: Types of PHAs [2].

***Corresponding author:** Amandeep Girdhar, School of Biotechnology, Rajiv Gandhi Proudyogiki Vishwavidhyalaya, State Technological University, Airport Bypass Road, Bhopal 426033, Madhya Pradesh, India, Tel: +918989206617

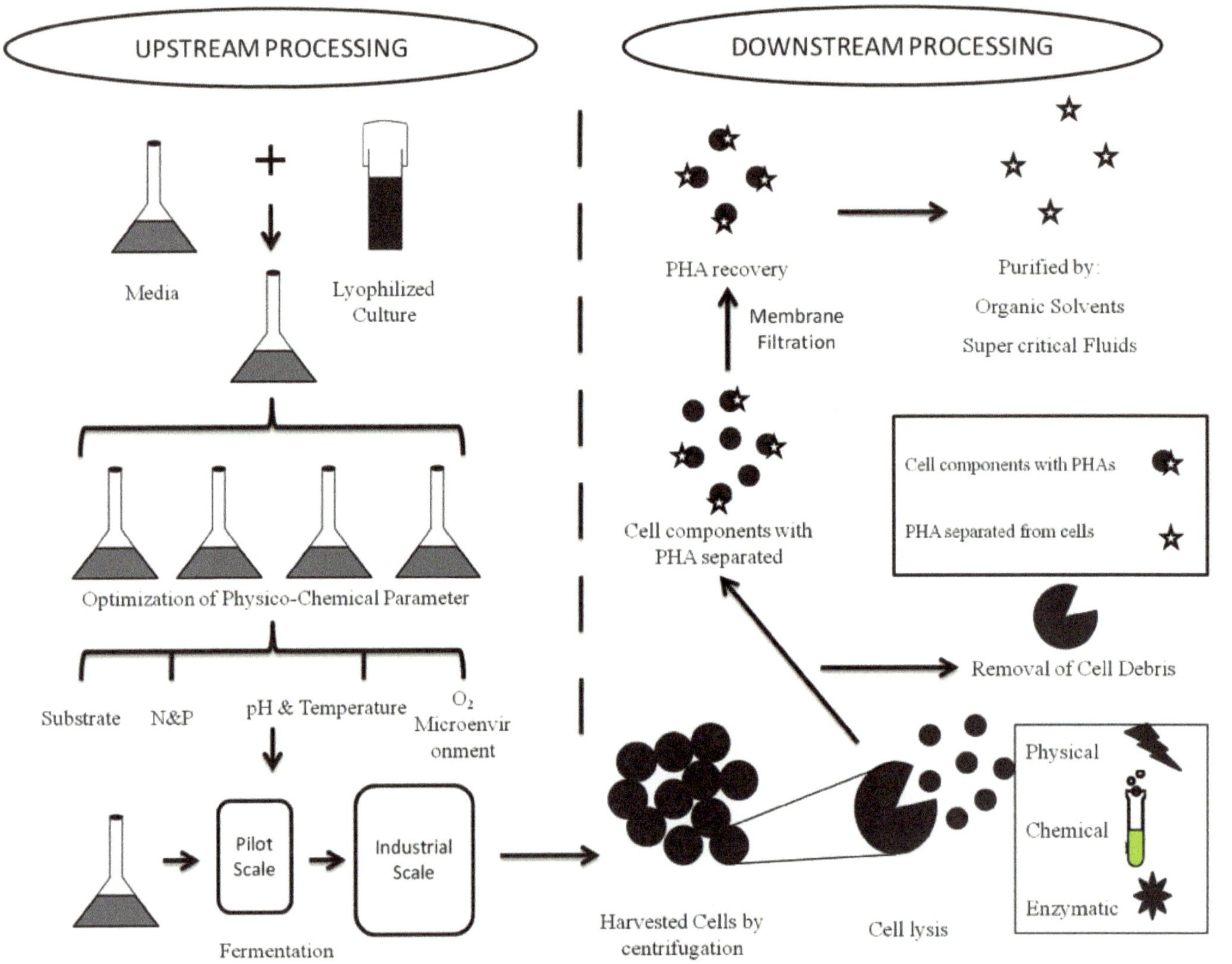

Figure 1: Biopolymer synthesis.

strains, the most commonly studied genera are *Bacillus sp, Alcaligenes sp, Pseudomonas sp, Aeromonas sp, Rhodopseudomonas sp, Halomonas sp*, Transgenic *Escherichia coli, and Burkholderia sp* [7,9]. Table 2 and Table 3 show carbon sources based feeding regimes used by different strains with the PHA Content. The various stress conditions trigger PHA production [11,12]; thus, variation in process parameters is required [24-26] to induce the biosynthesis.

Process parameters

Substrate: Carbon source is an important requisite but is not involved in induction of PHA biosynthesis, although is required for polymer production maximization [54,55]. Glucose is the preferred substrate at industrial scale, many alternatives have also been employed, like sucrose as stated in the research by Azhar et al. of Ain Shams University, Egypt with highest growth and polymer production by *Alcaligenes latus* [56]. Even a broad spectrum of substrates, namely, starch, sucrose, lactose, maltose, galactose, mannose, mannitol, fructose, glycerol, ethanol, lactic acid, malic acid, acetic acid and butyric acid have been used to compare with glucose [56]. In another report by Poonsuk Prasertsan, the halotolerant bacterial strains *Rhodobacter sphaeroides* has exhibited maximum growth and polymer production on acetate as a carbon source compared to glucose and fructose [25]. The conclusion deduced glucose, fructose and acetate as good substrates for cellular growth and polymer formation, but a combination of these substrates did not give good results [25].

Nitrogen and phosphate: A research aimed at the optimal requirement of phosphate and nitrogen in the media (specific or complex) to increase the production tested the growth of *Aeromonas hydrophila* on gluconate enriched MS media containing different phosphate concentrations as (Table 4). It was clearly inferred that a concentration of 11.66 mM phosphate in the media provided a good nutrient deficiency to the bacteria leading to a high PHA concentration [44]. This concentration of phosphate was 1/3 of the normal phosphate concentration used in the media for the bacterial growth. Table 4 illustrates the effect of limiting phosphate concentrations, with nitrogen deficiency in media, on PHA production [44].

Although the nutrient limitation is a prerequisite for PHA production, but, in a recent report on *Cupriavidus necator* in several nutrition media (Mineral medium, Bonnarme's medium, Mandels' medium and Luria broth Mineral medium, Bonnarme's medium, Mandels' medium and Luria broth) with 10g/l sodium glutamate not

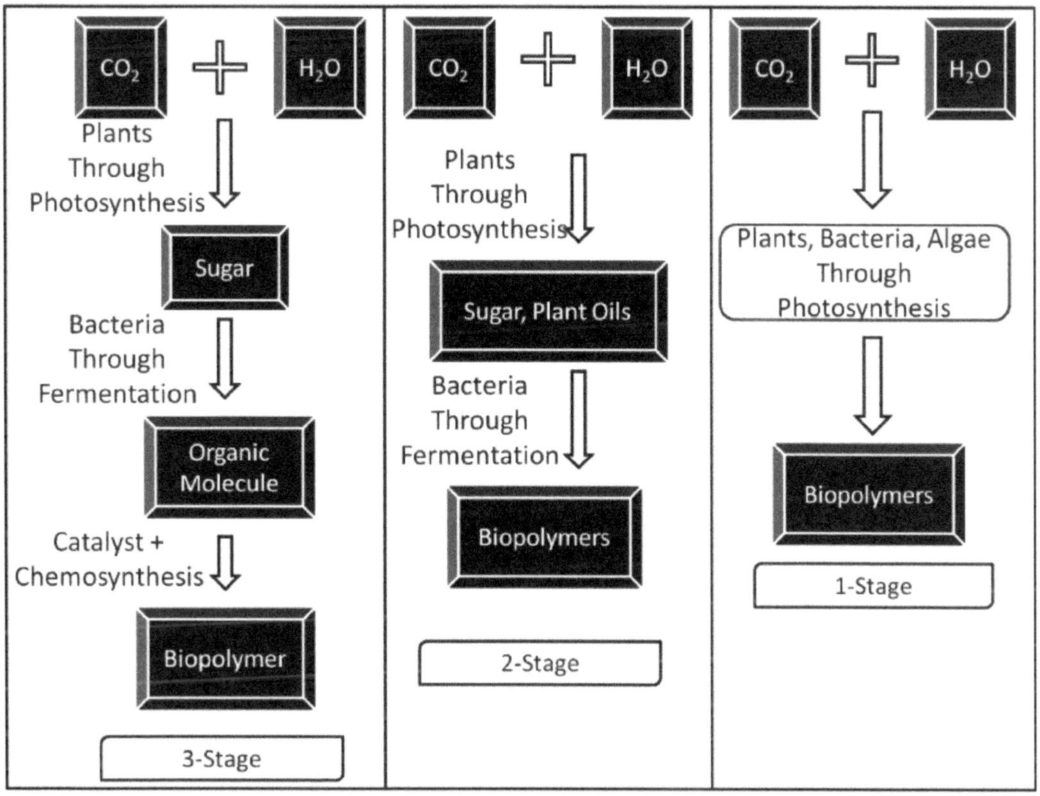

Figure 2: Systems for Biopolymer Synthesis.

only enhanced the overall productivity by 33 times but also decreased the need for limiting nutrients in media [57].

pH and temperature regimes: The pH and Temperature optima completely depend on the nature of organisms. According to Palleroni et al. [58] and recent studies done by Wei et al. [26], the optimal pH range for the growth of bacteria involved in the production of PHAs is 6.0 to 7.5 [26,58]. The temperature requirement for the optimal production is 30 to 37°C as reported by Yu-Hong Wei et al. [26]. These are the generalized range for fine bacterial growth.

Oxygen microenvironment: Oxygen limitation is regarded amongst the effective way of enhancing PHA accumulation, involves volumetric oxygen transfer coefficient (k_La), reducing (k_La) of the system significantly enhance the PHA yield of the bacterial cell, even at agitation speed of 300 rpm, the percentage PHA derivate was 29.8% whereas it was 50.7 % at 50 rpm [24]. Since K_La is lower at low agitation rate, indicates that increasing the oxygen stress enhances PHA production.

Modes of culture

Four modes of operations are used, these are Batch, Fed batch, Two stage and continuous; although Batch process has been the mode of choice for most of the research and commercial production.

Batch: The batch cultivation is performed from range starting from 100ml working volume at laboratory scale [16,28] to 15000L working volume at industrial scale [59]. *Alcaligenes latus* (BIOPOL, the first commercially produced biopolymer from *Alcaligenes latus* [60]), *Ralstonia eutropha*, *Aeromonas hydrophila*, *Burkholderia* sp. and *Pseudomonas putida* have been given equal importance for polymer production [9]. Maximum PHA production is reported between 96 to 120 hours in almost all species, where glucose and sucrose have been the substrate of choice (in shake flask batch mode *Alcaligenes latus* exhibited 45.96 % and 40.14 % PHB content, and in 2 L reactor it exhibited 56.59 % and 47.53% PHB content using sucrose and glucose as substrate, respectively), with ammonium sulfate as best nitrogen source [56,61]. A 2L batch operation yielded maximum PHB content after 80 hours of operation, after which the content decreased (probably due to hydrolysis during stationary phase), suggesting the critical importance of harvesting time [61].

Fed batch: Basic aim of using the fed-batch mode is to increase the biomass with respect to batch mode and is achieved by controlling the intermittent feeding to reduce the substrate inhibition without affecting the growth of the microbe(s). The basic causative agents of inhibition are the organic acids which in high concentration lead to the collapse of the transmembrane pH gradient of the bacterial cells and a low concentration supports the cell growth and hence PHA accumulation [62]. Two feeding rates for organic acid concentration variation were studied with constant and varying rate on *Ralstonia eutropha*. Three constant feeding rates of 1.2 g/l, 1.8 g/l and 2.5 g/l achieved greater biomass(RBM) and PHA content in case of slow feeding rate with zero residual acids, while it ceased at 24th and 36th hours respectively at intermediate feeding, with a residual acid of 2.02g/l and at 18th and 36th hour at fast feeding rate, with a residual acid of 4.02g/l suggesting that cell growth and production could only continue under a low level of residual acids in the medium, while PHAs accumulate at higher acid concentration. Thus it may be considered essential to feed acids during

Strain	Carbon Source Used	Polymer Yield	References
Aeromonas sp.	Municipal waste water	11.11%	[27]
	Palm oil mill effluent	22.97%	
	Glycerol	19.82%	
	Molasses	20.09%	
Alcaligenes sp.	Municipal waste water	41.11%	[27]
	Palm oil mill effluent	11.69%	
	Glycerol	2.33%	
	Molasses	0.00%	
Bacillus sp.	1:1 mixture of WBH and RBH	59%	[28]
	Municipal waste water	43.95%	[27]
	Palm oil mill effluent	58.925%	
	Glycerol	25.00%	
	Molasses	48.01%	
Bacillus cereus	Municipal waste water	38.07%	[27]
	Palm oil mill effluent	64.09%	
	Glycerol	0.00%	
	Molasses	23.94%	
Bacillus licheniformis	Municipal waste water	40.97%	[27]
	Palm oil mill effluent	62.96%	
	Glycerol	1.69%	
	Molasses	43%	
B. subtilis	Municipal waste water	46.71%	[27]
	Palm oil mill effluent	50.00%	
	Glycerol	18.92%	
	Molasses	16.13%	
Bacillus megaterium	Sugarcane molasses	43%	[15]
Burkholderia cepaca	Glycerol	31.35%	[29]
Chromobacterium sp.	Municipal waste water	25.00%	[27]
	Palm oil mill effluent	40.89%	
	Glycerol	11.86%	
	Molasses	27.85%	
Cupriavidus taiwanensis	Gluconic acid	72%	[26]
Escherichia coli	Xylose	59 ± 4%	[30]
	Sodium butyrate	7.81 ± 0.21%	[31]
Hydrogenophaga pseudoflava	Lactose	29.8 ± 3.0%	[13]
Haloarcula hispanica (previously deposited as Halobacterium hispanicum)	YE/Glucose	2.4%	[32]
Haloarcula marismortui	YE/Glucose	21%	[33]
Haloarcula **sp.**	Glucose	63%	[34]
Haloarcula japonica	Glucose	0.5%	[35]
Halobiforma haloterrestri	Butyric acid	40%	[36]
H. haloterrestris	YE	15%	[36]
Haloferax mediterranei	YE/Glucose	17%	[32]
H. mediterranei	Starch	6.48g/L	[37]
Haloferax gibbonsii	YE/Glucose	1.2%	[32]
Haloferax volcanii	YE/Glucose	7%	[32]
Haloterrigena hispanica	YE/Casamino acids	0.14%	[38,39]
Halopiger aswanensis	YE/Na acetate/Butyric acid	34%	[40,41]
Pseudomona oleovorans	Gluconic acid	0.9±0.1%	[26]
Pseudomonas putrefaciens	Corn cob	66.67%	[14]
Pseudomonas sp.	Municipal Waste Water	61.05%	[27]
	Palm oil mill effluent	60.08%	
	Glycerol	62.08%	
	Molasses	60.04%	
Proteus mirabilis	Municipal Waste Water	0.00%	[27]
	Palm oil mill effluent	0.00%	
	Glycerol	0.00%	
	Molasses	0.00%	
Ralstonia eutropha	Glucose	0.058	[42]
	Sodium gluconate	87.03%	[43]

Table 2: Data for carbon sources used by microorganisms for scl-PHA production.

Strain	Carbon Source Used	Polymer Yield	References
Aeromonas hydrophila	Gluconate	15%	[44]
Cupriavidus necator	Crude palm kernel oil (5g/L)	77 ± 3%	[22]
	Crude palm kernel oil (10g/L)	82 ± 2%	
	Jatropha oil	62 ± 3%	
	Crude palm oil	69 ± 6%	
	Palm olein	61 ± 4%	
	Soybean oil	65 ± 1%	
	Corn oil	63 ± 2%	
	Coconut oil	70 ± 3%	
Escherichia coli	Xylose	70 ± 1%	[30]
	Sodium octanoate	47.3 ± 5.0 %	[31]
	Decanoic acid	25.7 ± 0.9 %	
	10-Undecenoic acid	41.0 ± 8.0 %	
	Dodecanoic acid	28.6 ± 8.0 %	
Haloarcula hispanica	YE/Glucose	17.33 ± 0.04%	[45]
Halobacteriumnoricense	Triptone	0.08/0.03%	[46]
Halococcusdombrowskii	YE/HyCase	0.15/0.01%	[46]
Halococcussalifodinae	YE/HyCase	0.05/0.01%	[46]
Haloferax mediterranei	Glucose	27%	[47]
	ECS (Extruded Corn Starch)	38.7%	[47]
	ERB (Extruded Rice Bran) :ECS (1:8)	55.6%	[47]
	Bacto Casa amino acids/YE	18.21 ± 1.88%	[45]
	YE/Starch	24.88 ± 1.27 g/L	[45]
Hydrogenophaga pseudoflava	Whey (copolymer)	10.1 ± 0.9%	[13]
Pseudomonas putida	Octanoic acid	49.7 %	[48]
	Octanoate	21%	[49]
	Glucose	ND	
	Acetate	ND	
	Pyruvate	ND	
	Citrate	ND	
	Succinate	ND	
	Glucanoate	ND	
	fructose	ND	
	Dodecanoic acid	54.5 %	[48]
	Oleic acid	68.9 %	
	Xylose	20 %	[50]
	Nananoic acid	0.15g/g	[51]
	NanoicAcid:Glucose (0.8:1)	50.85%	
	Corn oil	17.5 ± 1.4%	[52]
	Glucose	19%	[44]

Pseudomonas oleovorans	Corn oil	15.7 ± 2.5 %	[52]
Pseudomonas chlororaphis	Corn oil	39.5 ± 0.8	[52]
Pseudomonas G101	Waste frying rapseed oil	20%	[53]
Natronobacterium gregoryi	YE/Casamino acids	0.1/0.3%	[46]
Sinorhizobiummeliloti	Rice bran hydrolysate	48.32%	[4]

Table 3: Data for carbon sources used by microorganisms for mcl-PHA production.

Concentration	Amount $Na_2HPO_4.12H_2O$ (g/l)	Amount KH_2PO_4 (g/l)	PHA concentration (%, w/w)
Normal phosphate (35mM)	9	1.5	4.6 ± 0.6
2/3 phosphate (23.35 mM)	6	1	5.5 ± 0.5
1/3 phosphate (11.66 mM)	3	0.5	10.2 ± 0.5

Table 4: Effect of limiting P and N concentrations on PHA production [44].

the cell growth phase, followed by gradually decreasing the flow rate to maintain a lower residual concentration of about 2 g/l. This strategy came out to be successful with the Dry Cell Weight (DCW) and PHA contents as 14.35g/L and 6.89g/L respectively at the end of 42nd hour [62]. pH-stat mode provide randomly variable rate of organic acids, which worked on the principle of acid adjustment/pH control and is successful if the pH of the system varies with the growth of the bacteria. The pH variation is adjusted by organic acids rather than HCl/NaOH leading to a random feeding of organic acids, which caused random feeding leading to less effective PHA production and DCW. In reference, a more careful feed rate control can further modulated for improvement in cell growth and PHA production.

Continuous: Continuous fermentation technology has got advantage of maintaining constant nutrient environment and is useful to study the effect of nutrient limitation on the bacterial growth and productivity in a quick and real time manner [63]. In fact, the documented reports are aimed at studying the effect of limiting the concentration of various nutrients. In a recent example, the metabolic response of *P. putida* KT2442 producing high levels of PHA under single- and multiple-nutrient-limited growth, with chemostat mode has been reported [64]. At the industrial level, continuous mode could prove very useful owing to continuous harvesting of the product, coupled with the ability to tweak the nutrient concentration as per the requirement; although the stringency of sterile environment and a check on mutability of the producing strain becomes obligatory under continuous mode, when compared to the ease of batch mode.

Two – stage batch culture as the best method for PHB production: The two-stage fermentation involves supply of nutrients to the culture in two different stages and composition. Thus, becomes handy for possessing the nutrient limitation, a prerequisite for PHA production [9]. In the first stage, culture is fed with simple carbon sources such as glucose or fructose and other essential nutrients to make sure that cells grow at high specific growth rate and lead to a higher cell biomass yield, whereas in the second stage, supply of essential nutrient such as nitrogen is limited as a trigger to initiate the biosynthesis [56]. A volumetric PHA productivity of 1.06 g L^{-1} h^{-1} was obtained by Two-stage continuous mode of *Pseudomonas oleovorans* culture [65].

Downstream Processing

All the bacterial species reported to produce polyhydroxyalkanoates, accumulate them intracellularly in the form of granules, as carbon and energy reserves [44,59]. It is thus apparent, that the downstream processing chain of PHAs begins from cell lysis of the biomass separated from the medium by centrifugation, so as to expose the cell trapped product [66]. There are many techniques used for the purpose, Table 5 describes a few of these.

Recovery

The PHAs are present in the cell debris after the cell lysis and centrifugation, which can be recovered with a good efficiency using different solvents like cold methanol, chloroform etc [77]. This leaves a dull white precipitate behind; further purification of the product is done by membrane filtration, which helps in segregating PHAs from other small biomolecules still present in the extract. A study with non-halogenated solvents for recovery from *R. eutropha* has shown to give high purity (upto 99%) and have proved potentiality of ethyl acetate and methyl isobutyl ketone to replace halogenated solvent, chloroform, with cost reduction benefits [78]. In a research, 0.1 μm ceramic tubular membranes have shown high separation efficiency owing to their chemical inertness and resistance to pH, temperature and concentration changes [77]. But polyethersulfone polymeric flat plate membrane modules are becoming the choices of the day because of their 5-10 times less cost than ceramic membranes and ability to retain even the smallest PHA granule. However, membranes like polyethersulfone come with an inherent disadvantage of limited life span and need to be replaced once in a year or two. It must be noted that the ultra-filtration carried out to separate the PHAs is operated in a cross flow regime, which avoids the blocking of the membranes and ensures continuous operation for long time [77]. The amount of product recovered in form of filtrate is then determined using Gas Chromatography to confirm

Classes	Techniques	Acting Principle	Microorganism	Reference
Physical/ MechanicalDisruption	Thermolysis	Disruption of Cell wall by the affect of ionic strength, pH and temperature with chelating agent		[67,68]
	Ultrasonication	Ultrasonic waves, followed by centrifugation	*Bacillus flexus*	[68]
	Bead Mill	Grinding cylinder containing beads made of wear resistant materials like glass, alumina, titanium carbide, zirconium oxide and zirconium silicate is driven by motor	*Alcaligenes latus*	[60,67,69]
	High Pressure Homogenizer	Disrupter fitted with a displacement pump monitors the pressure and a discharge valve to homogenize the solution pushed through pump	Gram Negative Bacteria	[70]
Chemical Disruption	Alkali Treatment	Exposure to basic pH (mild alkaline hydrolysis)	*Bacillus flexus*	[69]
	Detergent Solubilization	Detergents like SDS, CTAB, Triton X 100, Saponins, Tween 20 and Tween 80 etc, are used	*Ralstonia eutropha*	[67,71,72]
	Cell Wall Permeabilisation	Organic solvents like toluene, acetone, chloroform and ethylene carbonate are used, followed by non-solvent precipitation	*Bacillus cereus* SPV and *Cupriavidus necator*, respectively.	[73,74]
Enzymatic Disruption		Lytic enzymes in medium with detergent or chelating agent	*Cupriavidus necator*	[75,76]

Table 5: Cell Lysis techniques and acting principles.

Parameters	Synthetic Plastics	Biodegradable Plastics
Synthesis	Crude oil	Microorganism
	Flash distillation	Waste carbon sources
	Column distillation	Culturing
	Cracking (use of catalyst)	Biomass separation
	Polymerization	Polymer recovery
	Heat required for every step	Heat required for film preparation and drying
Degradation	Ethylene, CO_2, CH_4 production	Biological compounds are produced
	Toxic	Non-toxic
	Never degrade	Take days to months

Table 6: Factors deciding cost.

the percentage of recovery [43]. Quantification can be done by NMR [79], Flow Cytometry [80] etc.

Cost

The major factor of concern is the cost of the produced biomaterial i.e. PHA. The production cost is contributed by various factors. The process parameters are involved in upstream cost whereas the down streaming involves the cost of solvents used and the other purification strategies. The Table 6 illustrates parameters involved in cost of production and degradation through biological means. Processing cost in generally 20% of the total cost.

Conclusion

The growing environmental pollution is becoming a global concern, where synthetic plastics contribute to a great extent due to their non-degradability. The bioplastics can be regarded as a working alternative to conquer the conundrum. From all the classes of bioplastics, PHA has shown to have analogous properties with biodegradability. Thus, PHA can be looked forward as an alternate to synthetic plastics as they require natural sources for synthesis like Bacteria, that have come as an effective way for commercial production, *Ralstonia eutropha* is amongst the best producers. Being promiscuous in nature, their production involves certain process parameters to be optimized both at laboratory and industrial scale, where the optimum pH and temperature range depends on the microorganism used in the synthesis. Further the O_2 stress enhances the production. The mode of fermentation is responsible for influencing the biomass yield and accumulation of the product. Not to overlook the Cost factor, which can be reduced by various alternatives to novel carbon sources, but to attain a good yield and commercialization with these alternatives still remain a challenge.

References

1. Cavalheiro JMBT, de Almeida MCMD, Grandfils C, da Fonseca MMR (2009) Poly(3-hydroxybutyrate) production by Cupriavidus necator using waste glycerol. Process Biochem 44: 509-515.
2. Shrivastav A, Mishra SK, Pancha I, Jain D, Bhattacharya S, et al. (2011) Biodegradability studies of polyhydroxyalkanoate (PHA) film produced by a marine bacteria using Jatropha biodiesel byproduct as a substrate. World J Microb Biot 27: 1531-1541.
3. Ojumu TV, Yu J, Solomon Bo (2004) Production of Polyhydroxyalkanoates, a bacterial biodegradable polymer. African Journal of Biotechnology 3: 18-24.
4. Devi ES, Vijayendra SVN, Shamala TR (2012) Exploration of rice bran, an agro-industry residue, for the production of intra- and extra-cellular polymers by Sinorhizobium meliloti MTCC 100. Biocatalysis and Agricultural Biotechnology 1: 80-84.
5. Lemoigne M (1926) Produits de dehydration et de polymerisation de l'acide ß-oxobutyrique. Bull Soc Chim Biol 8: 770-782.
6. Zhu C, Chiu S, Nakas JP, Nomura CT (2013) Bioplastics from waste glycerol derived from biodiesel industry. J Appl Polym Sci 130: 1-13.
7. Setyawaty R, Setiadi T, Katayama-Hirayama K, Kaneko H, Hirayama K (2012) Polyhydroxyalkanoate (PHA) production from tapioca industrial wastewater treatment: Influence of operating conditions on PHA content. Sustain Environ Res 22: 123-127.
8. Jendrossek D (2009) Polyhydroxyalkanoate granules are complex subcellular organelles (carbonosomes). J Bacteriol 191: 3195-3202.
9. Khanna S, Srivastava AK (2005) Recent advances in microbial polyhydroxyalkanoates. Process Biochem 40: 607-619.
10. Verlinden RA, Hill DJ, Kenward MA, Williams CD, Radecka I (2007) Bacterial synthesis of biodegradable polyhydroxyalkanoates. J Appl Microbiol 102: 1437-1449.
11. Foster LJ (2007) Biosynthesis, properties and potential of natural-synthetic hybrids of polyhydroxyalkanoates and polyethylene glycols. Appl Microbiol Biotechnol 75: 1241-1247.
12. Kadouri D, Jurkevitch E, Okon Y, Castro-Sowinski S (2005) Ecological and agricultural significance of bacterial polyhydroxyalkanoates. Crit Rev Microbiol 31: 55-67.
13. Povolo S, Romanelli MG, Basaglia M, Ilieva VI, Corti A, et al. (2013) Polyhydroxyalkanoate biosynthesis by Hydrogenophaga pseudoflava DSM1034 from structurally unrelated carbon sources. N Biotechnol 30: 629-634.
14. Ogunjobi AA, Ogundele AO, Fagade OE (2011) Production of polyhydroxyalkanoates by Pseudomonas putrefaciens from cheap and renewable carbon substrates. Electronic Journal of Environmental, Agricultural and Food Chemistry 10: 2806-2815.
15. Chaijamrus S, Udpuay N (2008) Production and Characterization of Polyhydroxybutyrate from Molasses and Corn Steep Liquor produced by Bacillus megaterium ATCC 6748. Agricultural Engineering International: the CIGR Ejournal.
16. Baei MS, Najafpour GD, Younesi H, Tabandeh F, Eisazadeh H (2009) Poly(3-hydroxybutyrate) Synthesis by Cupriavidus necator DSMZ 545 Utilizing Various Carbon Sources. World Applied Sciences Journa 7: 157-161.
17. Solaiman DK, Ashby RD, Foglia TA, Marmer WN (2006) Conversion of agricultural feedstock and coproducts into poly(hydroxyalkanoates). Appl Microbiol Biotechnol 71: 783-789.
18. Omar FN, Rahman NAA, Hafid HS, Mumtaz T, Yee PL, et al. (2011) Utilization of kitchen waste for the production of green thermoplastic polyhydroxybutyrate (PHB) by Cupriavidus necator CCGUG 52238. African Journal of Microbiology Research 5: 2873-2879.
19. Sharmila T, Meenakshi SA, Kandhymathy K, Bharathidasan R, Mahalingam R, et al. (2011) Screening And Characterisation Of Polyhydroxybutyrate Producing Bacteria From Sugar Industry Effluents. World Journal of Science and Technology 1.
20. Verlinden RA, Hill DJ, Kenward MA, Williams CD, Piotrowska-Seget Z, et al. (2011) Production of polyhydroxyalkanoates from waste frying oil by Cupriavidus necator. AMB Express 1: 11.
21. Budde CF, Riedel SL, HÃ¼bner F, Risch S, PopoviÄ‡ MK, et al. (2011) Growth and polyhydroxybutyrate production by Ralstonia eutropha in emulsified plant oil medium. Appl Microbiol Biotechnol 89: 1611-1619.
22. Wong YM, Brigham CJ, Rha C, Sinskey AJ, Sudesh K (2012) Biosynthesis and characterization of polyhydroxyalkanoate containing high 3-hydroxyhexanoate monomer fraction from crude palm kernel oil by recombinant Cupriavidus necator. Bioresour Technol 121: 320-327.
23. Riedel SL, Bader J, Brigham CJ, Budde CF, Yusof ZA, et al. (2012) Production of poly(3-hydroxybutyrate-co-3-hydroxyhexanoate) by Ralstonia eutropha in high cell density palm oil fermentations. Biotechnol Bioeng 109: 74-83.
24. Pozo C, Martínez-Toledo MV, Rodelas B, González-López J (2002) Effects of culture conditions on the production of polyhydroxyalkanoates by Azotobacter chroococcum H23 in media containing a high concentration of alpechín (wastewater from olive oil mills) as primary carbon source. J Biotechnol 97: 125-131.
25. Sangkharak K, Prasertsan P (2008) Nutrient optimization for production of polyhydroxybutyrate from halotolerant photosynthetic bacteria cultivated under aerobic-dark condition. Electron J Biotechnol 11: 83-94.

26. Wei YH, Chen WC, Huang CK, Wu HS, Sun YM, et al. (2011) Screening and Evaluation of Polyhydroxybutyrate-Producing Strains from Indigenous Isolate Cupriavidus taiwanensis Strains. Int J Mol Sci 12: 252-265.

27. Sangkharak K, Prasertsan P (2012) Screening and identification of polyhydroxyalkanoates producing bacteria and biochemical characterization of their possible application. J Gen Appl Microbiol 58: 173-182.

28. Shamala TR, Vijayendra SV, Joshi GJ (2012) Agro-industrial residues and starch for growth and co-production of polyhydroxyalkanoate copolymer and Î±-amylase by Bacillus sp. CFR-67. Braz J Microbiol 43: 1094-1102.

29. Zhu C, Nomura CT, Perrotta JA, Stipanovic AJ, Nakas JP (2010) Production and characterization of poly-3-hydroxybutyrate from biodiesel-glycerol by Burkholderia cepacia ATCC 17759. Biotechnol Prog 26: 424-430.

30. Nduko JM, Matsumoto K, Ooi T, Taguchi S (2013) Effectiveness of xylose utilization for high yield production of lactate-enriched P(lactate-co-3-hydroxybutyrate) using a lactate-overproducing strain of Escherichia coli and an evolved lactate-polymerizing enzyme. Metab Eng 15: 159-166.

31. Tappel RC, Wang Q, Nomura CT (2012) Precise control of repeating unit composition in biodegradable poly(3-hydroxyalkanoate) polymers synthesized by Escherichia coli. J Biosci Bioeng 113: 480-486.

32. Kumar AS, Mody K, Jha B (2007) Bacterial exopolysaccharides--a perception. J Basic Microbiol 47: 103-117.

33. Han J, Lu Q, Zhou L, Zhou J, Xiang H (2007) Molecular Characterization of the phaECHm Genes, Required for Biosynthesis of Poly(3-Hydroxybutyrate) in the Extremely Halophilic Archaeon Haloarcula marismortui. Appl Environ Microbiol 73: 6058-6065.

34. Taran M, Amirkhani H (2010) Strategies of poly(3-hydroxybutyrate) synthesis by Haloarcula sp. IRU1 utilizing glucose as carbon source: Optimization of culture conditions by Taguchi methodology. Int J Biol Macromol 47: 632-634.

35. Nicolaus B, Lama L, Esposito E, Manca MC, Improta R, et al. (1999) Haloarcula spp able to biosynthesize exo- and endopolymers. J Ind Microbiol Biot 23: 489-496.

36. Hezayen FF, Tindall BJ, Steinbüchel A, Rehm BHA (2002) Characterization of a novel halophilic archaeon, Halobiforma haloterrestris gen. nov., sp. nov., and transfer of Natronobacterium nitratireducens to Halobiforma nitratireducens comb. nov. Int J Syst Evol Micr 52: 2271-2280.

37. Lillo JG, Rodriguez-Valera F (1990) Effects of Culture Conditions on Poly(beta-Hydroxybutyric Acid) Production by Haloferax mediterranei. Appl Environ Microbiol 56: 2517-2521.

38. Romano I, Poli A, Finore I, Huertas FJ, Gambacorta A, et al. (2007) Haloterrigena hispanica sp. nov., an extremely halophilic archaeon from Fuente de Piedra, southern Spain. Int J Syst Evol Microbiol 57: 1499-1503.

39. Donato PD, Fiorentino G, Anzelmo G, Tommonaro G, Nicolaus B, et al. (2011) Re-Use of Vegetable Wastes as Cheap Substrates for Extremophile Biomass Production. Waste and Biomass Valorization 2: 103-111.

40. Hezayen FF, Rehm BH, Eberhardt R, Steinbüchel A (2000) Polymer production by two newly isolated extremely halophilic archaea: application of a novel corrosion-resistant bioreactor. Appl Microbiol Biotechnol 54: 319-325.

41. Hezayen FF, Gutiérrez MC, Steinbüchel A, Tindall BJ, Rehm BH (2010) Halopiger aswanensis sp. nov., a polymer-producing and extremely halophilic archaeon isolated from hypersaline soil. Int J Syst Evol Microbiol 60: 633-637.

42. Moreno N, Gutiérrez I, Malagón D, Grosso V, Revelo D, et al. (2007) Bioprospecting and characterization of poly-bhydroxyalkanoate (PHAs) producing bacteria isolated from Colombian sugarcane producing areas. African Journal of Biotechnology 6.

43. Heinrich D, Madkour MH, Al-Ghamdi MA, Shabbaj II, Steinbüchel A (2012) Large scale extraction of poly(3-hydroxybutyrate) from Ralstonia eutropha H16 using sodium hypochlorite. AMB Express 2: 59.

44. Qiu YZ, Han J, Guo JJ, Chen GQ (2005) Production of poly(3-hydroxybutyrate-co-3-hydroxyhexanoate) from gluconate and glucose by recombinant Aeromonas hydrophila and Pseudomonas putida. Biotechnol Lett 27: 1381-1386.

45. Lu Q, Han J, Zhou L, Zhou J, Xiang H (2008) Genetic and biochemical characterization of the poly(3-hydroxybutyrate-co-3-hydroxyvalerate) synthase in Haloferax mediterranei. J Bacteriol 190: 4173-4180.

46. Legat A, Gruber C, Zangger K, Wanner G, Stan-Lotter H (2010) Identification of polyhydroxyalkanoates in Halococcus and other haloarchaeal species. Appl Microbiol Biotechnol 87: 1119-1127.

47. Huang TY, Duan KJ, Huang SY, Chen CW (2006) Production of polyhydroxyalkanoates from inexpensive extruded rice bran and starch by Haloferax mediterranei. J Ind Microbiol Biotechnol 33: 701-706.

48. Gumel AM, Annuar MS, Heidelberg T (2012) Biosynthesis and characterization of polyhydroxyalkanoates copolymers produced by Pseudomonas putida Bet001 isolated from palm oil mill effluent. PLoS One 7: e45214.

49. Escapa IF, del Cerro C, García JL, Prieto MA (2013) The role of GlpR repressor in Pseudomonas putida KT2440 growth and PHA production from glycerol. Environ Microbiol 15: 93-110.

50. Le Meur S, Zinn M, Egli T, Thöny-Meyer L, Ren Q (2012) Production of medium-chain-length polyhydroxyalkanoates by sequential feeding of xylose and octanoic acid in engineered Pseudomonas putida KT2440. BMC Biotechnol 12: 53.

51. Jiang X, Sun Z, Marchessault RH, Ramsay JA, Ramsay BA (2012) Biosynthesis and properties of medium-chain-length polyhydroxyalkanoates with enriched content of the dominant monomer. Biomacromolecules 13: 2926-2932.

52. Chung MG, Rhee YH (2012) Overexpression of the (R)-specific enoyl-CoA hydratase gene from Pseudomonas chlororaphis HS21 in Pseudomonas strains for the biosynthesis of polyhydroxyalkanoates of altered monomer composition. Biosci Biotechnol Biochem 76: 613-616.

53. Mozejko J, Wilke A, PrzybyÅ‚ek G, Ciesielski S (2012) Mcl-PHAs produced by Pseudomonas sp. Gl01 using fed-batch cultivation with waste rapeseed oil as carbon source. J Microbiol Biotechnol 22: 371-377.

54. Park SJ, Kim TW, Kim MK, Lee SY, Lim SC (2012) Advanced bacterial polyhydroxyalkanoates: towards a versatile and sustainable platform for unnatural tailor-made polyesters. Biotechnol Adv 30: 1196-1206.

55. Saxena A, Tiwari A (2011) Polyhydroxyalkonates: Green Plastics of the Future. International Journal of Biomedical and Advance Research 2: 356-367.

56. El-sayed AA, Abdel Hafez AM, Hemmat, Abdelhady M, Khodair TA (2009) Poduction of Polyhydroxybutyrate (PHB) Using Batch and Two-stage Batch Culture Strategies. Australian Journal of Basic and Applied Sciences 3: 617-627.

57. Berezina N (2013) Novel approach for productivity enhancement of polyhydroxyalkanoates (PHA) production by Cupriavidus necator DSM 545. N Biotechnol 30: 192-195.

58. Palleroni NJ, Palleroni AV (1978) Alcaligenes latus, a new species of hydrogen-utilizing bacteria. Int J Syst Bacteriol 28: 416-424.

59. Chen GGQ (2010) Plastics from Bacteria: Natural Functions and Applications (Microbiology Monographs). Springer USA.

60. Tamer IM, Moo-Young M, Chisti Y (1998) Disruption of Alcaligenes latus for Recovery of Poly(β-hydroxybutyric acid): Comparison of High-Pressure Homogenization, Bead Milling, and Chemically Induced Lysis. Ind Eng Chem Res 37: 1807-1814.

61. Azhar ESA, Abdelhady HM, Hafez AMA, Khodair TA (2009) Batch production of polyhydroxybutyrate (PHB) by Ralstonia eutropha and Alcaligenes latus using bioreactor different culture strategies. Journal of Applied Sciences Research 5: 556-564.

62. Yan Q, Du G, Chen J (2003) Biosynthesis of polyhydroxyalkanoates (PHAs) with continuous feeding of mixed organic acids as carbon sources by Ralstonia eutropha. Process Biochem 39: 387-391.

63. Hoskisson PA, Hobbs G (2005) Continuous culture--making a comeback? Microbiology 151: 3153-3159.

64. Poblete-Castro I, Escapa IF, Jäger C, Puchalka J, Lam CM, et al. (2012) The metabolic response of P. putida KT2442 producing high levels of polyhydroxyalkanoate under single- and multiple-nutrient-limited growth: highlights from a multi-level omics approach. Microb Cell Fact 11: 34.

65. Jung K, Hazenberg W, Prieto M, Witholt B (2001) Two-stage continuous process development for the production of medium-chain-length poly(3-hydroxyalkanoates). Biotechnol Bioeng 72: 19-24.

66. Preethi R, Sasikala P, Aravind J (2012) Microbial production of polyhydroxyalkanoate (PHA) utilizing fruit waste as a substrate. Research in Biotechnology 3: 61-69.

67. Sivasankar B (2005) Biosperations: Principles and Techniques. PHI Learning New Delhi India.

68. Divyashree MS, Shamala TR, Rastogi NK (2009) Isolation of polyhydroxyalkanoate from hydrolyzed cells of Bacillus flexus using aqueous two-phase system containing polyethylene glycol and phosphate. Biotechnology and Bioprocess Engineering 14: 482-489.

69. Kunasundari B, Sudesh K (2011) Isolation and recovery of microbial polyhydroxyalkanoates. express polym lett 5: 620-634.

70. Diels AM, Michiels CW (2006) High-pressure homogenization as a non-thermal technique for the inactivation of microorganisms. Crit Rev Microbiol 32: 201-216.

71. Choi J, Lee SY (1999) Efficient and economical recovery of poly(3-hydroxybutyrate) from recombinant Escherichia coli by simple digestion with chemicals. Biotechnol Bioeng 62: 546-553.

72. Yang YH, Brigham C, Willis L, Rha C, Sinskey A (2011) Improved detergent-based recovery of polyhydroxyalkanoates (PHAs). Biotechnol Lett 33: 937-942.

73. Valappil SP, Misra SK, Boccaccini AR, Keshavarz T, Bucke C, et al. (2007) Large-scale production and efficient recovery of PHB with desirable material properties, from the newly characterised Bacillus cereus SPV. J Biotechnol 132: 251-258.

74. Fiorese ML, Freitas F, Pais J, Ramos AM, de Aragão GMF, et al. (2009) Recovery of polyhydroxybutyrate (PHB) from Cupriavidus necator biomass by solvent extraction with 1,2-propylene carbonate. Engineering in Life Sciences 9: 454-461.

75. Kathiresan Y, Aroua MK, Ramachandran KB (2011) Chemical Characterization of medium-chain-length polyhydroxyalkanoates (PHAs) recovered by enzymatic treatment and ultrafiltration. J Chem Technol Biot 82: 847-855.

76. Kapritchkoff FM, Viotti AP, Alli RC, Zuccolo M, Pradella JG, et al. (2006) Enzymatic recovery and purification of polyhydroxybutyrate produced by Ralstonia eutropha. J Biotechnol 122: 453-462.

77. Yasotha K, Aroua MK, Ramachandran KB, Tan IKP (2006) Recovery of medium-chain-length polyhydroxyalkanoates (PHAs) through enzymatic digestion treatments and ultrafiltration. Biochem Eng J 30: 260-268.

78. Riedel SL, Brigham CJ, Budde CF, Bader J, Rha C, et al. (2013) Recovery of poly(3-hydroxybutyrate-co-3-hydroxyhexanoate) from Ralstonia eutropha cultures with non-halogenated solvents. Biotechnol Bioeng 110: 461-470.

79. Cavalheiro JM, Raposo RS, de Almeida MC, Cesário MT, Sevrin C, et al. (2012) Effect of cultivation parameters on the production of poly(3-hydroxybutyrate-co-4-hydroxybutyrate) and poly(3-hydroxybutyrate-4-hydroxybutyrate-3-hydroxyvalerate) by Cupriavidus necator using waste glycerol. Bioresour Technol 111: 391-397.

80. Saranya V, Poornimakkani, Krishnakumari MS, Suguna P, Binuramesh C, et al. (2012) Quantification of intracellular polyhydroxyalkanoates by virtue of personalized flow cytometry protocol. Curr Microbiol 65: 589-594.

Supramolecular Structures of Oil Systems as the Key to Regulation of Oil Behavior

Tukhvatullina AZ[1]*, Barskaya EE[1], Kouryakov VN[2], Ganeeva YM[1], Yusupova TN[1] and Romanov GV[1]

[1]*AE Arbuzov Institute of Organic and Physical Chemistry of Kazan Scientific Center of Russian Academy of Sciences, Kazan, Russia*
[2]*Oil and Gas Research Institute of Russian Academy of Sciences, Moscow, Russia*

Abstract

The paper is devoted to investigation of supramolecular structures formed by high-molecular components of oil – asphaltenes and waxes. Petroleum asphaltenes that are characterized by their ability to self-organization can be considered as a polydisperse system. The aggregation of A1and A2 asphaltenes both in toluene solution and in diluted heavy oils was studied by dynamic light scattering technique. The stabilizing role of resins was confirmed. Due to the fact that the supramolecular structures in oils can be also formed by high molecular microcrystalline waxes, their influence as well as A1 and A2 asphaltenes on wax crystallization process was evaluated using optical microscopy. The effect of the different molecular structures of asphaltenes (A1 and A2) on the rheological properties of crude oils was also observed. Self-organization of asphaltenes directly in heavy high-viscosity oils with high content of resins was investigated by varying the ratios oil:precipitant. It was shown that extremely stable supramolecular structures in heavy oils had been formed. The influence of ultrasonic action for the destruction of stable supramolecular structures and the reduction of oil viscosity were considered.

Also petroleum asphaltenes can be the indicators of many processes that occur in the reservoir. We have shown the possibility of using the gradient of asphaltenes, as well as their structure, to confirm the influx of deep light hydrocarbons in carbonate reservoirs. Thus, all obtained results of investigation of structure formation processes will let regulate the macroscopic properties of heavy oils in the future and identify the processes occurring in the reservoir.

Keywords: Supramolecular structures; Disperse system; Asphaltenes; Asphaltene aggregation

Introduction

According to the modern concepts, the oil is a complex multi-component disperse system. The theory of oil disperse systems [1-4] supposes that the liquid phase compounds are oriented around the supramolecular structures forming solvation layers. The disperse particles are characterized by a complex structure in this case: the core formed by high-molecular components of oil (asphaltenes, microcrystalline waxes) and the solvation shell surrounding the core and consisting of resins. The disperse media in such model is presented by hydrocarbons of oils. At present, the focused attention is paid to the most high-molecular-mass, polar oil components-asphaltenes [5-30]. Asphaltenes are characterized by their ability to self-organization (or self assembly), and therefore the presence of a hierarchy of structures in oil media: molecules-nanoaggregates-clusters of nanoaggregates [5-9] (Figure 1). That is why asphaltenes can be attributed to the objects of supramolecular chemistry.

Petroleum asphaltenes can be considered as a complex polydisperse system of molecules in which subcomponents with significantly different characteristics can be distinguished as the result of the fractionation procedure [27]. At present, two asphaltene molecule types-"continental" (A1 fraction) and "archipelago" (A2 fraction) are adopted as the basic models (Figure 2) [10-15,26]. It should be noted that in crude oils asphaltene aggregates of both types exist in various proportions. As the asphaltene fractionation procedure is rather time and labor consuming, it is not possible to isolate absolutely pure fractions of asphaltenes but concentrates enriched in asphaltenes of one or another structure [26,27].

The main difference between A1 and A2 asphaltenes is their structure determining different ability to aggregate. A1 asphaltene molecules even at low concentrations form aggregates, followed by the introduction of A2 asphaltene molecules into these units [13]. Thus, A1 and A2 asphaltene molecules in oils form a disperse phase where the core is presented by stacks of the "continental" type of asphaltenes and the periphery rich in the "archipelago" type of asphaltenes [14,15].

In recent scientific literature, there are many studies on the self-organization of asphaltenes in solutions of solvents [28-30]. However, the study of the asphaltenes behavior directly in the oil system is of a much greater interest, since regularities obtained in solutions do not reflect the nature and completeness of the processes that occur in real oils. Therefore we have studied aggregation of A1 and A2 asphaltenes both in toluene solution and in diluted heavy oils with concentrations that are close to real oils by dynamic light scattering technique. The concentration of solutions is also limited by the sensitivity of instruments. The effect of the different molecular structures of asphaltenes was also shown by measuring rheological properties of crude oils.

***Corresponding author:** Tukhvatullina AZ, AE Arbuzov Institute of Organic and Physical Chemistry of Kazan Scientific Center of Russian Academy of Sciences, Kazan, Russia

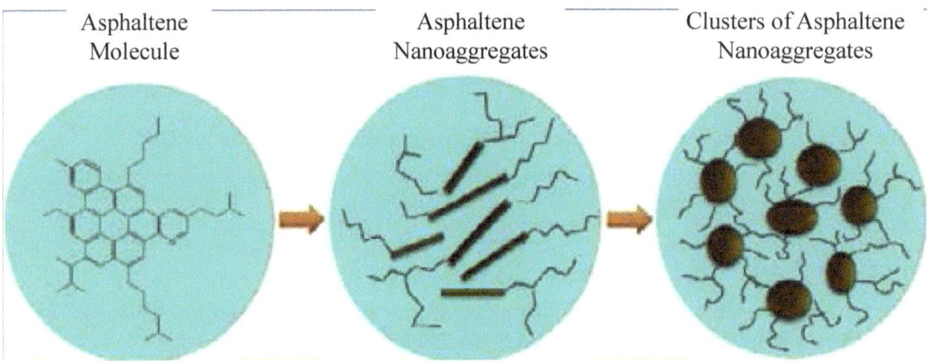

Figure 1: A modified Yen model (or Yen-Mullins model of asphaltenes) [5].

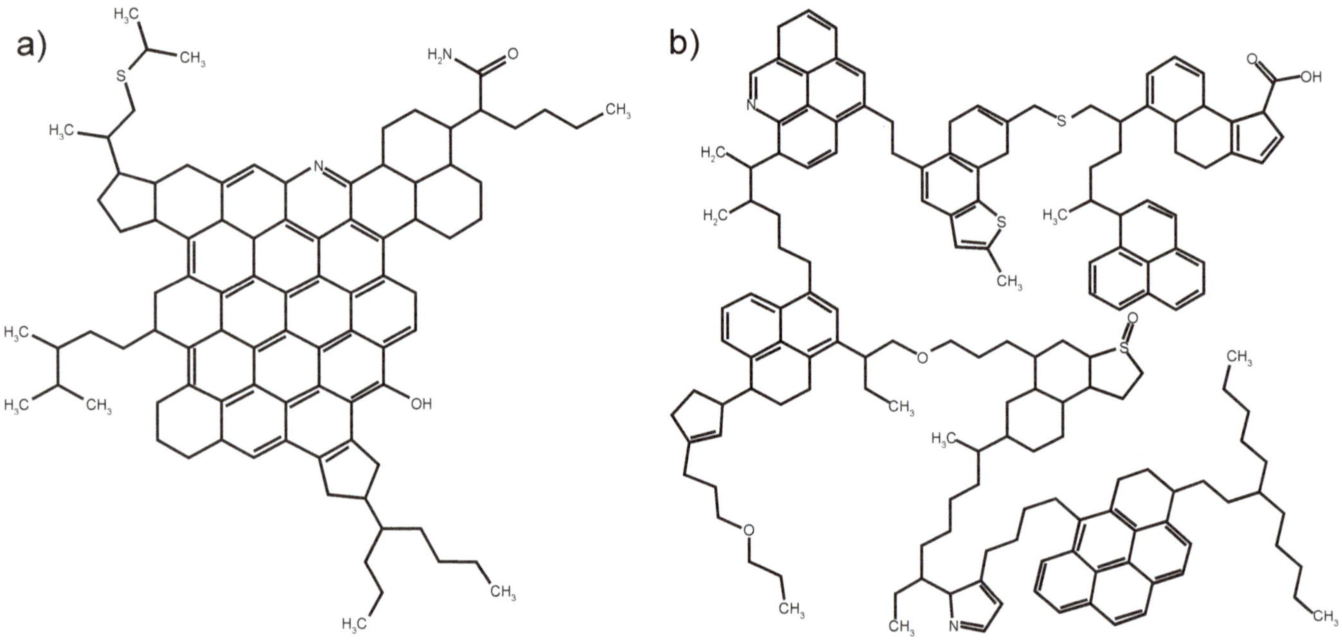

Figure 2: Model of asphaltene molecules: a) the "continental" type, and b) the "archipelago" type [11].

Method

The objects of study were heavy high-viscosity crude oils from carbonates of the Akanskoe field of Tatarstan Republic, Russia. It should be noted that oil samples from one of the wells have been specially selected from the upper (top) and bottom parts of the reservoir [31].

Asphaltenes under investigation were precipitated from the crude oils with a 40-fold excess of nhexane followed by separation into three asphaltene fractions according to their ability to form the aggregates [12]. It is assumed that the first asphaltene fraction that is precipitated under the minimal addition of the precipitant, is a concentrate of A1 asphaltene molecules (the "continental" type), while the third fraction, which remains in solution, presented by A2 asphaltene molecules (the "archipelago" type). The intermediate fraction is a mixture of A1 and A2 asphaltene molecules [12].

The asphaltene aggregation was studied by photon correlation spectroscopy (dynamic light scattering). Toluene-heptane system used is a widely accepted model system for studying the properties of the asphaltenes [32-34].

The photomicrographs of oils and model systems were obtained using Olympus CX41 microscope with a digital photomicroscopy system based on Olympus E420 digital camera and ImageScope M software. Photomicrographs were obtained in transmitted light between crossed polaroids at 100-400 x magnifications.

Thermal analysis (TA) of asphaltenes and their fractions was performed with a MOM Q-1500D differential thermal analyzer (Hungary) over the temperature range of 20-1000°C at an oven heating rate of 10°C/min. The chemical composition of asphaltenes was characterized by the mass loss values at different stages of thermal oxidative degradation (20-410°C, 410-530°C and 530-700°C) [35].

The presence of waxes in asphaltenes was fixed by high-temperature gas-liquid chromatography technique (GLC) on a Perkin-Elmer chromatograph with a flame-ionization detector in the mode of temperature programming from 20 to 360°C.

Electron Paramagnetic Resonance (EPR) spectrum of asphaltenes and their fractions were obtained at room temperature with a spectrometer type SE/X-2544. The content of stable free radicals (IR) was estimated by the amplitude of a single line in the center of the EPR spectrum. Complex viscosity measurements were carried out using Rheotest RN4.1 (MessgeräteMedingen GmbH) over temperature range 20-80°C.

Oil processing by ultrasound was performed using an ultrasonic apparatus "Alena" [36] with the frequency of 22 kHz. The intensity of ultrasound was regulated from 10 to 100 W/cm^2. Exposure time was from 1.5 to 20 minutes [37].

Results

One of the most promising techniques to monitor in real time the kinetics of asphaltene aggregation and measure a changing size of the aggregates is the dynamic light scattering (DLS) [32-34]. It allows determining the value of the threshold concentration of precipitant that characterizes so called "zero rate of aggregation" -the stability threshold of system. The aggregation of initial asphaltenes and their fractions were studied in toluene solutions, diluted heavy oils and model systems by dynamic light scattering technique. Initial asphaltenes were obtained from the crude oils from the top and bottom parts of the reservoir. In order to identify the role of asphaltene molecules-A1 and A2, the model mixtures from deasphalted oil with different asphaltene fractions were prepared.

The kinetics of asphaltene aggregation in toluene solutions with asphaltene concentration of 0.1 g/l is described by the diffusion-limited mechanism, i.e. aggregation rate is mainly determined by diffusion of particles. The changes in the aggregates sizes are shown in Figure 3. The rate of aggregation increases significantly with increasing of precipitating agent volume. For each volume of precipitant aggregate, size increases with time according to a power law.

The values of the stability threshold were determined for the majority of solutions of asphaltenes and their fractions with a concentration of 0.1 g/l and 5 g/l (Table 1). The solutions with the concentration of asphaltenes 5 g/l in toluene were investigated at installation with the Infrared (IR) laser. The wavelength of IR laser was 980 nm; the wavelength of red laser was 632 nm. In the solutions of A2 asphaltenes with concentration of 0.1 g/l the formation of aggregates was not observed up to a precipitant concentration of 90% (by weight). The aggregation of A2 asphaltenes was fixed only for solutions with a concentration of 5 g/l and to a much lesser extent than in solutions of A1 asphaltenes even with a concentration of 0.1 g/l.

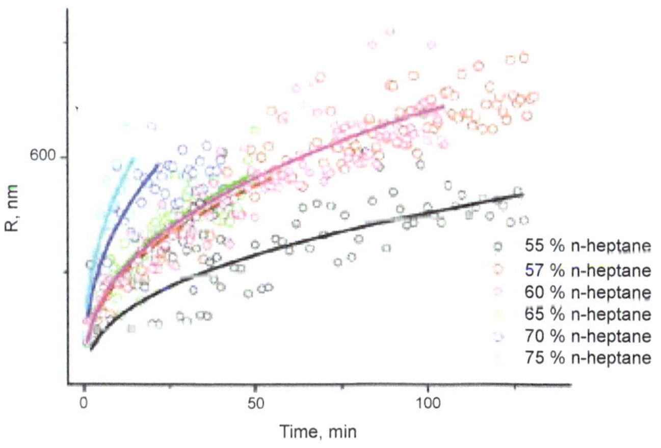

Figure 3: Aggregation kinetics of 0.1g/l asphaltene solution for various heptane volumes.
R-the size of aggregates, nm

No.	Sample	The threshold concentration of heptane,% wt. (for 0.1g/l asphaltenes/ toluenesolutions)	The threshold concentration of heptane,% wt. (for 5g/l asphaltenes/toluene solutions)
1	Initial asphaltenes (top part)	46	40
2	Initial asphaltenes (bottom part)	48	46.5
3	Asphaltenes A1 (top part)	43	25
4	Asphaltenes A1 (bottom part)	28	20.2
5	Asphaltenes A2 (top part)	-	69
6	Asphaltenes A2 (bottom part)	-	55

Table 1: The threshold concentrations of heptane for 0.1g/l and 5g/l solutions of asphaltenes and their fractions in toluene.

In toluene solutions with a low asphaltene concentration the stability threshold of asphaltenes of oils from top and bottom parts of the carbonate reservoir is almost the same despite the different content of A1 and A2 fractions (which differ by their stability) in initial asphaltenes. Initial asphaltenes of oil from the bottom of the reservoir are enriched in A1 asphaltenes (64.6%) and depleted in A2 asphaltenes (8.1%) while asphaltenes from the top of the reservoir contains less A1 asphaltenes (49.4%) and twice the amount of A2 asphaltenes (15.7%) [31]. The rest part of asphaltenes is presented by the intermediate fraction-a mixture of A1 and A2 asphaltenes [12]. This may be stipulated by the fact that asphaltenes in both cases exist only as nanoaggregates in solutions with a low asphaltene concentration.

Further research was devoted to investigation of asphaltene aggregation in diluted heavy oils as well as model systems with the concentration of 125 g/l by weight in toluene. Model systems are the mixtures of deasphalted oils (maltenes) supplemented with 5%(15%) of different asphaltene fractions (A1 or A2). Solutions with the concentration of 500 g/l by weight in toluene were also prepared for crude oils (ratio 1:1 by weight). The threshold stability of oil systems was defined by titration.

Thus, the method of dynamic light scattering showed a greater capacity of asphaltenes enriched in A1 molecules for aggregation in the oil systems. With increasing concentration of A1 asphaltenes from 5 to 15% the threshold stability is not so much reduced as it could be expected, while the viscosity increases from 1020 mPs-s to 7185 mPs-s (the shear rate was 11s-1.)

Probably, the increasing concentration of A1 asphaltenes leads to the formation of coagulation structures which impede their further additional aggregation of asphaltenes.

An interesting fact is that the stability threshold of crude oils is only slightly dependent on the volume of added toluene (Table 2). This suggests a slight influence of the solvent proportion (1:1 or 1:4) on the process of asphaltene aggregation, particularly the threshold concentration of heptane, despite the fact that viscosity changes are essential.

It was found that aggregation of asphaltenes and A1 and A2 fractions in toluene solution has a higher rate in comparison with the aggregation in diluted oils and model systems (Table 2), confirming the stabilizing role of components of oil systems. According to the literature [19-22], resins are the stabilization factor of asphaltenes in

No.	Sample	The concentration of solution, g/l	The threshold concentration of n-heptane, % wt.
1	Oil (top part)	500	60.8
2	Oil (bottom part)	500	57.5
3	Oil (top part)	125	61.4
4	Oil (bottom part)	125	59.7
5	Maltenes + A1 asphaltenes (5%) (top part)	125	48.1
6	Maltenes + A1 asphaltenes (15%) top part)	125	43.3
7	Maltenes + A2 asphaltenes (5%) (top part)	125	77.4
8	Maltenes + A1 asphaltenes (5%) (bottom part)	125	44.9
9	Maltenes + A2 asphaltenes (5%) (bottom part)	125	69.5

Table 2: The threshold concentration of heptane for solutions of oils and model systems.

crude oils, due to their peptizing action. However, there is virtually no experimental data quantifying the role of resins.

To study the effect of resins on the stability of the hydrocarbon fluids the kinetics of asphaltene aggregation in toluene-heptane system was investigated during addition of resins with three different concentrations: 0.2 g/l, 0.6 g/l and 1 g/l. The concentration of asphaltenes in all cases was 0.1 g/l. For ease of comparison, the kinetics of growth of aggregates was measured at concentrations of n-heptane equal to 60, 65, 70, 75 and 80 weight percent. In all cases the same power-law dependence of the radius on time as that in the absence of resins was observed, i.e. for solutions with asphaltene concentration of 0.1 g/l the diffusion limited aggregation is observed.

Figure 4 shows the measurement results of the radius of asphaltene aggregates in the solutions containing different amounts of resins at a fixed (70 weight percent) concentration of n-heptane. The results obtained for other n-heptane concentrations (60, 65, 75 and 80 weight percent) are quite similar.

It was shown that an increase in resins concentration leads to the decrease of the rate of aggregates growth as well as their amount. Furthermore, the introduction of resins into the asphaltene solution increases the threshold concentration of n-heptane. In the field of hydrocarbon solutions instability with respect to precipitation of heavy fractions, addition of resins leads to a significant slowdown in the aggregation reaction. Thus, the stabilizing role of resins is substantiated.

Due to the fact that the supramolecular structures in oils can be also formed by high molecular microcrystalline waxes, their influence as well as influence of asphaltene fractions on wax crystallization process was evaluated using optical microscopy. Figure 5 shows photomicrographs of oils and model systems. Observation between crossed polaroids allows to characterize the distribution and size of the wax crystallites in the samples.

Maltenes (deasphalted oil) from the oil of the top part of the reservoir contain wax crystallites that are large enough compared with the original oil (Figure 5). This can be explained by the procedure of maltenes separation resulting in removing gasoline fractions that serve as a solvent of waxes, and asphaltenes inhibiting the wax crystallization process from crude oils. In maltenes from the oil of the bottom of the reservoir the crystal growth is not observed due to the fact that the oil is enriched in microcrystalline waxes with carbon atoms number from 40 to 60 (C40-C60), which are likely to disperse macrocrystalline waxes (C16-C40) [31]. The content of microcrystalline waxes in the oil from the top of the reservoir is substantially lower but the content of macrocrystalline waxes is higher. The dispersion of the crystalline

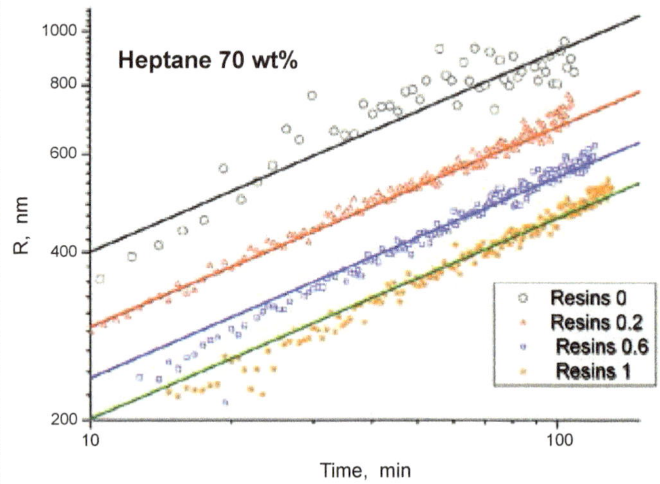

Figure 4: Aggregation kinetics of asphaltene solution for various resins amount.
R-the size of aggregates, nm

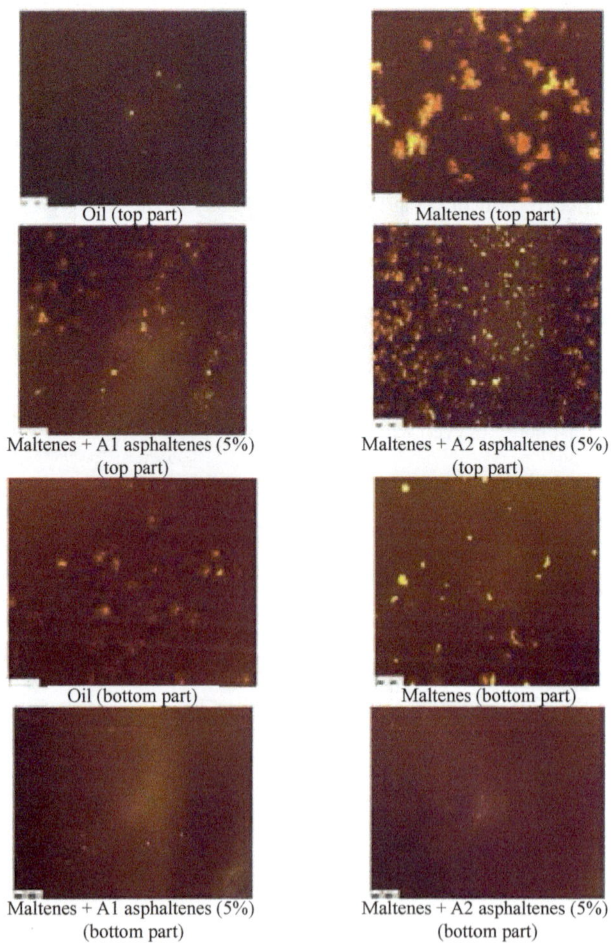

Figure 5: Photomicrographs of oils and model systems obtained by optical microscope.

phase can also be observed when asphaltenes (A1 and A2 fractions) are added to maltenes. This finding confirms the role of asphaltenes as the inhibitors of wax crystallization [23,24]. Moreover, the ntroduction of A2 asphaltenes in maltenes leads to more intensive dispersion.

Self-organization of asphaltenes directly in heavy high-viscosity oils with high content of resins was investigated by varying the ratios oil:precipitant [38]. It was found that the quantitative content of precipitating asphaltene aggregates in heavy oils from carbonate reservoirs is not dependent on the concentration of the first portion of the precipitant-n-alkane (10, 20, 30%) (Table 3). The thermal analysis data (Δm^3-mass loss values at very high temperatures-530-700°C) [35] and EPR spectroscopy data (content of stable free radicals) showed that a decrease in the share of the first portion of the precipitant leads to reducing the content of polycondensed 8 aromatic structures in the asphaltene aggregates (Figures 6 and 7) due to co-precipitation of waxes according to GLC data (Figure 8). The purest separation of asphaltenes is observed during a conventional technique of asphaltene precipitation in a 40-fold excess of n-alkane.

The effective technique to fix the formation/destruction of supramolecular structures in oils is rheology method. The capacity of high-viscosity oils from carbonate reservoirs to structure forming at higher temperatures was shown. In order to identify the effect of asphaltene molecules-propagators (A1) and terminators (A2) of aggregation on the oil structure, the study of model mixtures from deasphalted oil with different asphaltene fractions was carried out.

The phase transition temperatures were identified (Figure 9). At low temperatures (20-30°C) according to the high values of the activation energy of viscous flow oil systems are characterized by coagulation structure (so called the bound-dispersed state). Since investigated oil systems contain waxes, the phase transition at 30-60°C can be attributed to the process of destruction of the crystalline phase.

Sample	The ratio oil:precipitant upon addition the first portion of the precipitant									
	1:40	1:30			1:20			1:10		
		I fr.,%	II fr.,%	Σ,%	I fr.,%	II fr.,%	Σ,%	I fr.,%	II fr., %	Σ,%
Crude oil	11.0-12.0	11.9	0.1	12	12	0.2	12.2	11.5	0.5	12

Table 3: The content of asphaltenes obtained from crude oils by two-step precipitation and standard precipitation techniques.

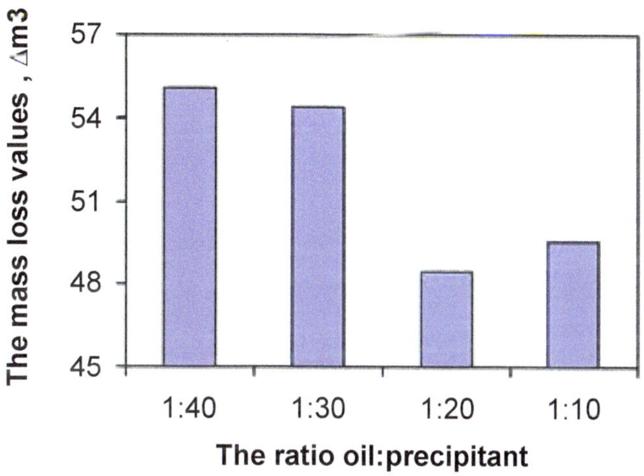

Figure 6: The thermal analysis data of the first asphaltene fractions.

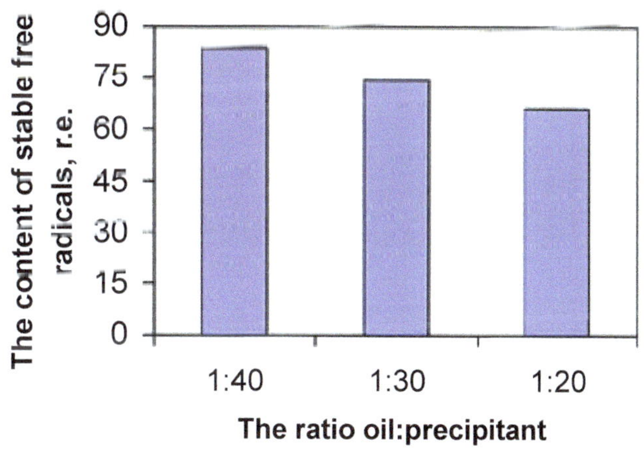

Figure 7: The EPR spectroscopy data of the first asphaltene fractions.

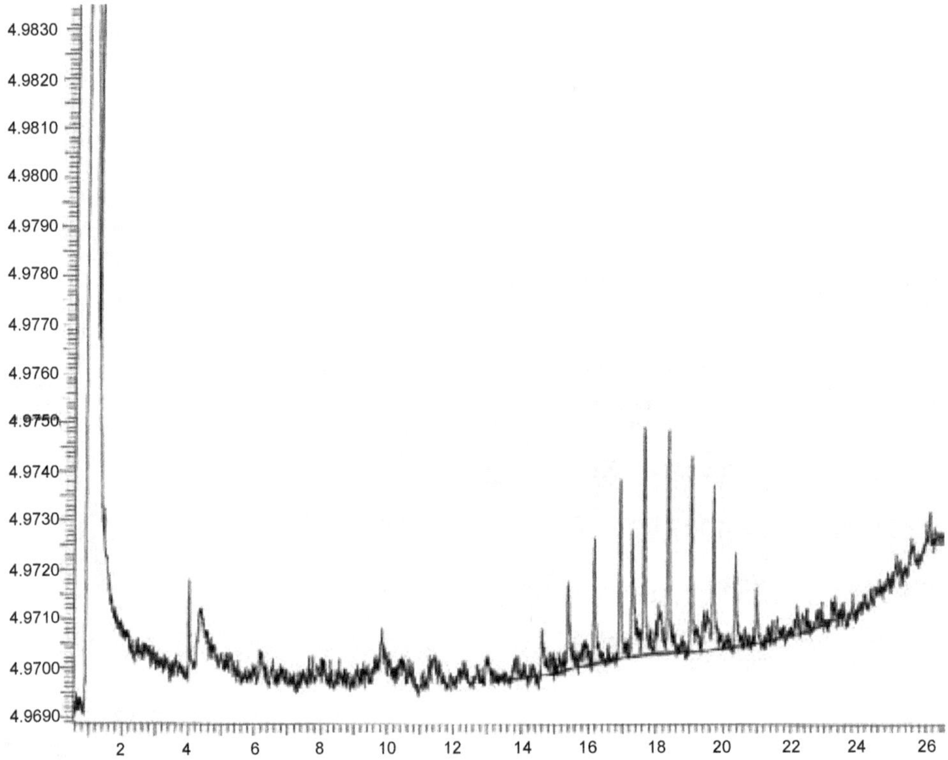

(a)

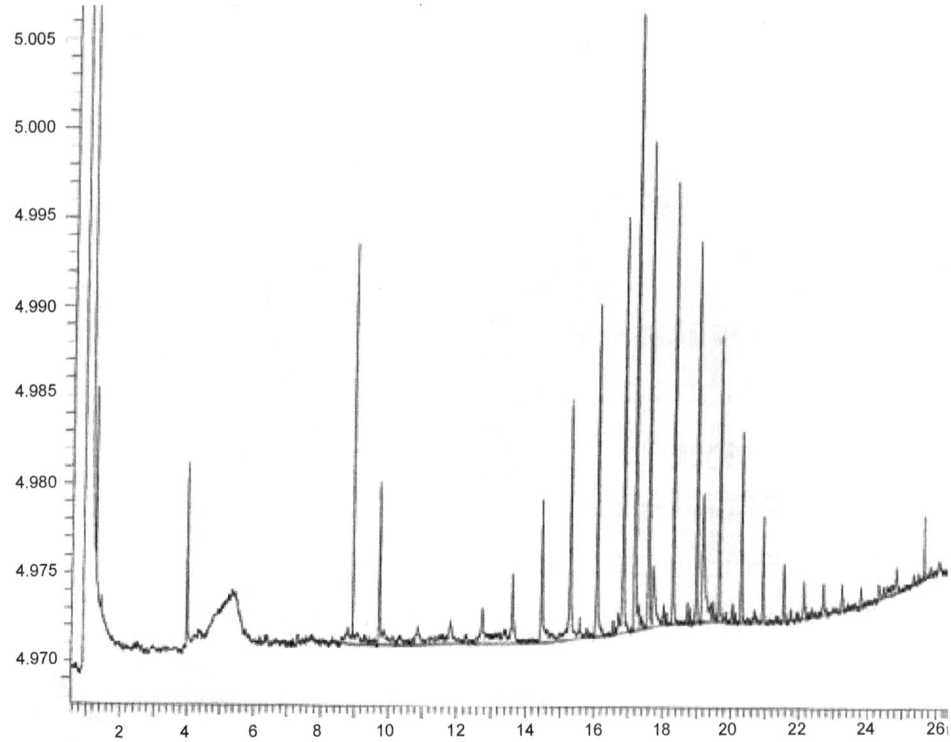

(b)

Figure 8: Chromatograms of asphaltenes separated varying ratio oil:precipitant: a) 1:10, b) 1:40.

At a temperature of 60°C or higher there is a transition into the freely dispersed state with low values of the activation energy of viscous flow.

It was found that A1 asphaltenes contribute to the formation of the most stable bound-dispersed structure of oils, which greatly increases their viscosity. A2 asphaltenes disperse waxes greater than A1 asphaltenes, preventing the formation of the large crystals (Figure 5). Comparative analysis of the oil residue after distilling of the low-boiling fractions and maltenes with A1 asphaltenes it was shown that the stabilizing effect of resins in the absence of A2 asphaltenes is less pronounced.

Since the stable supramolecular structures are formed in heavy oils, it is necessary to apply technologies that affect directly to the core of the oil dispersed systems in order to reduce their viscosity. One of such technologies is the ultrasonic action [39-43]. The regularities of the influence of ultrasonic treatment on the structure of heavy oil with high content of resins were identified. It was found that ultrasound treatment leads to a decrease of oil viscosity (Figure 10), which is associated with changes in their dispersion structure [37].

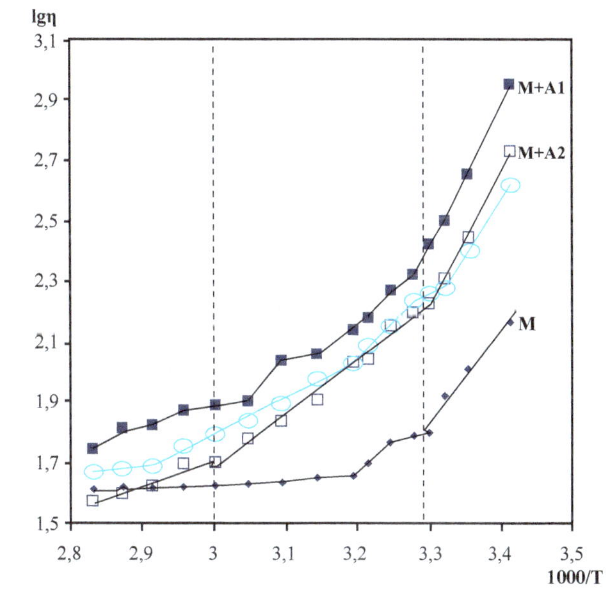

Figure 9: Viscosity-temperature dependence of the maltenes (M) and model systems (M+A1, M+A2).

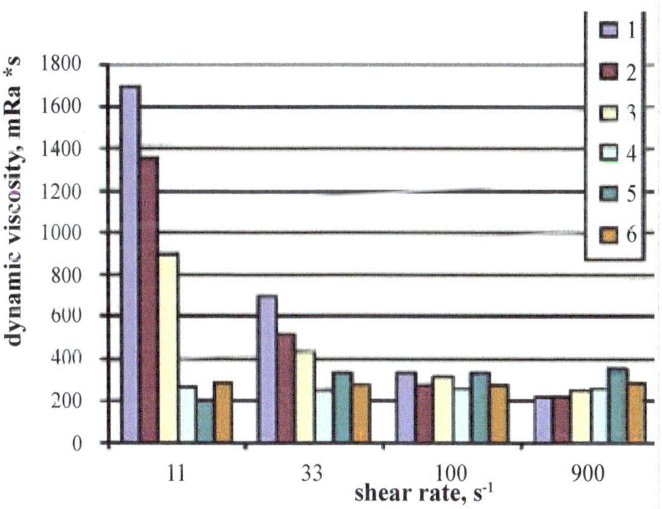

Figure 10: Change of the viscosity of the highly resinous oil before (1) and after sonication, depending on the intensity: 2-10 W/cm²; 3-25 W/cm²; 4-30 W/cm²; 5-50 W/cm²; 6-100 W/cm²

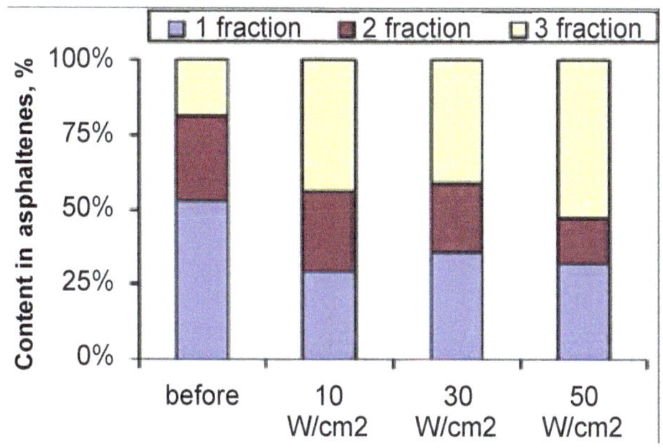

Figure 11: Change of the fractional composition of the asphaltenes from the highly resinous oil before and after sonication, depending on the intensity.

The destruction of A1 and A2 asphaltene aggregates by sonication is proved by comparative data of asphaltenes fractionation from oil before and after sonication (Figure 11). The high content of A1 asphaltenes before ultrasound treatment can be explained by segregation of stable aggregates of A1 and A2 asphaltenes. After processing it is possible to isolate a more pure fraction of A1 asphaltenes according to EPR data (the content of stable free radicals: before sonication is 80 r.e., after sonication–more than 90 r.e.), the amount of which is markedly reduced. This effectively reduces the oil viscosity.

Asphaltenes are the indicators of many processes that occur in the reservoir. Previously [31] we have shown the possibility of using the gradient of asphaltenes, as well as their structure, to confirm the influx of deep light hydrocarbons in carbonate reservoirs.

Thus, all obtained results of investigation of structure formation processes will let regulate the macroscopic properties of heavy oils in the future and identify the processes occurring in the reservoir.

Acknowledgements

Authors would like to thank the RFBR Foundation for financial support (grants I.D. 12 03-00487 and research project No. 12-05-31345) and the Academy of Sciences of the Republic of Tatarstan (grant No. 03-47/2013 (G)).

References

1. Syunyaev RZ, Safieva RZ (2012) Oil disperse systems: softness, nanostructure, hierarchy, phasic behavior. Georesources 3: 39-40.

2. Syunyaev ZI, Syunyaev RZ, Safieva RZ (1990) Oil disperse systems. Khimiya, Moscow, Russia.

3. Safieva RZ (1998) Physical chemistry of crude oil: Psysicochemical foundations of petroleum refining technology. Khimiya, Moscow, Russia.

4. Tumanyan BP (2000) Scientific and Applied Aspects of the Theory of Petroleum Disperse Systems. Tekhnika, OOO TUMAgrupp, Moscow, Russia.

5. Mullins OC (2010) The Modified Yen Model. Energy Fuels 24: 2179-2207.

6. Mullins OC, Seifert DJ, Zuo JY, Zeybek M (2013) Clusters of Asphaltene Nanoaggregates Observed in Oilfield Reservoirs. Energy Fuels 27: 1752-1761.

7. Mullins OC, Betancourt SS, Cribbs ME, Dubost FX, Creek JL, et al. (2007) The colloidal structure of crude oil and the structure of oil reservoirs. Energy Fuels 21: 2785-2794.

8. Zuo JY, Mullins OC, Freed D, Zhang D, Dong C, et al. (2011) Analysis of downhole asphaltene gradients in oil reservoirs with a new bimodal asphaltene distribution function. J Chem Eng Data 56: 1047-1058.

9. Mullins OC, Zuo JY, Freed DE, Elshahawi H, Cribbs ME, et al. (2010) Downhole fluid analysis coupled with novel asphaltene science for reservoir evaluation. 51th Annual logging symposium, SPWLA, USA.

10. Pacheco-Sanchez JH, Alvarez-Ramirez F, Martinez-Magadan JM (2004) Morphology of aggregated asphaltene structural models. Energy Fuels 18: 1676-1686.

11. Rahimi H, Nazar ARS (2010) Asphaltene aggregates fractal restructuring model, a population balance approach. Energy Fuels 24: 1088-1093.

12. Tojima M, Suhara S, Imamura M, Furuta A (1998) Effect of heavy asphaltene on stability of residual oil. Catalysis Today 43: 347-351.

13. Acevedo S, Garcia LA (2012) Changes of diameter distribution with temperature measured for asphaltenes and their fractions A1 and A2. Impact of these measurements in colloidal and solubility issues of asphaltenes. Energy Fuels 26: 1814-1819.

14. Acevedo S, Castro A, Vasquez E, Marcano F, Ranaudo MA (2010) Investigation of Physical Chemistry Properties of Asphaltenes Using Solubility Parameters of Asphaltenes and Their Fractions A1 and A2. Energy Fuels 24: 5921-5933.

15. Marcano F, Flores R, Chirinos J, Ranaudo MA (2011) Distribution of Ni and V in A1 and A2 Asphaltene Fractions in Stable and Unstable Venezuelan Crude Oils. Energy Fuels 25: 2137-2141.

16. Buckley J (2012) Asphaltene deposition. Energy Fuels 26: 4086-4090.

17. Liao Z, Zhou H, Graciaa A, Chrostowska A, Creux P, et al. (2005) Adsorption/ occlusion characteristics of asphaltenes: some implication for asphaltene structural features. Energy Fuels 19: 180-186.

18. Derakhshesh M, Bergmann A, Gray MR (2013) Occlusion of polyaromatic compounds in asphaltene precipitates suggests porous nanoaggregates. Energy Fuels 27: 1748-1751.

19. Rogel E (2008) Molecular thermodynamic approach to the formation of mixed asphaltene-resin aggregates. Energy Fuels 22: 3922-3929.

20. Rogel E (2011) Effect of Inhibitors on Asphaltene Aggregation: A Theoretical Framework. Energy Fuels 25: 472-481.

21. Murgich J, Abanero JA (1999) Molecular recognition in aggregates formed by asphaltene and resin molecules from the Athabasca oil sand. Energy Fuels 13: 278-286.

22. Ortega-Rodriguez A, Cruz SA, Gil-Villegas A, Guevara-Rodrigues F, Lira-Galeana C (2003) Molecular view of the asphaltene aggregation behavior in asphaltene-resin mixtures. Energy Fuels 17: 1100-1108.

23. Kriz P, Anderson SI (2005) Effect of asphaltenes on crude oil wax crystallization. Energy Fuels 19: 948-953.

24. Oh K, Deo M (2009) Characteristics of wax gel formation in the presence of asphaltenes. Energy Fuels 23: 1289-1293.

25. Garcia MC, Carbognani L (2001) Asphaltene-paraffin structural interactions. Effect on crude oil stability. Energy Fuels 15: 1021-1027.

26. Ganeeva YM, Yusupova TN, Romanov GV (2011) Asphaltene nano-aggregates: structure, phase transitions and effect on petroleum systems. Russ Chem Rev 80: 1034-1050.

27. Akhmetov BR, Evdokimov IN, Eliseev NY (2002) Some features of the supramolecular structures in petroleum media. Chemistry and Technology of Fuels and Oils 4: 266-270.

28. Mohammadi AH, Richon D (2008) Estimating Onset of Precipitation of Dissolved Asphaltene in the Solution of Solvent + Precipitant Using Artificial Neural Network Technique. The Open Thermodynamics Journal 2: 82-88.

29. Castillo J, Gutierrez H, Ranaudo M, Villarroel O (2010) Measurement of the Refractive Index of Crude Oil and Asphaltene Solutions: Onset Flocculation Determination. Energy Fuels 24: 492-495.

30. Castillo J, Hung J, Fernandez A, Mujica V (2001) Nonlinear optical evidences of aggregation in asphaltene–toluene solutions. Fuel 80: 1239-1243.

31. Yusupova TN, Ganeeva YM, Tukhvatullina AZ, Romanov GV, Muslimov RKh, et al. (2012) Composition of Oils in Vereiskian and Bashkirian Carbonate Reservoirs of the Akanskoe Oilfield in the Republic of Tatarstan. Petroleum Chemistry 4: 215-220.

32. Yudin IK, Nikolaenko GL, Gorodetskii EE, Kosov VI, Melikyan VR, et al. (1998) Mechanisms of asphaltene aggregation in toluene-heptane mixtures. Journal of Petroleum Science and Engineering 20: 297-301.

33. Yudin IK, Anisimov MA (2006) Dynamic light scattering monitoring of asphaltene aggregation in crude oils and hydrocarbon solutions. Asphaltenes, Heavy Oils and Petroleomics. Springer, New York, USA.

34. Burya YG, Yudin IK, Dechabo VA, Kosov VI, Anisimov MA (2001) Light-scattering study of petroleum asphaltene aggregation. Appl Opt 40: 4028-4035.

35. Yusupova TN, Petrova LM, Mukhametshin RZ, Romanov GV, Foss TR, et al. (1999) Distribution and Composition of Organic Matter in Oil- and Bitumen-cotaining Rocks in Deposits of Different Ages. J Therm Anal Calorim 55: 99-107.

36. http://u-sonic.ru/en/print/228

37. Barskaya EE, Tukhvatullina AZ, Ganeeva YM, Yusupova TN (2012) Influence of ultrasonication on the dispersed structure of the crude oils. 3rd International Conference on Elastic wave effect on fluid in porous media. Moscow, Russia.

38. Ganeeva YM, Yusupova TN, Morozov VI, Romanov GV (2013) Self-organization behavior of asphaltene molecules in heavy oils with varying the oil: precipitant ratio. Petrol Chem 53: 220-224.

39. Kirillov LB, Pivovarova N, Vlasov GV, Milyaev AP, Radyuk RG (2011) Development of new methods for the preparation of primary hydrocarbon. Res Innov 5: 96-98.

40. Klokova TP, Volodin YA, Glagoleva OF (2006) Effect of ultrasound on the colloidal-disperse properties of petroleum systems. Chemistry and Technology of Fuels and Oils 42: 43-46.

41. Volkova GI, Prozorova IV, Anufriev RV, Yudina NV, Mullakaev MS, et al. (2012) Ultrasonic treatment of oils for improvement of viscosity-temperature characteristics. Petroleum Refining and Petrochemistry 2: 3-6.

42. Loskutova YV, Prozorova IV, Yudina NV, Rikkonen SV (2004) Change in the rheological properties of high-paraffin petroleums under the action vibrojet magnetic activation. Journal of Engineering Physics and Thermophysics 77: 1034-1039.

43. Guoxiang Y, Xiaoping L, Pingfang H, Xuan S (2010) Desalting and dewatering of crude oil in ultrasonic standing wave field. Journal of Petroleum Science and Engineering 70: 140-144.

Surveying a Diverse Pool of Microalgae as a Bioresource for Future Biotechnological Applications

Gisela Jakob[1,2], Juliane Wolf[1,2], Tony VL Bui[1], Clemens Posten[3], Olaf Kruse[4], Evan Stephens[1,2], Ian L Ross[1,2] and Ben Hankamer[1,2]*

[1]Institute for Molecular Bioscience, The University of Queensland, Queensland, Australia
[2]Solar Biofuels Research Centre, The University of Queensland, Queensland, Australia
[3]Institute of Life Science Engineering, Bioprocess Engineering, University of Karlsruhe, Karlsruhe, BW D-76131, Germany
[4]Department of Biology, Algae Biotechnology & Bioenergy, University of Bielefeld, Bielefeld, NRW D-33501, Germany

Abstract

Resource limitation is an escalating concern given human expansion and development. Algae are increasingly recognised as a promising bioresource and the range of cultivated species and their products is expanding. Compared to terrestrial crops, microalgae are very biodiverse and offer considerable versatility for a range of biotechnological applications including the production of animal feeds, fuels, high value products and waste-water treatment. Despite their versatility and capacity for high biomass productivity on non-arable land, attempts to harness microalgae for commercial benefit have been limited. This is in large part due to capital costs and energy inputs remaining high, the necessity of identifying 'suitable' land with proximal resource and infrastructure availability and the need for process and strain optimisation. Microalgae represent a relatively unexplored bioresource both for native and engineered strains. Success in this area requires (1) appropriate methods to source and isolate microalgae strains, (2) efficient maintenance of motherstocks, (3) rapid strain characterisation and correct matching of strains to applications, (4) ensuring productive and stable cultivation at scale, and (5) ongoing strain development (breeding, adaptation and engineering). This article illustrates a survey and isolation of over 150 local microalgae strains as a bioresource for ongoing strain development and biotechnological applications.

Keywords: Microalgae; Bioresource; Motherstocks

Introduction

The global population is projected to increase from its current level of ~7 billion up to ~9 billion by 2050 [1]. This, together with unprecedented levels of lifestyle change in developing countries and policies designed to alleviate poverty (though global effect on addressing hunger appears to have recently stalled [2]), is by 2050 forecast to result in the requirement of ~70% more food [3] and ~50% more fuel [4], as well as ~50% more fresh water [5] and an increasing amount of chemical feedstocks. To supply these resources while simultaneously reducing global CO_2 emissions requires a transition away from fossil fuels, and towards renewable systems. The scale of this challenge should not be underestimated, given the urgent need for a very significant CO_2 emission reduction in this decade if we are to stay within the so called 'safe limit' (2°C) defined by the Intergovernmental Panel on Climate Change [6]. This is an ambitious target given recent claims that 80% of remaining fossil fuels must be left in the ground to prevent progressing past this threshold [7].

Fuel, food and water resources are all inextricably connected within our production-consumption cycles. For example, high levels of fertiliser use and water desalination are already required to support our existing population and will likely have to increase to provide food and water security. This in turn requires increased fuel consumption. More efficient means for utilising biological systems as sustainable bioresources to produce food, fuel chemical feedstocks and high value products are becoming increasingly important as consuming ancient fossil fuels becomes more controversial, and the necessity of CO_2 emission reductions becomes more widely represented in global policy.

Microalgae production systems are positioned at the nexus of these challenges as many species have high efficiencies relative to conventional crops in terms of using solar energy to drive the conversion of CO_2 to biomass (stored chemical energy). This biomass can subsequently be used to produce a broad range of downstream products. It has been widely stated that microalgae have the advantage that they can be produced on a proportion of non-arable land (non-arable land is ~25% of global surface area vs. ~3% arable land area [8,9]) and in many cases can use saline and waste water streams. This theoretically opens up the opportunity to extend global photosynthetic capacity beyond arable lands and assist with a transition from the current *food vs fuel* position [10,11] to a more sustainable *'food and fuel'* future. However the simplicity of the concept has not progressed to commercial reality despite a significant international research effort. This is primarily due to the many interconnected challenges of optimising biology and engineering parameters for high efficiency production and integrating these into commercially viable systems. Newly emerging strategies for high efficiency microalgae production [12,13] may contribute significantly to a food and fuel future but they are not the panacea that some have promoted. Opposing opinions that microalgal production systems lack the appropriate production strains suitable to overcome the challenges of economic and environmental sustainability for competitively priced biofuel production may be valid at the present time, but such arguments are insubstantial given the early stage of technology maturity, the rapid ongoing development in the field currently, and the large microalgae biodiversity (~350,000 species)

***Corresponding author:** Ben Hankamer, Institute for Molecular Bioscience, The University of Queensland, Queensland, Australia

and advanced genetic engineering techniques that can be tapped for strain optimisation [14-16]. Exploiting such a large biological resource is clearly an advantage but also presents a considerable undertaking, and high-throughput processes for strain isolation and maintenance are certainly required to increase the efficiency of traditionally laborious methods. This article describes the establishment of native Australian microalgae collections in terms of bioresource potential, and summarises the purification and cryopreservation protocols developed to efficiently isolate over 150 native strains from a range of water sources for ongoing strain development in a broad range of applications.

Founding A Microalgae Strain Library

International microalgae collections such as the Culture Collection of Algae and Protozoa (CCAP), Culture Collection of Algae at Göttingen University (SAG) and the University of Texas (UTEX) algae collection already offer a valuable resource for the provision of microalgae reference, research and breeding stocks. However international strain collections have their limitations and would benefit from augmentation with complementary local native strain collections which can offer a number of advantages. First, indigenous species are less likely to trigger local quarantine regulations (e.g. some imported strains are considered invasive 'weed' species or contain compounds undesirable for introduction into natural ecosystems). Second, indigenous species are generally more adapted to local climate conditions (e.g. light and temperature) and local biology (e.g. competitors and predators). Third, if correctly maintained and preserved (e.g. cryopreserved) wild type collections of indigenous species can be prevented from adapting to laboratory conditions (i.e. low selection pressures) which over time can result in a loss of culture robustness and suitability for large scale outdoor mass cultivation. Fourth, many strain collections are encumbered with intellectual property restrictions which specific local strain collections can avoid, although governments, national parks and private land owners can exert certain rights over commercially interesting strains isolated in such owned areas. The establishment of a phenotypically broad collection of local strains provides both a motherstock suited for further strain development and optimisation, and an improved understanding of competitor species that can invade aspiring monocultures of local or imported species. The aim of this article is to assist others with the establishment of similar local collections.

Methods

Capturing a broad range of phenotypic diversity from natural water sources requires collection from a broad range of environmental conditions. In this study, saline and fresh water sources, as well as photoautotrophic and mixotrophic environments were sampled. Sampling from extreme environmental conditions is possible and can reveal extremophile species which continue to yield significant potential. 'Moderate' extremophiles like *Dunaliella* or some *Tetraselmis* strains (growing in hyper saline ponds) or *Arthrospira* (growing in alkaline ponds) are relatively easy to cultivate using these methods, but 'extreme' extremophiles generally require more advanced facilities (e.g. 60°C cultivation systems) that are not discussed here.

Isolation of strains from water samples is indelibly influenced by the isolation process design, and furthermore both passive analytical screens (e.g. productivity and compositional monitoring) and active biological response screens (e.g. selection pressure applied through cultivation) can be used to guide the strain selection processes and the subsequent development of databases of strain characteristics. A flow diagram of the strategies used for microalgae isolation is shown in Figure 1.

In the strategy presented here the collection of crude water samples was followed by microscopic analysis (Figure 1 Native water samples) and subsequent incubation of the sampled species both in 'sterile source water' (to maintain species diversity) and in 'nutrient enriched water' samples supplemented with artificial medium for selection of the most adaptable species (Figure 1 Pretreatment). Following incubation several isolation techniques were employed including micromanipulation (Figure 1 Microman.), fluorescence activated cell sorting (Figure 1 FACS) and dilution (Figure 1 Dilution). Once isolated the method of choice for long-term storage was cryopreservation (Figure 1 Cryo) while serial cultivation on agar plates and in liquid media (Figure 1 Serial) was used for storage of sensitive strains. These isolates were identified via 18S [17,18] and 16S ribosomal sequencing [19] in conjunction with morphological classification (Figure 1 Identification) [20,21]. They were subjected to further screening to improve cultivation conditions and identify species for specific traits of interest (Figure 1 Screening) and to evaluate commercial cultivation capacity (Figure 1 Scale-up) to assist with strain selection and development for specific biotechnological applications. Each method step is described below.

Water samples

500 mL samples were collected from a broad range of local water sources in the east and south of Australia (Table 1). At the location site, samples were taken between the water surface and 10 cm depth. Samples from biofilms on plant and rock surfaces were also obtained. Microscopic analysis (Nikon Ti-U fitted with a Nikon Digital Sight DS-U2, 5mp colour head; 200x and 400x magnification) was performed prior to further treatment to record microorganism diversity and provide an initial basis for morphological classification (e.g. Figure 1 Native water samples).

Pre treatment

Sterile source water cultivation: The 'sterile source water' strategy was used to maintain maximum biodiversity. Although original water samples were non-sterile, the source water was sterilised (0.2 μm Supor® Membrane Syringe Filter, Acrodisc® 32 mm, Pall Life Sciences) to produce a natural water supply for subculture. Sterile technique was practiced throughout the purification process to preserve initial biodiversity and prevent further contamination. The microalgae were cultivated (100 rpm, C10 Platform Shaker, New Brunswick Scientific; illumination at 10 to 100 μE $m^{-2}s^{-1}$ cool white fluorescent light, relative to cell density) to increase the microalgae concentration.

Nutrient enriched water based cultivation: In this scenario water samples were enriched with nutrients to favour the selection of strains capable of fast nutrient uptake and fast growth. For nutrient enrichment, TP medium (TAP media [22] without acetate) was added to base water at a 1:3 enrichment ratio with subsequent cultivation for 4-7 days. Following initial enrichment and isolation, strains were transitioned to a range of fully artificial media including TP, TP +250 mM NaCl, TP +500 mM NaCl, TP + vitamins (3.9 μM thiamine, 7.5 nM cyanocobalamin, and 0.16 μM biotin, and these same vitamin concentrations were maintained as constant for all vitamins included media in this work, denoted as +V), TAP+V, 3NBBM+V [23], BG11+V for cyanobacteria [23], and DM+V for diatoms [23]. TAPY (TAP + 0.35% yeast extract) was used to encourage growth of contaminating microorganisms to confirm establishment of axenic cultures. Reagents were supplied by *Sigma-Aldrich*, *Chem-Supply* and *Amresco*.

Figure 1: Flow chart of the isolation, maintenance and analysis connected to the establishment of a mid size microalgae strain collection for biotechnological applications.

Isolation

Fluorescence activated cell sorting (FACS): FACS offers a rapid isolation technique to purify microalgae from the original sample or from contaminants. FACS has become increasingly popular in freshwater and marine ecology studies [24-26], and for these isolation procedures [27,28] due to the efficacy and high throughput aspects of this process. Success in this approach relies on several factors including the algal cell density and composition of the sample. Dominating species are more likely to be successfully obtained, and therefore the algal diversity of purified cells can be compromised. The size and shape of individual algae cells also has an influence on the success rate of sorting, and the survival rate differs from species to species because of sensitivity to physical stress. Fragile diatoms for example had lower survival rates than chlorophytes. For FACS analysis 5 mL samples of the *sterile source water* and *nutrient enriched water* based cultivations were pre-filtered (40 μm, Nylon Cell strainer, BD Falcon) into a FACS

Collection sites	Salinity	Collected samples	Isolation technique	Non-axenic isolates	Axenic isolates	Acetate utilising isolates	Isolates lost	Isolates cryopreserved	Identified isolates
Australia, QLD, Brisbane, rain water tank	fresh	2	TAP Enrichment + FACS	12	12	12	0	12	*Chlorella* sp., *Scenedesmus* sp., *Desmodesmus* sp., *Desmodesmus intermedius*
Australia, QLD, Brisbane, Nursery 1	fresh	3	TAP Enrichment + FACS	12	12	10	2	2	*Chlorella* sp., *Chlorella sorokiniana*, *Micractinium pusillum*
Australia, QLD, Brisbane, Nursery 2	brackish/ fresh	3	TAP Enrichment + FACS	18	18	14	4	14	*Chlorella* sp., *Chlorella sorokiniana*, *Micractinium pusillum*, *Scenedesmus* sp.
Australia, QLD, Brisbane, Nursery 3	fresh	1	TAP Enrichment + FACS	6	6	6	1	2	*Chlorella* sp., *Micractinium* sp.
Australia, NSW, rainforest waterfall	fresh	1	TAP Enrichment + FACS, Micro-manipulation	15	15	15	0	6	*Chlorella* sp., *Chlorococcum* sp., *Chlamydomonas* sp., *Desmodesmus* sp.
Australia, QLD, SE Townsville, port, seaside pond	salt	1	Micro-manipulation, FACS	2	2	1	2	0	No confirmed identifications
Australia, QLD, Townsville, river outlet	salt	4	Micro-manipulation, FACS	27	20	18	3	8	*Chlorella* sp., *Chlorella sorokiniana*,
Australia, QLD, NE Townsville, pond	salt	1	Micro-manipulation, FACS	13	4	4	9	0	*Chlorella* sp.
Australia, QLD, Townsville, lake	brackish	2	FACS in Liquid	9	3	5	3	0	*Chlorella* sp., *Micractinium* sp., *Navicula pelliculosa* sp.
Australia, QLD, Gold coast, fish tank	fresh	2	Dilution, FACS in Liquid	18	11	4	3	7	*Stichococcus* sp., *Merismopedia* sp., *Elakatothrix* sp., , *Ankistrodesmus* sp., *Chlorella* sp.
Australia, QLD, UQ, pond	fresh	3	Dilution, Micro-manipulation	9	5	9	0	3	*Chlorella* sp., *Scenedesmus* sp.
Australia, QLD, D'Aguilar, river	fresh	1	Micro-manipulation	7	7	0	1	3	*Ankistrodesmus* sp., *Chlorella* sp., *Scenedesmus abundans*
Australia, QLD, Central coast lake (1)	fresh	1	Micro-manipulation	3	1	0	0	0	*Chamydomonas* sp., *Chlorella* sp.
Australia, QLD, Central coast lake (2)	fresh	1	Micro-manipulation	4	1	0	1	0	*Euglena* sp., *Chlamydomonas* sp., *Chlorella* sp.
Australia, SA, Waikerie, Murray River	fresh	1	Micro-manipulation	5	3	0	1	0	No confirmed identifications
Australia, NSW, Yanga, storm water	fresh	1	Micro-manipulation	4	1	0	1	0	*Anabaena* sp., *Staurastrum* sp., *Coleastrum* sp., *Nannochloris* sp.
Australia, QLD, Goondiwindi, creek	fresh	1	Micro-manipulation	3	0	0	0	0	*Aulacoseira* sp., *Closterium* sp.,

Table 1: Statistical analysis of algae isolation success from crude water samples. Collection sites, water characteristics, the number of water samples and strains isolated to increasing levels of purity are provided, together with their ability to utilise acetate, storage characteristics and species identification.

tube and analysed in a BD FACS Aria unit (BD Biosciences). The samples were then probed with a laser to detect individual 'events' corresponding to specific particles (e.g. algae cells or bacteria). The resultant dot plots present individual algae cells as population clusters (Figure 2a) which can be analysed in terms of parameters such as forward and side scatter (which represent cell size and granularity). In addition chlorophyll fluorescence was monitored (488 nm excitation wavelength, 695 ± 40 nm transmitting filter) to distinguish between bacteria and dead/stressed algal cells (low fluorescence) and healthy algae cells (high fluorescence). This is achieved through the application of gating thresholds (Figure 2a delineated regions) which define different subpopulations based on size and fluorescence (e.g. P1-P6).

In order to maximise species diversity it is important to select cells and discrete regions to avoid oversampling dominant species, a process that is simplified by FACS.

Single or multiple events (e.g. individual or multiple cells) with different cell size and chlorophyll content were sorted into 96 well plates at a gating setting of one or more events per well (Figure 2b), containing 150 μL of solid agar media, 150 μL of liquid media, or 150 μL solid agar media topped with 50 μL liquid media (media as defined above). After sorting, microalgal growth was monitored via inverted microscope (Figure 2c), and success rates were ranked for each strain on the basis of colony formation and contamination status. Using a setting of three sorting events per well into liquid media yielded the highest success rate of single species recovery (>63%). Using a lower events/well setting or sorting on solid agar media resulted in a lower success rate (<10%) and less diverse algae populations constant with [29] who also reported a relationship between sorting success, culture media and culture vessel size of the micro well plate.

Micromanipulation: Micromanipulation is a laborious but powerful technique which allows the manual targeting of specific cells within a complex mixture. This is useful for sensitive strains and to increase biological diversity. Individual target cells were identified by microscopy (Olympus BX 41, 100x magnification) and extracted with a micromanipulator MM33 (Maerzhauser Wetzlar). Replaceable glass capillaries (Drummond Scientific, length 3.5", outer diameter 1.14 mm, inner diameter 0.53 mm) were used to select and transfer the cell into either *sterile source water* or *nutrient enriched water* as defined above. Individual cells were directly selected from these water samples. An alternative strategy involved spreading the microalgae sample (50 μl) onto agar and selecting cells after they had settled [29].

Dilution technique: Dilution either in liquid or solid media can be used as an alternative technique to resolve and purify individual algae strains. Achieving effective dilution on solid media [30] involves streaking of a small volume of the original sample onto agar plates (TP or original sterile water source media) with an inoculation loop in a three- or four phase streaking pattern. Plates were then incubated (conditions as above) until colonies appeared (some originating from a single isolated cell) which could then be manipulated individually. Re-streaking was repeated until pure cell colonies were observed. In parallel, liquid serial dilution was performed using 96 well plates. Enriched as well as untreated water samples were serially diluted (4:1) through 48 wells filled with 500 μl of the appropriate medium. Samples were incubated under low light conditions (~50 μE m^{-2} s^{-1} cool white fluorescent light) and examined daily (Nikon Ti-U inverted microscope).

Maintenance

Enrichment and maintenance of established isolates: Established isolates were enriched further with artificial media and incubated in larger volumes (10 mL) to increase cell number and concentration. For some microalgal isolates a stepwise increase of the concentration of artificial medium was found to be beneficial and was applied, with growth monitored microscopically and by optical density (OD_{750}) measurements. For long term storage triplicate samples of each isolate were cryopreserved using 3-5 x 10^6 cells per cryo-vial using a refined two-step freezing protocol developed for microalgae [31]. The final volume (1 mL containing 6.5 % DMSO and 0.2 M sucrose (*Sigma-Aldrich, Chem-Supply*)) was stored at -80°C for at least 4 hours before being transferred to -196°C for long-term storage in liquid N_2 vapour phase. Strains that could not be efficiently cryopreserved were maintained through serial cultivation using both liquid and solid media.

Analysis

Screening: Screening for desirable properties is an ongoing process that can be repeated once a microalgal collection has been established. The isolates can be re-screened for a variety of applications, and where breeding is not possible, rapidly advancing methods for engineering microalgae can enable further advancement. The screening characteristics used here are therefore illustrative only. The principles, however are universal – very specific screens are usually time consuming so early rapid screening for indicative traits can be utilised first, followed by specific screening on a smaller subset of parameters. The isolates obtained in this work were initially screened on the basis of biomass productivity, and have already been subjected to a rigorous set of secondary screens and this work will be reported in the near future.

Identification: Only a subset of ~20% of strains, which performed well in early screens, were selected for full identification (though this is clearly flexible). Identification consisted of morphological investigation (Olympus BX42 and Nikon Ti-U, 200x and 400x magnification) [20,21]

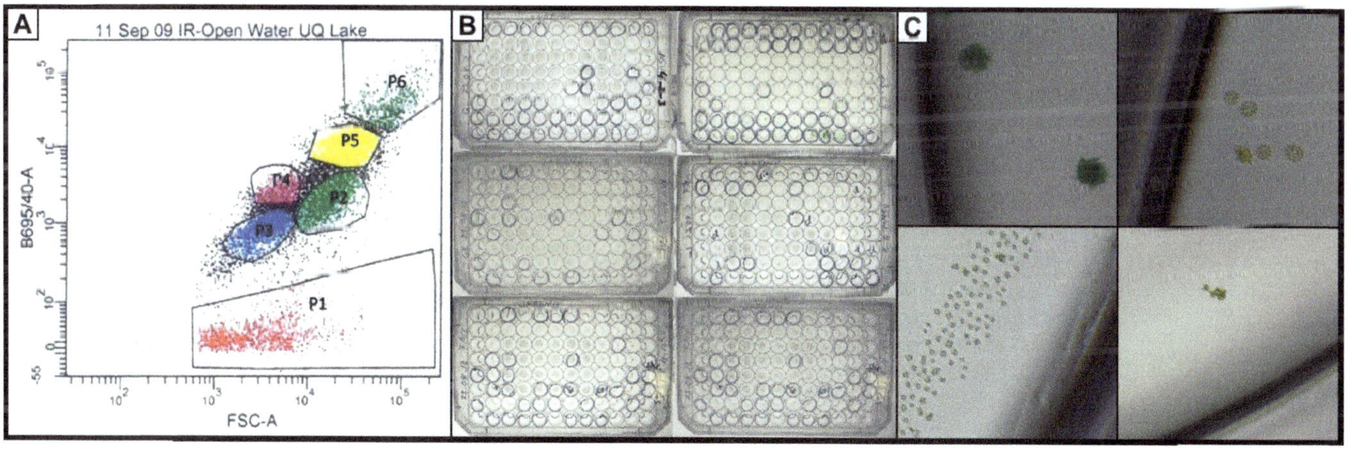

Figure 2: FACS isolation process. (A) FACS dot plot based upon laser excitation (y-axis) and forward scatter (x-axis) of cell mixtures which facilitates population analysis and selection of defined regions (P1-P6) for sorting; (B) 96 well plates containing purified algae; (C) microalgal isolate growth in individual wells monitored at 100x magnification.

Figure 3: Micromanipulation setup. Cells are imaged on agar plates. The insert shows individual cells at the tip of the needle used for aspiration (needle ID 0.53 mm).

and molecular classification by rDNA analysis. For the latter, DNA was isolated according to [32] though a 10 min sonication step was required to break open the cell walls of numerous wild type strains. Both18S and 16S ribosomal DNA analysis was performed. The amplification of 18S rDNA and its sequencing was outsourced to the Australian Genome Research Facility (AGRF). The analysis of 16S rDNA was performed in house using two 'universal' primers [19] that specifically target cyanobacteria and eukaryotic photosynthetic plastids. PCR amplicons were sequenced at AGRF. Sequences were aligned using nucleotide BLAST (NCBI, http://blast.ncbi.nlm.nih.gov/Blast.cgi) against the 'nucleotide collection (nr/nt)' database.

Results and Discussion

To establish this subset of our local microalgae strain library, water samples were sourced from 17 locations which included rainforest water sources, creeks, ponds and rivers, brackish/saline seaside ponds and river outlets, as well as artificial systems such as rainwater tanks, plant nurseries and fish tanks (Table 1). Clearly the biodiversity recovered from any biodiscovery program is dependent upon the biodiversity of the original water samples. The number of species recovered from a given water sample may be related to the trophic status of the source (i.e. oligo-, meso- or eutrophic) [33]. It has been previously reported [34] that oligotrophic conditions can have a higher level of species biodiversity for algae and while eutrophic water sources may contain more algae, species biodiversity is often lower as fewer species tend to dominate. Our observations supported this, and isolation processes yielded more isolates from eutrophic water sources.

In this study the 17 locations sampled, initially yielded a total of 167 non-axenic isolates. Of these ~95% were Chlorophytes (e.g. *Chlorella, Chlorococcum, Scenedesmus* and *Chlamydomonas*), ~4% cyanobacteria (e.g. *Anabaena* and *Merismopedia*) and diatoms (e.g. *Navicula*), and ~1% were unidentified cell types. 104 strains were recovered using the *sterile source water* approach and 63 were obtained using the *nutrient enrichment water* method. Although source and treatment specific, the fact that the *sterile source water* approach generally yielded a greater biodiversity suggests that it may be the better standard method for the establishment of bio-diverse local microalgae culture collections. For the isolation of fast growing strains higher levels of nutrient enrichment were beneficial.

120 of the initial non-axenic strain samples were recovered using FACS, 42 using micromanipulation, and 5 by dilution. This clearly shows the benefit of using FACS as a platform for developing local microalgae strain collections as it can sort and dispense over 500 events per hour, particularly if augmented with strains isolated using micromanipulation to increase biodiversity. Using a combination of FACS and micromanipulation yielded an average of approximately 10 ±7 strains per water sample. Using a FACS setting of 3 sorting events instead of 1 per well resulted in only slightly higher bacterial contamination levels, but increased the success rates of recovering algal cell isolates. Despite this it was noted that the survival rate of sorted algae cells rose when 3 events per well were used and so this is suggested as a sensible starting point for FACS purification. It was also noted that the use of 96 well plates instead of 384 well plates improved species recovery, with 150 µL solid agar media topped with 50 µL liquid media being the preferred media configuration.

Of the 167 non-axenic isolates, 121 were purified to the axenic level. This was confirmed by microscopic investigation of cultures supplemented with acetate and yeast extract as a carbon source to encourage heterotrophic growth and demonstrating the absence of contamination. Antibiotic treatment in some cases was able to assist with the production of axenic cell lines, but in many cases proved toxic to the algae themselves and so was of limited utility. Overall approximately 90% of the 121 axenic strains were purified from bacteria simply by using FACS or through repeated subcultivation on carbon-free agar media. The remaining 46 non-axenic isolates could not be successfully purified from contaminating bacteria. This may indicate the presence of either strong adhesion of the bacteria to the algae cells or the presence of endogenous bacteria. The observation that most of the non-axenic algae cultures visibly exhibited a white biofilm around the cells, suggests that the former was predominantly the case. Furthermore certain species having complex shapes (e.g. constricted symmetrical arrangements, spiral twisted, colonial or filamentous) such as the Chlorophyte *Staurastrum* proved more difficult to purify from bacterial contamination. Whether these strong interactions between the bacteria and algae are simply physical or represent a form of symbioses remains to be established, however it is commonly noted in our open pond trials and by others that in healthy and relatively stable raceway pond systems many bacteria and algae can coexist effectively. Indeed one benefit to their presence may be that the bacteria use the dissolved oxygen in the culture produced through the photosynthetic reactions of microalgae. The importance of this is that dissolved oxygen levels become increasingly inhibitory to algae photosynthetic processes. A further benefit of bacterial interactions might be the synthesis of essential vitamins required by certain algae (e.g. Vitamin B12 [35]), as well as some other beneficial compounds [36].

Of the 121 axenic cultures 57 were successfully cryopreserved using the method of Bui et al (47% success rate) [31]. Strains having a diameter of 3 to 50 µm were effectively recovered although some of the very large strains proved difficult, acceptable this step of the process would clearly benefit from improvement. Critical parameters include the optimisation of light level as high light can result in oxidative damage, as well as in the optimisation of nutrient conditions for specific strains.

Of the 64 strains that could not be cryopreserved 24 were lost during serial subcultivation. One reason for this is that the standard media used may not be sufficiently specific to the needs of individual species. Ongoing research is therefore required to optimise media composition.

Initial species identification was based on morphological classification but was refined through ribosomal sequencing. Ribosomal sequence analysis can be based on 18S rDNA and 23S rDNA analysis (derived from the nuclei of eukaryotes), or on 16S rDNA analysis (derived from chloroplasts and mitochondria) present in both eukaryotic microalgae and prokaryotic cyanobacteria. In this study 18S rDNA analysis was used as the primary rDNA analysis method but was supplemented with 16S rDNA analysis, contributing to the expansion of this resource. The 18S rDNA sequencing approach has the advantage that corresponding databases (e.g. NCBI) are more advanced than those for 16S rDNA, making it possible to achieve a higher quality of strain identification. Furthermore the 18S rDNA approach can currently enable identification to the species level in many cases. In practice our analysis typically yielded sequence identities of >95% but less than 100%, suggesting that while closely related to some strains in the online database, many of these wild isolates have not been previously catalogued. Exact matches occurred at low frequency and in some cases two or more hits with a similar identity greater than 95% were noted. Theoretically the combined use of 18S and 16S rDNA sequence analysis may facilitate improved identification and could also resolve the origin of specific plastids within a given species, contributing not only to species identification but the evolutionary relationships between specific nuclear and plastid genomes.

Conclusion

In this paper we have demonstrated a streamlined process for microalgae recovery from a broad range of water sources and used this to conduct a mid-scale survey of species native to Australian waters. Typically the water sources collected yielded ~10 strains of microalgae per sample, of which approximately half could be effectively cryopreserved to minimise maintenance costs and genetic drift, with most of the remainder being amenable to traditional subculture. Through the use of rDNA sequence analysis and morphological examination the resultant isolates were identified, either to the genus or species level providing a solid basis to assist the international research community with the establishment of multiple local strain collections to maximise microalgae species recovery as a breeding stock for cell lines beneficial for a wide range of biotechnological applications including the production of food, fuel, chemical feedstocks, high value products and for applications for wastewater treatment and bioremediation. Although there are already large international algae collections, the benefit of local strains collections include the establishment of robust, well adapted and locally derived breeding stocks that are often without the IP encumbrance associated with commercial strain collections. These can be used for the development of improved cell lines for a wide range of biotechnological applications. At a time when the global population is expanding from ~7 to ~9 billion people by 2050 and food, fuel and water demands are predicted to increase by 70%, 50% and 50% respectively the importance of establishing such diverse stocks becomes apparent. The ongoing exploration of the diversity of microalgal biology is already yielding advances in high performance wild types with commercial potential and genetic characteristics that could enable improvements for engineered strains. Initial screens focused on biomass productivity as a primary criteria (being a critical economic driver for commercialisation) but ongoing strain development will require further screens for a range of other useful characteristics including oil composition and profile, predator resilience, flocculation and other traits that enhance harvestability (e.g. floatation or sedimentation), and capacity for wastewater systems and bioremediation.

Acknowledgements

The authors would like to thank Eugene Zhang and Lou Brillault for their support in maintaining the microalgae collection. BH gratefully acknowledges the financial support of the Australian Research Council (LP0883380) as well as from a National and International Research Alliance Project (NIRAP) grant supported by Queensland State Government and project partners (KBR, Neste Oil, Siemens, Cement Australia, The University of Queensland, University of Bielefeld and Karlsruhe Institute of Technology).

References

1. Population Reference Bureau (2009) World Population Data Sheet.
2. Food and Agriculture Organization of the United Nations (2012) Economic growth is necessary but not sufficient to accelerate reduction of hunger and malnutrition. The State of Food Insecurity in the World.
3. Food and Agriculture Organization of the United Nations (2009) How to Feed the World in 2050.
4. International Energy Agency (2010) World Energy Outlook 2010.

5. OCED (2012) OECD Environmental Outlook to 2050: The Consequences of Inaction.

6. Stocker TF (2013) The Closing Door of Climate Targets. Science 339: 280-282.

7. Climate Commission (2013) The Critical Decade 2013: Climate Change Science, Risks and Responses.

8. Food and Agriculture Organization of the United Nations (2010) FAO Statistical Yearbook 2010.

9. Central Intelligence Agency (2013) The World Factbook.

10. Pimentel D, Marklein A, Toth MA, Karpoff MN, Paul GS, et al. (2009) Food Versus Biofuels: Environmental and Economic Costs. Hum Ecol 37: 1-12.

11. Gomiero T, Paoletti MG, Pimentel D (2010) Biofuels: Efficiency, Ethics, and Limits to Human Appropriation of Ecosystem Services. J Agric Environ Ethics 23: 403-434.

12. Vasudevan V, Stratton RW, Pearlson MN, Jersey GR, Beyene AG (2012) Environmental Performance of Algal Biofuel Technology Options. Environ Sci Technol 46: 2451-2459.

13. Stephens E, Ross IL, Hankamer B (2013) Expanding the microalgal industry--continuing controversy or compelling case? Curr Opin Chem Biol 17: 444-452.

14. Norton TA, Melkonian M, Andersen RA (1996) Algal biodiversity. Phycologia 35: 308-326.

15. Larkum AW, Ross IL, Kruse O, Hankamer B (2012) Selection, breeding and engineering of microalgae for bioenergy and biofuel production. Trends Biotechnol 30: 198-205.

16. Georgianna DR, Mayfield SP (2012) Exploiting diversity and synthetic biology for the production of algal biofuels. Nature 488: 329-335.

17. Huss VAR, Frank C, Hartmann EC, Hirmer M, Kloboucek A, et al. (1999) Biochemical Taxonomy and Molecular Phylogeny of the Genus Chlorella Sensu Lato (Chlorophyta). J Phycol 35: 587-598.

18. de-Bashan LE, Trejo A, Huss VA, Hernandez JP, Bashan Y (2008) Chlorella sorokiniana UTEX 2805, a heat and intense, sunlight-tolerant microalga with potential for removing ammonium from wastewater. Bioresour Technol 99: 4980-4989.

19. Stiller JW, McClanahan (2005) Phyto-specific 16S rDNA PCR primers for recovering algal and plant sequences from mixed samples. Molecular Ecology Notes 5: 1-3.

20. Entwisle TJ, Sonneman JA, Lewis SH (1997) Freshwater Algae in Australia A Guide to Conspicuous Genera. Sainty & Associates Australia.

21. Hallegraeff GM, Bolch CJS, Hill DRA, Jameson I, Leroi JM, et al. (2010) Algae of Australia: Phytoplankton of Temperate Coastal Waters. ABRS, Canberra & CSIRO Publishing Melbourne, Canberra, Melbourne 432.

22. Gorman DS, Levine RP (1965) Cytochrome f and plastocyanin: their sequence in the photosynthetic electron transport chain of Chlamydomonas reinhardi. Proc Natl Acad Sci USA 54: 1665-1669.

23. CCAP (2012) Culture Collection of Algae and Protozoa.

24. Reckermann M (2000) Flow sorting in aquatic ecology. Sci Mar 64: 235-246.

25. Cellamare M, Rolland A, Jacquet S (2010) Flow cytometry sorting of freshwater phytoplankton. J Appl Phycol 22: 87-100.

26. Crosbie ND, Teubner K, Weisse T (2003) Flow-cytometric mapping provides novel insights into the seasonal and vertical distributions of freshwater autotrophic picoplankton. Aquat Microb Ecol 33: 53-66.

27. Andersen RA (2005) Algal Culturing Techniques. Academic Press London UK.

28. Pienkos, PT (2008) Update on Algal Biofuel Research at NREL. 2nd Algae Biomass Summit. Seattle, WA, USA.

29. Sieracki M, Poulton N, Crosbie, N (2005) Automated Isolation Techniques for Microalgae, in Algal Culturing Techniques, Anderson RA Elsevier Academic Press London UK 101-116.

30. Anderson RA, Kawachi M (2005) Traditional Microalgae Isolation Techniques. Algal Culturing Techniques Anderson RA Elsevier Academic Press London 83-100.

31. Bui TVL, Ross IL, Jakob G, Hankammer B (2013) Impact of Procedural Steps and Cryopreservation Agents in the Cryopreservation of Chlorophyte Microalgae.

32. Melanie O (2009) Chloroplasts as bioreactors : high-yield production of active bacteriolytic protein antibiotics.

33. Schenk PM, Thomas-Hall SR, Stephens E, Marx UC, Mussgnug JH, et al. (2008) Second Generation Biofuels: High-Efficiency Microalgae for Biodiesel Production. BioEnergy Research 1: 20-43.

34. Rawson DS (1956) Algal Indicators of Trophic Lake Types. Limnol Oceanogr 1: 18-25.

35. Croft MT, Lawrence AD, Raux-Deery E, Warren MJ, Smith AG (2005) Algae acquire vitamin B12 through a symbiotic relationship with bacteria. Nature 438: 90-93.

36. Gonzalez LE, Bashan Y (2000) Increased Growth of the Microalga Chlorella vulgaris when Coimmobilized and Cocultured in Alginate Beads with the Plant-Growth-Promoting Bacterium *Azospirillum brasilense*. Appl Environ Microbiol 66: 1527-1531.

The Application of Acrylic Acid as Precipitation Inhibitor for ASP Flooding

Khaled Abdalla Elraies* and Shuaib Ahmed Kalwar

Faculty of Geoscience and Petroleum Engineering. Petroleum Engineering Department, Universiti Teknologi Petronas, Malaysia

Abstract

Alkaline-Surfactant-Polymer (ASP) flooding has shown incredible successes for enhancing oil recovery for both sandstone and carbonate reservoirs. However, the main constraint of ASP flooding in carbonate reservoirs is the presence of undesired minerals either within the reservoir rock or reservoir brine. These minerals could react with the added chemicals to form their insoluble salts as precipitations. In this paper, the performance of the acrylic acid was evaluated in the presence of sodium metaborate as an alkaline, alpha olefin sulfonate as a surfactant and AN-125 SH as a polymer. The effect of various acrylic acid concentrations on alkalinity, interfacial tension reduction and polymer viscosity were investigated using hard brine with a total salinity of 59,940 ppm. Fluid-fluid compatibility test indicates that acrylic acid has the potential to prevent any precipitation when hard brine is used. The acrylic acid to alkali ratio of 0.6:1 was found to be the optimum ratio for keeping the solution without precipitations for 30 days at 80°C. It was also observed that the combination of ASP with acrylic acid has a positive effect on interfacial tension and solution viscosity. This makes the new system more flexible for offshore application in which hard brine or sea water could be used to prepare ASP slug without any negative effects.

Introduction

Alkali-surfactant-polymer (ASP) flooding has been proven as the most effective and easiest method in sandstone reservoirs [1,2]. The successful results in sandstone formation have widened this method to carbonate reservoirs as well [3]. It is found that about 60% of the world's oil reserves and 40% of the world's gas reserves exist in carbonate rocks [4]. Having been potentially attractive, these rocks exhibit immense variation in properties that include porosity, permeability, and flow mechanism. These rocks can be naturally fractured and preferentially oil-wet [4,5]. In addition, carbonate formations contain large amounts of calcium and magnesium ions in the form of calcite ($CaCO_3$) and dolomite. These rocks also have small quantities of calcium sulphate minerals that are anhydrite and gypsum [4,6]. The main constraint of ASP flooding in carbonate reservoir is the formation of precipitation. Alkali reacts with divalent cationic minerals, resulting in an insoluble salts known as precipitations. These precipitates can cause significant formation damage and pore plugging in the reservoir formation [6].

The purpose of ASP flooding is to increase oil production by reducing the residual oil saturation after waterflooding [7]. For the displacement of oil in the porous media and reducing the capillaries of petroleum reservoir rocks, it is necessary to reduced interfacial tension between oil and ASP slug to an ultra-low interfacial tension [8]. Mobility control is also one of the most important mechanisms for oil recovery with respect to ASP flooding [9]. Many papers focus on the study of interfacial tension and mobility control for the purpose of practical applications.

In this paper, alkaline-surfactant-polymer formulation was developed by incorporating acrylic acid to form in-situ inhibitor to overcome the precipitation problems caused by reservoir minerals and brine compositions. Acrylic acid has the ability to react with sodium ions present in the brine to forms sodium acrylate. The produced sodium acrylate accumulates on surfaces of the divalent cations (calcium and magnesium ions) to prevent them from precipitating [10]. Sodium metaborate was blended as an alkali with acid to generate more acrylate. Sodium metaborate is a novel alkali that has high capability to tolerate multivalent ions. Its borate ion reacts with these cationic minerals and forms amorphous crystals that are effectively dissolvable in aqueous solution [11]. When acrylic acid and sodium metaborate were mixed in the presence of synthetic hard brine, the combination works more effectively to sequester these minerals from any precipitation [12]. The main advantage of this combination involves the use of hard brine without the need of softening the injection water for ASP flooding application.

Experimnetal Methodology

Materials description

Crude oil used in this study, was collected from Dulang oil field, Malaysia. The density of the crude oil was 0.799 g/cc (80°C) and the API gravity was 35.5°C. The crude oil viscosity is 0.96 cp at 80°C.

Sodium metaborate was supplied by Rio Tinto Minerals. Acrylic acid was supplied by Sigma Aldrich. C_{20-24} Alpha Olefin Sulfonate (AOS) was supplied by Stepan Chemical Co. AN-125 SH copolymer was supplied by SNF Floerger. Synthetic hard brine with a total salinity of 59,940 ppm was used to prepare all solutions and main compositions of the hard brine are given in Table 1.

Apparatus and Methods

Fluid-fluid compatibility test

The first step for this test was to prepare samples of various sodium metaborate concentrations using the synthetic brine. Purpose of this test was to evaluate the compatibility of the sodium metaborate with the brine. Next, the compatibility of different acrylic acid to alkali weigh ratios was performed. The aim of this test was to find an optimum ratio,

***Corresponding author:** Khaled Abdalla Elraies, Faculty of Geoscience and Petroleum Engineering. Petroleum Engineering Department, Universiti Teknologi Petronas, Malaysia, Tel: +605 368 7037; Fax: +605 365 5670

Components	Concentration
Sodium, ppm	20,445
Calcium, ppm	2523
Magnesium, ppm	239
Chloride, ppm	36,733

Table 1: Composition of synthetic hard brine.

at which no precipitation exists. In the third step, samples of surfactant and polymer with and without acid-alkali were prepared in order to investigate the compatibility of surfactant and polymer with the hard brine. All the compatibility tests were conducted at 80°C for a period of time and observation were made by visual evaluation with the formation of solids being indicative of incompatible fluids

Interfacial tension test

Interfacial Tension (IFT) between Dulang crude oil and various solutions were measured by spinning drop method. Model SVT 20 spinning drop tensiometer equipped with video camera was utilized to determine the IFT at 80°C using hard brine.

Viscosity measurement test

A polymer stock solution with 5000 ppm of AN-125 SH was prepared with deionised water. The stock solutions were mixed slowly with a magnetic stirrer for 1 to 2 days to properly hydrate the dry powdered polymer. Stock solutions were then filtered with 1.2 μm filter paper using OFIT model filter press under a constant pressure of 15 psig. The solutions were collected in a graduated cylinder, and the time was recorded after every 10 ml collected. The collected data from filtration was then used to calculate filtration ratio by using equation 1. A filtration ratio less than 1.2 is considered excellent.

$$Filtration_{ratio} = \frac{\Delta t_{80-100ml}}{\Delta t_{20-40ml}} \quad (1)$$

Where $\Delta t_{20\text{-}40\ ml}$ is the time to filter equal volumes of polymer solution near the beginning (20 to 40 ml) and $\Delta t_{20\text{-}40\ ml}$ is the time to filter equal volumes of polymer solution near the end (80-100 ml). High pressure high temperature Viscometer (OFIT 1100 model) was then used to determine the polymer viscosity at different shear rates. All viscosity measurements were performed at 80°C.

Results and Discussions

Alkali-brine interaction

The sodium metaborate employed in this study was incompatible with the synthetic hard brine. Although, sodium metaborate showed some resistance against water hardness at the ambient temperature, precipitations were formed immediately at 80°C. This is attributed to the interaction of bicarbonate group with divalent metal cations to form precipitates of carbonates. Magnesium hydroxide can also be generated to for precipitations at high pH. Table 2 shows the results of incompatibility of sodium metaborate with different concentrations ranging from 0.2% to 1.0 wt% during 15 at 80°C

Acrylic acid and alkali interaction

Based on the previous compatibility test, sodium metaborate was not compatible with the synthetic brine. Hence, the alkali and acrylic acid were combined with the brine to form in-situ inhibitor and to keep the solutions clear without any precipitations. Different acrylic acid to alkali weight ratios was screened to evaluate the performance of the in-situ precipitation inhibitor at 80°C. Table 3 summarizes the results for different acid to alkali weight ratio using sodium metaborate.

The performance of in-situ inhibitor was very effective in preventing Ca^{++} and Mg^{++} precipitations over time. Figure 1 shows the compatibility of different acid-alkali weight ratios after 30 days at 80°C. It was also observed that the efficiency of the inhibitor increased as the acid to sodium metaborate weight ratio was increased. For instance, with a small acid to alkali weight ratio, the produced inhibitor (sodium acrylate) was insufficient to prevent the precipitations. All the acid was neutralized by the added alkali and sodium ions present in the brine. However, when a high acid to alkali weight ratio was used, only the required amount of acid was neutralized by the added alkali. The inhibitor adsorbs on the active growth sites of the Ca^{++} and Mg^{++}

Days	Sodium Metaborate wt%								
	0.2	0.3	0.4	0.5	0.6	0.7	0.8	0.9	1.0
	pH								
	8.96	8.94	9.15	9.18	9.23	9.38	9.72	9.98	10.19
0	*clo*	*clo*	*clo*	*clo*	*clo*	*ppt*	*ppt*	*ppt*	*ppt*
1	*ppt*	*Ppt*	*ppt*	*ppt*	*ppt*	*ppt*	*ppt*	*ppt*	*ppt*
5	*ppt*	*Ppt*	*ppt*	*ppt*	*ppt*	*ppt*	*ppt*	*ppt*	*ppt*
10	*ppt*	*Ppt*	*ppt*	*ppt*	*ppt*	*ppt*	*ppt*	*ppt*	*ppt*
15	*ppt*	*Ppt*	*ppt*	*ppt*	*ppt*	*ppt*	*ppt*	*ppt*	*ppt*

Table 2: Occurrences of precipitation by sodium metaborate during 15 days at 80°C.

Days	Sodium Metaborate, 1 wt%					
	Acrylic Acid, wt%					
	0.2	0.3	0.4	0.5	0.6	0.7
	pH					
	9.59	8.94	8.40	7.61	6.81	6.09
1	*cld*	-	-	-	-	-
5	-	-	-	-	-	-
10	*ppt*	*ppt*	*ppt*	*ppt*	-	-
20	*ppt*	*ppt*	*ppt*	*ppt*	-	-
30	*ppt*	*ppt*	*ppt*	*ppt*	*ppt*	-

ppt denotes precipitation, - denotes clear solution, cld denotes cloudy

Table 3: Summary of the acid-alkali compatibility test during 30 days at 80°C.

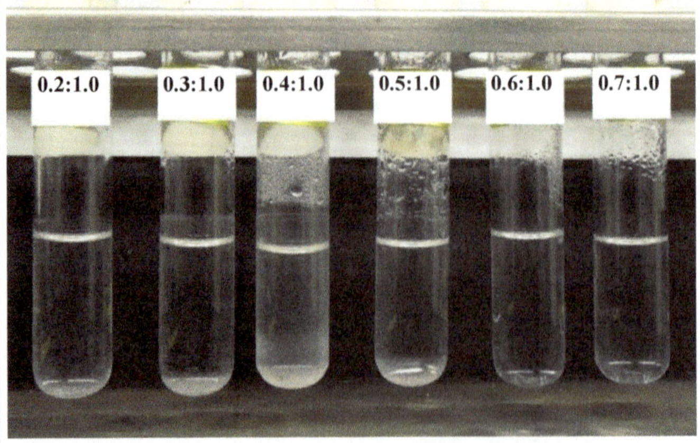

Figure 1: Performance of the in-situ inhibitor for preventing precipitation after 30 days at 80°C.

surfaces to prevent them from precipitating. By considering pH into account, the acid to alkali weight ratio of 0.6:1 was selected as optimum ratio for this case.

Acid-alkali-surfactants-polymer interaction

The compatibility of the acid-alkali-surfactant-polymer with the hard brine was investigated under different scenarios. AOS surfactant alone was found ineffective with the brine and formed precipitation or cloudiness at ambient temperature. Copolymer AN-125 SH was compatible with the brine and with the blend of acid-alkali-surfactant for 30 days at 80°C.

The combination of acid-alkali-surfactant-polymer was also found to be very compatible with the hard brine. No precipitations were generated for 30 days at 80°C. This shows the effectiveness of the in-situ inhibitor for creating such environment where all chemicals are blended without incompatibility issues. The results for the compatibility tests including acid-alkali-polymer and acid-alkali-surfactant-polymer are given in Table 4. It should be noted that acid-alkali weight ratio of 0.6:1 was applied whenever alkali is used.

Interfacial tension measurements

The effect of precipitation inhibitor in the IFT was investigated using different acid-alkali concentrations and 0.8% surfactant concentration. As shown in Figure 2, the IFT between crude oil and aqueous solution was significantly decreased by the presence of the acid and alkali in the system. When there was no acid and alkali in the system, 0.8% surfactant concentration reduced the IFT from about 14.9 mN/m to 0.64 mN/m. More reduction in the IFT was observed as the acid and alkali concentrations were increased to 0.48% and 0.8% respectively. Beyond this point, a slight increase in the IFT was detected due to the excess concentration of sodium acrylate. This will cause the salinity to increase, resulting in a subsequence increase in the IFT. Thus, with the presence of 0.8% surfactant, the optimum concentration for acid and alkali was found as 0.48%, 0.8% respectively. Using the optimum concentrations, interfacial tension can be reduced from 14.9 mN/m to 0.401 mN/m.

Viscosity measurements

The effect of salinity and hardness on the shear viscosity of polymer was investigated using 2000 ppm of AN-125 SH with distilled water, soft brine with 5.2% of NaCl and hard brine with 5.99%. All measurements were conducted at 80°C and the results for all cases are given in Figure 3. It can be seen from the Figure 4 that NaCl and divalent ions have a minor effect on the polymer. Although, hardness concentrations were extremely high in the hard brine, only a slight reduction in polymer viscosity was observed. This illustrates that the AN-125 SH polymer has high tolerance towards monovalent and divalent ions and is also thermally stable to control viscosity.

	Polymer AN-125 SH, 2000 ppm				
	Alpha olefin sulfonate, wt%				
Days	0.0	0.0	0.4	0.6	1.0
	Acid to alkali concentrations, wt%				
	0 : 0	0.6:1.0	0.4:0.7	0.6:1.0	0.8:1.3
1	-	-	-	-	-
15	-	-	-	-	-
30	-	-	-	-	-

- denotes clear solution

Table 4: Summary of acid-alkali-surfactant-polymer interaction test during 30 days at 80°C.

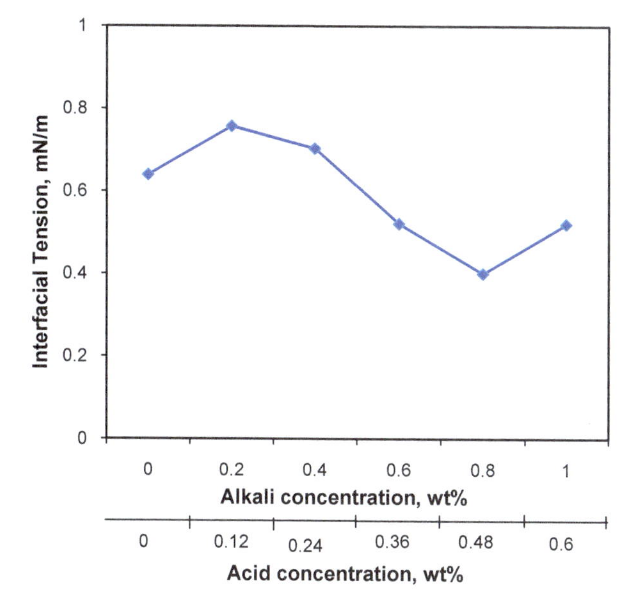

Figure 2: IFT between crude oil and various acid-alkali concentrations in the presence of 0.8% surfactant.

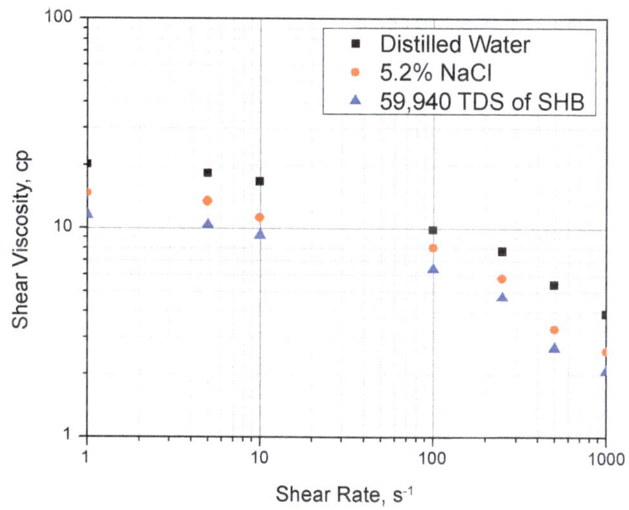

Figure 3: Effect of salinity and hardness on shear viscosity of the polymer.

Figure 4 shows the impact of various acid-alkali concentrations on the polymer viscosity at 80°C. All solutions were prepared with the synthetic hard brine using 2000 ppm of polymer. The acid-alkali weight ratio of 0.6:1 was employed for all samples to keep the solutions clear without precipitations. As shown in Figure 4, the polymer viscosity was not affected much by the presence of the different acid and alkali concentrations. For instance, the shear viscosity was 9.3 cp at 10 s^{-1} shear rates when the sample was prepared by only hard brine while the viscosity was reduced to only 7.6 cp in the presence of 0.12% and 0.2% of acid and alkali respectively at the same shear rate. This reduction might be attributed to the presence of alkali and acid. Alkali reduces viscosity because of charge shielding mechanism, while acid decreases pH which also affects polymer viscosity.

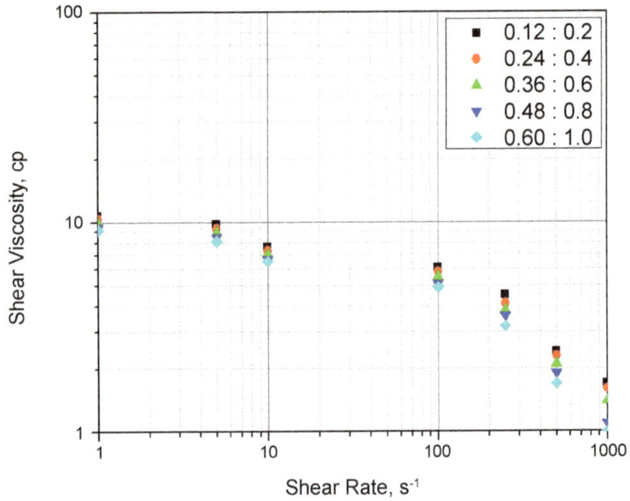

Figure 4: Effect of various acid-alkali concentrations on shear viscosity.

Conclusions

The present study shows that sodium metaborate and Alpha olefin sulfonate are significantly affected by the presence of divalent metal cations. Precipitations were immediately generated when they are mixed with the hard brine used in this study. However, AN-125 SH copolymer showed excellent properties towards monovalent and divalent ions and no precipitations were seen for 30 days at 80°C.

The combination of acid and alkali was very effective in preventing any precipitations in the presence of surfactant and polymer. No precipitation was detected for 30 days at 80°C when the optimum acid-alkali ratio of 0.6:1 was employed. Moreover, the generated in-situ inhibitor has insignificant impact on the interfacial tension and shear viscosity of the polymer.

The benefit of this new system is the use of hard brine instead of softened water while maintaining the desired chemicals activities. Thus, the new acid, alkali, surfactant and polymer AASP formulation can have favourable outcomes and can be cost-effective agent for chemical EOR especially for carbonate reservoirs.

Acknowledgement

The authors would like to thank the MOR-EOR at Universiti Teknologi PETRONAS for supporting this work. This research work was fully sponsored by YUTP grant no 0153AA-A69. We also thank RIO Tinto Minerals, Stepan Chemical Co. and SNF Floerger for providing some of the materials for this project.

References

1. Shutang G, Huabin L, Hongfu L (1995) Laboratory Investigation of Combination of Alkaline-Surfactant-Polymer for Daqing EOR.SPE Reservoir Eng 10: 194-197.
2. Pratap M, Gauma MS (2004) Field Implementation of Alkaline-Surfactant-Polymer (ASP) Flooding : A maiden effort in India. SPE Asia Pacific Oil and Gas Conference and Exhibition, 18-20 October 2004, Perth, Australia.
3. Manrique EJ, Muci VE, Gurfinkel ME (2007) EOR Field Experiences in Carbonate Reservoirs in the United States. SPE Res Eval & Eng 10: 667-686.
4. Akbar M, Vissapragada B, Alghamdi AH, Allen D, Herron M, et al. (2000) A Snapshot of Carbonate Reservoir Evaluation. Oilfield Rev 12: 20-41.
5. Mahroos FA, Zubari H (2009) Challenges of Producing Middle Eastern Oilfileds During the Maturation Stage- A Paradigm Shift. SPE Middle East Oil and Gas Show and Conference, 15-18 March 2009, Bahrain, Bahrain.
6. Levitt D, Dufour S, Pope G, Morel D, Gauer P (2012) Design of an ASP flood in a High-Temperature, High-Salinity, Low-Permeability Carbonate. International Petroleum Technology Conference, 7-9 February 2012, Bangkok, Thailand.
7. Al-Mjeni R, Arora S, Cherukupalli P, Wunnil, Edwards J, et al. (2010) Has the Time Come for EOR? Oilfield Rev 22: 16-35.
8. Zhao Z, Bi C, Li Z, Qiao W, Cheng L (2006) Interfacial tension between crude oil and decylmethylnaphthalene sulfonate surfactant alkali-free flooding systems. Colloid Surface A 276: 186-191.
9. Lee S, Kim DH, Huh C, Pope GA (2009) Development of a Comprehensive Rheological Property Database for EOR Polymers. SPE Annual Technical Conference and Exhibition, 4-7 October 2009, New Orleans, Louisiana USA.
10. Elraies KA, Tan IM, Awang M, Fathaddin MT (2010) A New Approach to Low-Cost, High Performance Chemical Flooding System. SPE Production and Operations Conference and Exhibition, 8-10 June 2010, Tunis, Tunisia.
11. Flaaten AK, Nguyen QP, Zhang J, Mohammadi H, Pope GA (2010) Alkaline/Surfactant/Polymer Chemical Flooding Without the Need for Soft Water. SPE J 15: 184-196.
12. Elraies KA, Ahmed S (2012) A New Strategy for Minimizing Precipitations during ASP Flooding in Carbonate Reservoirs. World Academy of Science, Engineering and Technology 72: 1527-1529.

22

Enhanced Biodegradation of Mobil Oil Hydrocarbons by Biosurfactant Producing Bacterial Consortium in Wheat and Mustard Rhizosphere

Rajesh Kumar[1]*, Ram Naresh Bharagava[1], Manish Kumar[1], Sanjeev K Singh[2] and Govind K[3]

[1]*Department of Environmental Microbiology, School for Environmental Sciences, Babasaheb Bhimrao Ambedkar University, Vidya Vihar, Rae Bareli Road, Lucknow, Uttar Pradesh, India*

[2]*Division of Eco-Restoration, National Environmental Engineering Research Institute, Nehru Marg, Nagpur, India*

[3]*Department of Microbiology, College of Basic Science and Humanities, Pantnagar, Uttarakhand, India*

Abstract

Hydrocarbon fuels are one of the most common global environmental pollutants which cannot be easily degraded owing to their hydrophobic nature. The present study involves the degradation of hydrocarbons by biosurfactant producing bacterial consortia. In this study, results showed 73.66% and 75.80% degradation of 2% mobil oil hydrocarbons in contaminated soil by biosurfactant producing bacterial consortium with wheat (*Triticum aestivum*) and mustard (*Brassica juncea*) crops, respectively. Therefore, it indicates that the developed bacterial consortium are capable for the effective degradation of mobil oil hydrocarbons in wheat and mustard rhizosphere and hence can be employed effectively for the degradation of mobil oil hydrocarbons in oil contaminated soils. Phthalate esters formed during the degradation can be used for industrial applications like PVC softening.

Keywords: Phthalate esters; Bioremediation; Biosurfactants; Hydrocarbon degradation; Fluorescent pseudomonas

Introduction

Hydrocarbons are the hydrophobic organic chemicals having toxicity, persistence and negative influence on living organisms and account for 90-95% or more of the total contaminant mass, exhibit limited solubility in groundwater and tend to partition to the soil matrix [1]. The soil pollution by hydrocarbons like mobil oil is mainly responsible for the changes in physical and chemical properties of soils which result in adverse effects on plant growth parameters [2,3]. Moreover, bioremediation is an efficient, economic and versatile alternative to the physico-chemical treatment methods mainly with bacterial consortia or individually [4,5]. This bioremediation process of hydrocarbon removal occurs through two distinct, but interrelated processes of biodegradation and microbial uptake, which enhance the hydrocarbon degradation by increasing the substrate bioavailability and by increasing the hydrophobicity of surfaces, by mobilization, solubilization or emulsification [6-8].

Biodegradation is a "microbially driven chemical transformation of organic compounds" [9] whereas microbial uptake is the direct removal of contaminants by adsorbing the compounds onto membrane surface or absorbing compounds through cell membrane. Many microbial species including fluorescent pseudomonads are reported to excrete a variety of totally or partially extracellular, amphipathic biosurfactant molecules that reduces surface tension and facilitates the uptake of hydrocarbons [10,11] and can be employed for the biodegradation and removal of mobil oil hydrocarbons from contaminated soil [1]. However, fluorescent pseudomonad (gram-negative, non-spore forming, motile, rod shaped and versatile bacteria) have been reported to promote plant growth in rhizosphere either directly by producing plant growth regulators and by increasing the plant uptake of some micro and macro elements from the rhizosphere [12,13] or indirectly through biological control of pathogens or induction of host defense mechanisms [14]. This Plant Growth Promoting Rhizobacteria (PGPR) contributes in lowering the abiotic stresses on plants and therefore considered to be an important tool for the phytoremediation technology [3,15].

In this study, a consortium containing fluorescent pseudomonads and *Bacillus sp.* was developed and used for the degradation and removal of mobil oil hydrocarbons from soil planted with wheat (*Triticum aestivum var.* 2565) and mustard (*Brassica juncea var.* Kranti) plants.

Materials and Methods

Collection of hydrocarbon contaminated soil/sludge samples

The soil/sludge samples were collected from the different oil contaminated sites: Haldia (22°03'37.65"N latitude and 88°06'35.09"E longitude), Jhansi (22°26'38.84"N latitude and 78°34'03.37"E longitude) and Pantnagar (29°01'20"N latitude and 79°29'15"E longitude), India. These soil/sludge samples were used for the isolation of plant growth promotory and biosurfactants producing bacterial strains.

Isolation and characterization of plant growth promoting and biosurfactant producing bacterial strains

The test strains were isolated from soil/sludge samples collected from different sites by serial dilution method on King's B agar media at 30°C in 48 hours. These isolated test strains were characterized morphologically, biochemically as per the standard procedures [16] and for plant growth promoting properties such as siderophore [17], ammonia, indole acetic acid [18], HCN production [19] and phosphate solubilization [20]. In addition, the test strains were also screened for biosurfactant production on Mineral Salt Medium (composition in g/l, $NaNO_3$ 2.5; K_2HPO_4 1.0; KH_2PO_4 0.5; $MgSO_4$ (anhydrous) 0.5; KCl 0.1; $FeSO_4$ 0.01; $CaCl_2$ 0.01; Na_2HPO_4 5.67; $MnSO_4$ 0.002; NH_4NO_3 0.39 and dextrose 30). The pH of medium was maintained at 7.0, followed

***Corresponding author:** Rajesh Kumar, Department of Environmental Microbiology, School for Environmental Sciences, Babasaheb Bhimrao Ambedkar University, Vidya Vihar, Rae Bareli Road, Lucknow, Uttar Pradesh, India

by incubation at 30°C for 48 hours under shaking flask condition at 150 rpm on a rotary shaker. The promising test strains for the above properties were characterized molecularly on the basis of 16S rRNA gene sequence analysis and the phylogenetic tree was constructed by blasting the data (FASTA format) using the software *Mega4* to know the relatedness between the isolated test strains.

Bioassay for biosurfactant production and its characterization by FTIR analysis

The biosurfactants produced by the test strains on Mineral Salt Medium were characterized on the basis of different properties *viz;* foaming, surface tension (after centrifugation at 10,000 rpm for 20 minutes) by De Nuoy Ring Detachment Method [21], emulsification index [22], drop collapse test and haemolytic activity. The biosurfactants were extracted by diethyl ether extraction method. Seventy two hours grown test cultures on Mineral Salt medium at 30°C under shaking flask conditions at 150 rpm were centrifuged at 10,000 rpm for 20 minutes at 4°C, pH of the supernatant was adjusted to 2.0 with 6 N HCl and allowed to stand overnight at room temperature. The biosurfactant was extracted with 2 volumes of diethyl ether in a separating funnel with vigorous shaking for 60 minutes on a shaker. The upper organic layer was collected and transferred to pre-weighed dry beakers. Diethyl ether was allowed to evaporate for 24 hours and the samples were dried in an oven at 60°C for one hour, light brown coloured biosurfactant obtained was subjected to FTIR analysis. The samples were grounded with KBr in the ratio of 100:1 (100 mg KBr and 1mg sample) in an agate motor to make the pellets by using a Hydraulic press (CAP 15T) at a pressure of ten tons. Finally, the grinded samples were analysed by using FTIR (Bruker, Vertex 70).

Development of hydrocarbon emulsifying and plant growth promoting bacterial consortia

The promising hydrocarbon emulsifying and plant growth promoting bacterial strains were selected to develop consortia on the basis of compatibility studies. By using agar diffusion technique on King's B solid medium, strains were studied for compatibility. Those showing compatibility were selected for making consortia. The developed consortia were again checked for compatibility in liquid media (50 ml MSM broth) in presence and absence of hydrocarbons (mobil oil 2%). The absorbance was taken at periodic interval of 24 hours from 0 hours upto 120 hours in spectrophotometer (Beckman DU 340B model) and finally, based on growth pattern, three consortia, C1 (H1A, H1B, H1C), C2 (J2A, P2A, J1A) and C3 (J1A, J1B, H1C) were selected for the *in situ* bioremediation of mobil oil hydrocarbons in wheat (*Triticum aestivum*) and mustard (*Brassica juncea*) rhizosphere.

Pot experiment

The experimental soil was mixed, allowed to dry to get water content less than 1%, sieved with 2 mm sieve and autoclaved at 121°C for 1hr for three consecutive days. From this autoclaved soil, 1.5 kg of soil was transferred to each pot and mixed with 30g (2%) of mobil oil hydrocarbons. Different consortia were added to the experimental pots (50 ml i.e. ~ 1×10^7cells ml^{-1}). Suitable controls without hydrocarbons and/or bacterial consortia were kept as control. Surface sterilized seeds of wheat *(Triticum aestivum)* and mustard (*Brassica juncea*) were sown at a depth of 2.5 cm. The pots were kept under observation for germination (seedling germination percentage) and plant health parameters (i.e. seedling survival, average plant height, chlorophyll content using Spadmeter optisciences 2000, plant biomass, average shoot height, average stem dry weight with fruit, average seed weight, average root weight and average yield per pot). The data obtained was analysed using statistical software. This experiment was carried out for 120 days in poly house at room temperature.

Characterization of residual hydrocarbons

Extraction of residual hydrocarbons from wheat and mustard rhizosphere: Approximately 5g of soil sample from mobil oil contaminated soil was taken initially (0 days, control) and after cultivation of wheat and mustard at 120 days. To these soil samples, 15 ml of acetone was added followed by 1hr shaking and the residue obtained was filtered and washed again twice with acetone, followed by filtration and concentration on a rotatory vaccum evaporator at 50°C and analyzed for the presence of residual hydrocarbons using FTIR. 2.6.2 Gas Chromatography Mass Spectrophotometry (GC-MS) analysis of extracted residual hydrocarbonsFor GC-MS analysis, the concentrated samples were dissolved in 1 ml of hexane. The quantification of hydrocarbon compound was done by high resolution gas chromatography using QP-2010 gas chromatograph (Shimadzu) equipped with a HP-DB 5MS column (60 m × 0.25 mm, 0.25 µm film thickness) coupled with mass spectrometer detector (HP 5972).

Results and Discussion

Characteristics of biosurfactant producing plant growth promotory bacterial strains

In this study, initially, twenty three bacterial strains were isolated, out of which, seven compatible and promising strains for biosurfactant production and plant growth promotory properties were selected for further studies. The test strains H1A, H1B and H1C were isolated from oily sludge samples collected from Haldia; J1A, J1B from sludge samples and J2A from oily soil samples collected from Jhansi while test strain P2A from soil of Pantnagar Industrial area.

The morphological and biochemical characteristics of the isolated bacterial strains are represented in Table 1. All the test strains were found rod shaped and Gram negative except J2A, which was Gram positive and resembled *Lysinibacillus sphaericus*. Further, the molecular analysis of these bacterial strains showed that most of them belongs to genus *Pseudomonas* (homology 97-99%), only one bacterial strain (J2A) belongs to genus *Bacillus* (Figure 1). In addition, the 16S rRNA nucleotide sequences obtained from test strains were also submitted to gene bank and accession number allotted were JX100389, JX149542, JX149543, JX100388, JX149548, JX149545 and JX149546 for the isolated test strains H1A, H1B, H1C, J1A, J1B, J2A and P2A, respectively.

The selected test strains showed various plant growth promotory properties such as siderophore, indole acetic acid, ammonia production, phosphate solubilization, HCN production as shown in Table 2. The isolated test strains were categorized as excellent (++++), good (+++), fair (++), poor (+) based on zone size/optical density measurements amongst the test strains. On the basis of these plant growth promotory properties, it is evident that the isolated strains were capable to solubilize phosphate and sequester iron from soil and provide to plant cell under stress conditions. Moreover, our results are also supported by similar findings as reported by various authors [23,15]. Further, out of these isolated bacterial strains, the Indole Acetic Acid (IAA) production was best shown by the strains H1A and J1A that can enhance the various stages of plant growth [3,24].

Characteristics of biosurfactant produced by isolated test strains

Biosurfactant production and result of its various properties are

Sr No.	Characteristics	Test Strains						
		H1A	H1B	H1C	J1A	J1B	J2A	P2A
1	Gram reaction	-ve	-ve	-ve	-ve	-ve	+ve	-ve
2	Chromogenesis on King's B agar	Green fluorescence	Green fluorescence	Green fluorescence	Green fluorescence	Green fluorescence	Pale Green fluorescence	Green fluorescence
3	Oxidase test	+ve	+ve	+ve	+ve	+ve	+ve	+ve
4	Catalase test	+ve	+ve	+ve	+ve	+ve	+ve	+ve
5	Gelatin liquefaction	+ve	+ve	+ve	+ve	+ve	+ve	+ve
6	Starch hydrolysis	+ve	+ve	+ve	-ve	-ve	+ve	+ve
7	Casein hydrolysis	+ve	+ve	+ve	+ve	+ve	+ve	+ve
8	Indole	-ve	-ve	-ve	-ve	-ve	+ve	-ve
9	Methyl red	-ve	-ve	-ve	-ve	-ve	-ve	-ve
10	Vogus Prausker	-ve	-ve	-ve	-ve	-ve	-ve	-ve
11	Citrate utilization test	+ve	+ve	+ve	+ve	-ve	+ve	+ve

Table 1: Morphological and biochemical characteristics of the test strains.

Test Strains	Siderophore Production	IAA Production	NH_3 Production	PO_4^{2-} Solubilization	HCN Production
H1A	++++	+++	+++	+++	+++
H1B	++	++	++	+++	+++
H1C	+++	++	++	+++	+++
J1A	++++	+++	+++	++++	+++
J1B	+++	++	+++	+++	+++
J2A	++	+	–	+	–
P2A	+++	++	++	+++	+++

(++++) excellent, (+++) good, (++) fair, (+) poor.

Table 2: Plant Growth Promotory Properties of the isolated bacterial test strains.

Test strains	Properties of biosurfactants				
	Foaming	Emulsification index (E_{24})	Drop collapse test (+ indicates surfactant production)	Type of hemolysis	Biosurfactant yield (g l^{-1})
H1A	+	2.95 ± 0.004	+	β-hemolysis	1.2 ± 0.049
H1B	+	2.94 ± 0.036	+	β-hemolysis	1.4 ± 0.081
H1C	+	8.83 ± 0.020	+	β-hemolysis	2.6 ± 0.067
J1A	+	7.65 ± 0.044	+	β-hemolysis	1.0 ± 0.098
J1B	+	8.23 ± 0.004	+	β-hemolysis	3.6 ± 0.026
J2A	+	30.59 ± 0.012	+	α-hemolysis	1.6 ± 0.077
P2A	+	11.18 ± 0.032	+	β-hemolysis	2.0 ± 0.044

Table 3: Biosurfactant properties of the test strains.

summarized in Table 3. All the test strains were found to produce foam in the mineral salt medium. There was a reduction in the surface tension by the test strain surfactants (Figures 2a-2e). Surface tension measured by tensiometer at different time intervals (0, 24, 48, 72, 96 and 120 h) of 7 test strains in MSM broth showed that H1A, H1B, and H1C reduced surface tension from 72.8 to 25.13, 25.71 and 25.1 dynes cm^{-1}, respectively and other test strains J1A, J1B, J2A and P2A reduced surface tension from 72.8 to 26-29 dynes cm^{-1}. Growth pattern also showed an increase upto 72 hours of incubation period in all the cases, beyond which the growth stabilised as measured by the turbidity in MSM broth. However, in case of test strains H1A and H1B, there was a decline in the growth phase beyond 72 hours (Figures 2a-2e).

The mobil oil hydrocarbons emulsification by the biosurfactants varied from 30.59 (J2A) to 2.94 (H1B) as shown in Table 3. All the test isolate were found positive for drop collapse test (drop collapsed after spreading). In addition, beta and alpha haemolysis were shown by surfactants with J2A surfactant showing alpha haemolysis while surfactants produced by other strains showed beta haemolysis. Biosurfactant yield using diethyl ether method ranged from 1.0 (Strain J1A) to 3.6 (Strain J1B) g l^{-1}. Further, the biosurfactants were characterized by using FTIR analysis.

FTIR analysis of biosurfactant produced by the test strains

The IR spectra of biosurfactants produced by all the test strains showed the presence of different bonds as shown in (Figures 3a-3g). However, there are certain bonds which are common like $-CH_2$ stretching frequency from 2848 (H1B) to 2927 (P2A), C-C stretching vibration (1045 1147) and C-H deformation frequency (1378-1461). Moreover, the biosurfactant from the test strain J1B did not show C-H deformation frequency. Further, P-H bonding (2356-2372) is shown by H1A, H1C and P2A only, while secondary amide (3359) by J1A and P2A. The differences in the peaks can be used for differentitating the biosurfactants produced by the test strains as well as a marker for identifying the biosurfactants.

Effect of developed bacterial consortium on seed germination and plant growth parameters.

In the present study, seed germination was recorded as 80 and 60% in wheat and mustard, respectively on soft agar while in case of pot experiment there was a delay in seed germination in presence of 2% mobil oil hydrocarbons. However, the seed germination percentage was better in presence of bacterial consortium over the control. Further, it was found that in case of wheat plant, consortium C3 was better at

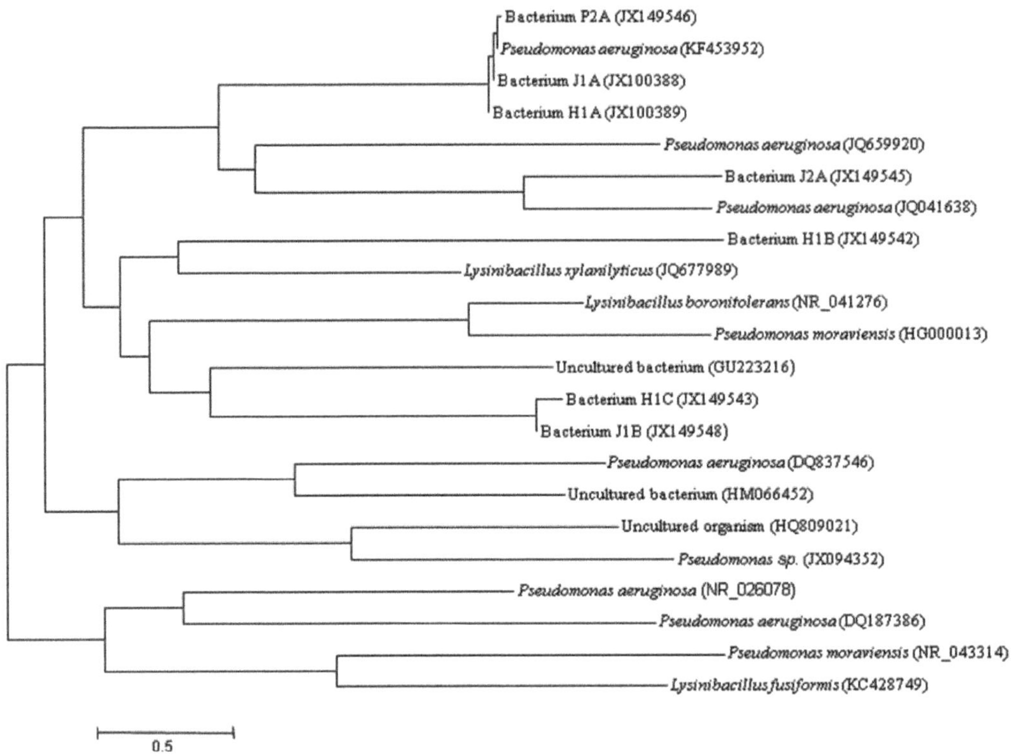

Figure 1: Phylogenetic tree: Relationship amongst test strains by neighbor joining method.

10 DAS while by 15 DAS, consortium C2 was found to be better in seed germination than the other two consortiums C1 and C3. While, in case of mustard plants, consortium C3 was better in seed germination over the other two consortiums at 10 and 15 days after sowing. Similar findings regarding delay in seed germination have been also reported by Okolo et al. [25].

Further, in case of plant growth parameters such as average plant height, the order for the effectiveness of the developed bacterial consortium in case of wheat plant was C1>C3>C2, C3>C1>C2 and C3>C1>C2 at a time interval of 30, 60 and 90 Days After Sowing (DAS), respectively, while in case of mustard plants, it was C1>C2>C3, C2>C1>C3 and C2>C1>C3 in the same time interval, respectively. The chlorophyll content was found to be more in presence of consortium C3 and C2 for wheat and mustard plants, respectively at 45 days after sowing. Moreover, the average root weight, average stem weight with spikes and average seed weight was recorded more in case of consortium C3 as compared to other two consortium at 120 DAS in case of wheat plants. But, in case of mustard plants, consortium C1 was found more effective for average root weight and average seed weight whereas for average stem weight with pods, consortium C2 was found more effective as compared to other two consortiums.

All the plant growth parameters, in presence of consortia were found to be better compared to the control plants grown in absence of 2% mobil oil hydrocarbons. Similar findings have been also reported earlier by Akinola et al. [26] and Merkl et al. [27] while studying the effect of petro-hydrocarbons on plant growth parameters. In some of the parameters like seed germination and plant height, there was a delay in performance of the consortia and as the time passed, consortia performed better and even better than plants without mobil oil hydrocarbons.

GC-MS analysis

The GC-MS analysis showed that the compounds appearing in the control sample at 0 DAS in different retention time were mostly absent from both the samples after 120 days of treatment. It indicates that most of the compounds were degraded by the developed bacterial consortia and utilized as carbon and energy sources. Further, the peaks were characterized for the type of metabolite produced during the biodegradation of mobil oil hydrocarbons in wheat and mustard rhizospheric soil. In case of wheat with consortium C3, some new peaks were observed at Retention Time (RT) 24.354 and identified as 1,2-benzenedicarboxylic acid, dioctyl ester while in case of mustard with C1 consortium, two new peaks were observed at Retention Time (RT) 24.342 and 30.721, identified as1,2-Benzenedicarboxylic acid, mono (2-ethylhexyl) ester and 4a,7,7,10a-tetramethyl-dodecahydro-benzo[f] chromen-3-ol respectively (Figure 4). The average% degradation of mobil oil (2%) contaminated soil in C3M2W was 73.66% and in C1M2M was 75.80%.

Conclusion

Hence, in this study, we concluded that the bacterial consortium C3 (J1A, J1B, H1C) and consortium C1 (H1A, H1B and H1C) were capable to enhance degradation of 2% mobil oil hydrocarbons in wheat (*Triticum aestivum*) and mustard rhizosphere, respectively. Moreover, consortium C3 degraded mobil oil hydrocarbons in wheat rhizosphere with an efficiency of 73.66% while consortium C1 degraded hydrocarbons with an efficiency of 75.80% in mustard rhizosphere in 150-180 days. Sathish Kumar et al. [28] also reported enhanced degradation of 1% crude oil by a mixed bacterial consortium consisting of Bacillus, Corynebacterium, Pseudomonas HPS2 and BPS1 strains as compared to individual strains, however, the percentage degradation was only

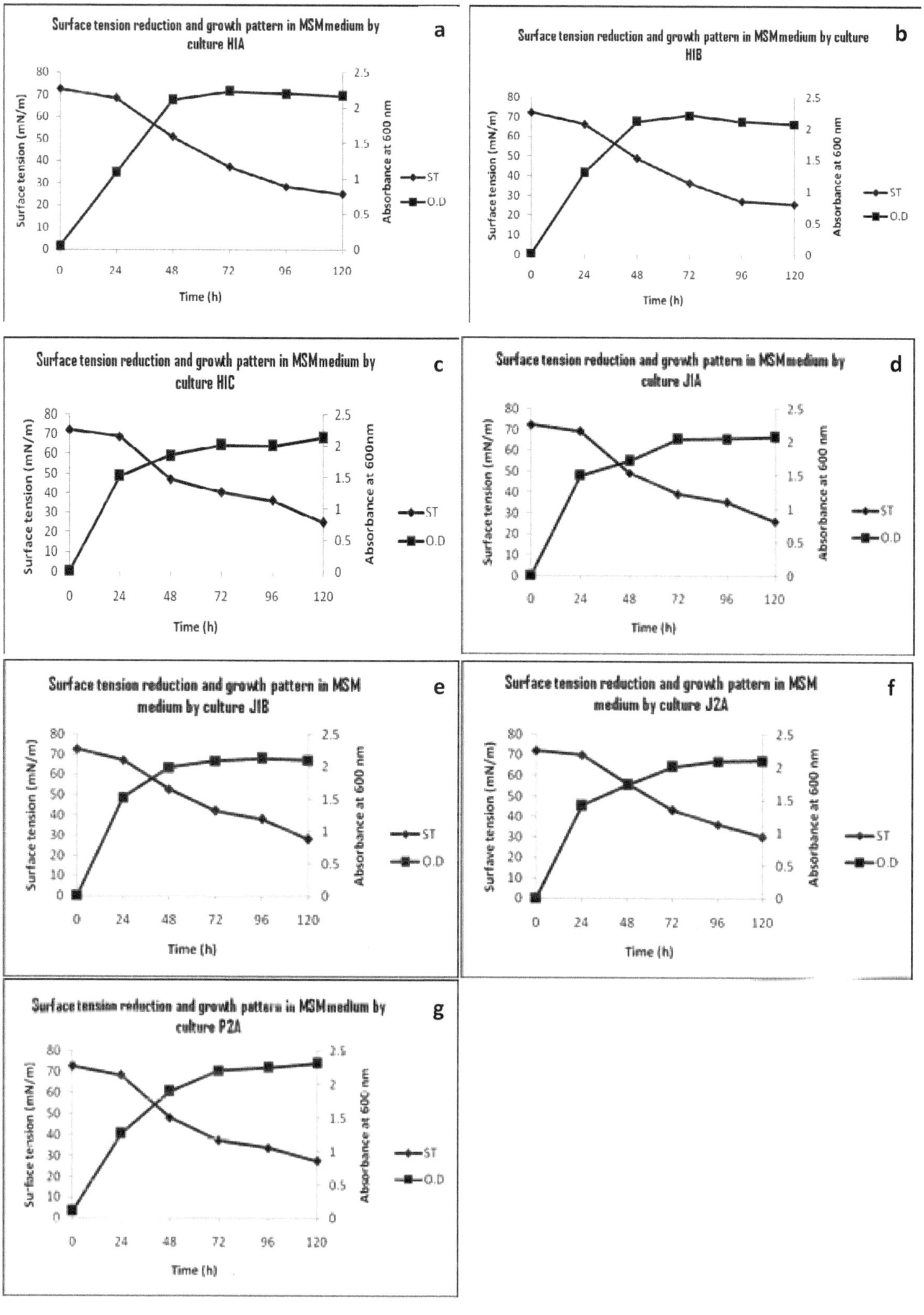

ST: Surface Tension
OD: Optical Density for Growth Measurement

Figure 2: Surface Tension reduction by crude biosurfactant produced by the isolates (a) H1A, (b) H1B, (c) H1C, (d) J1A, (e)J1B, (f) J2A, (g) P2A.

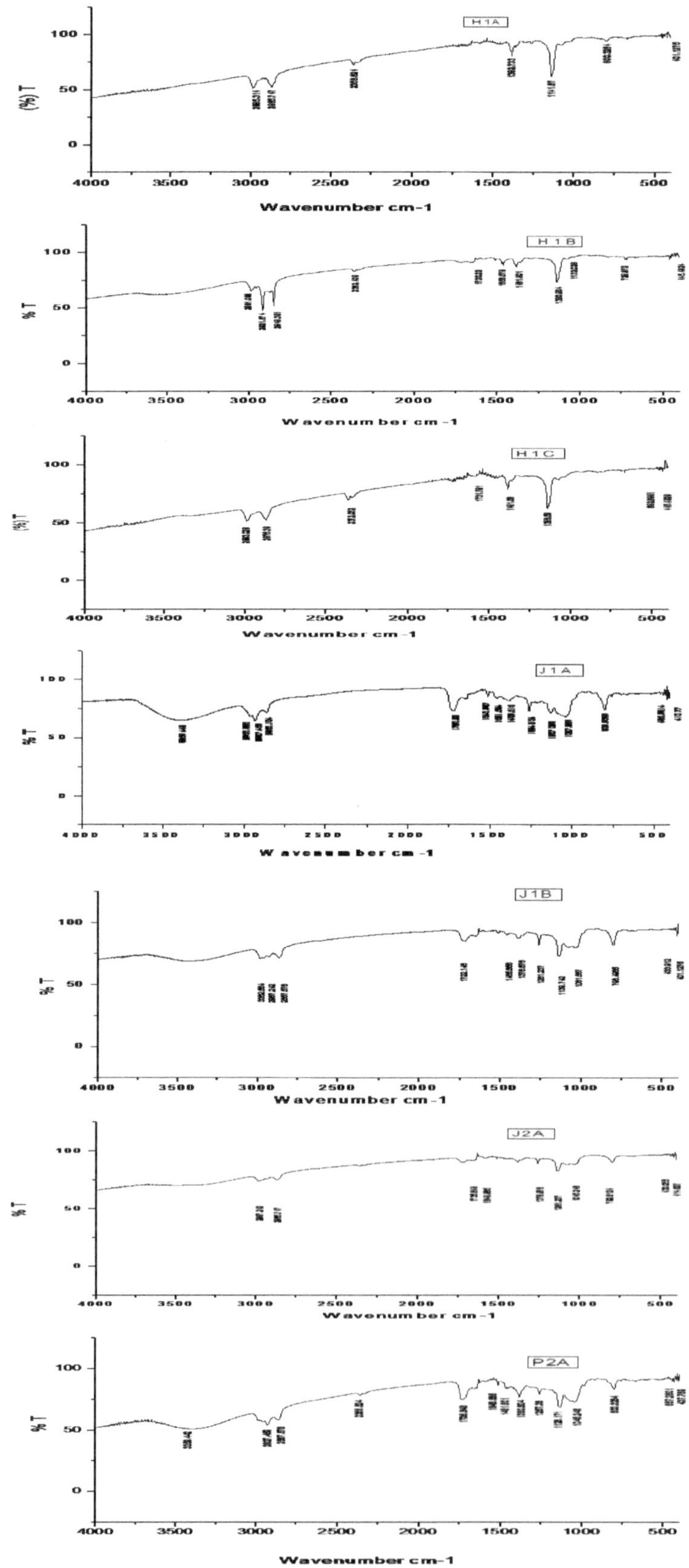

Figure 3: FTIR spectra of biosurfactant produced by various test strains in MSM media amended with dextrose carbon source.

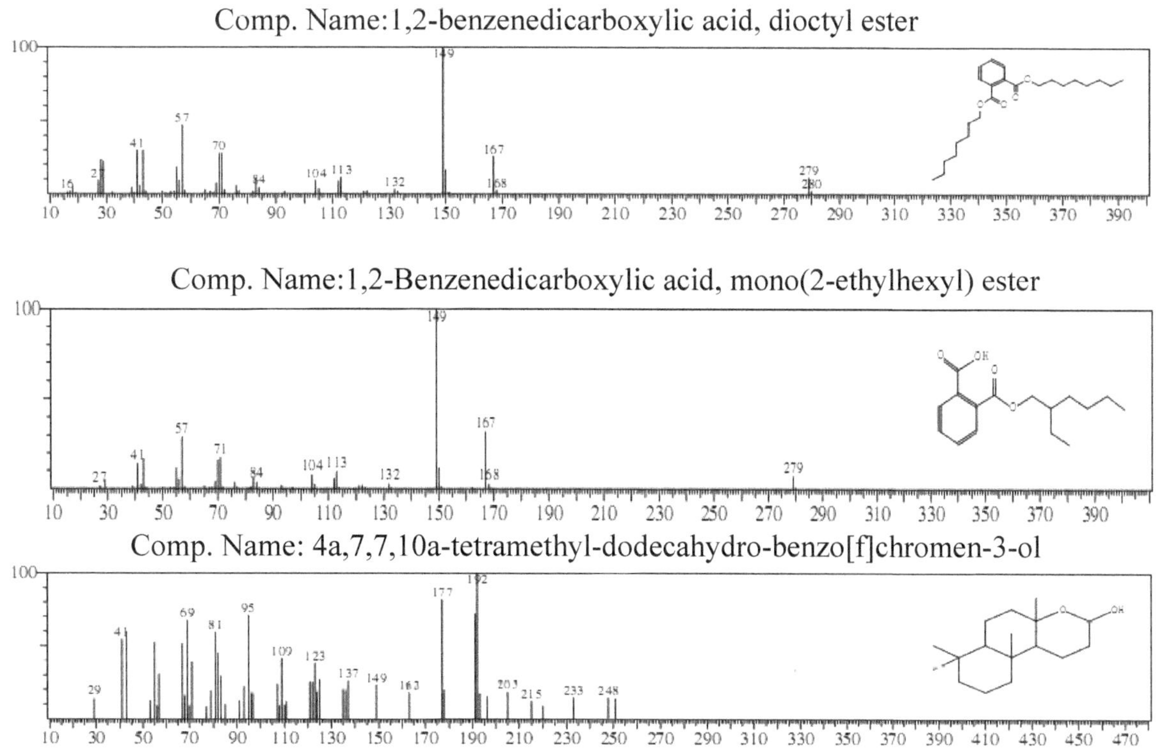

Figure 4: New peaks observed after degradation of 2% diesel hydrocarbons and identification of new compound in wheat rhizospheric soil (A) and mustard rhizospheric soil (B).

77%. Tapis massa oil degradation by bacterial consortium from crude oil refinery ground water has also been reported [29]. The phthalate esters [1,2-benzenedicarboxylic acid and 1,2-Benzenedicarboxylic acid, mono (2-ethylhexyl) ester] formed during the degradation of mobil oil (Figure 4) can be used as plasticizer or for softening of PVC material as they are less toxic to mammalian health.

So, the consortium used in the present study besides biodegradation of mobil oil hydrocarbons (bioremediation) from contaminated sites, can also be used in the synthesis of phthalate esters for industrial applications.

Acknowledgment

The author of this paper wishes to acknowledge Indian National Science Academy, New Delhi for providing financial assistance to carry out this work.

References

1. Pacwa-PÅ‚ociniczak M, PÅ‚aza GA, Piotrowska-Seget Z, Cameotra SS (2011) Environmental applications of biosurfactants: recent advances. Int J Mol Sci 12: 633-654.
2. Glick BR (2003) Phytoremediation: synergistic use of plants and bacteria to clean up the environment. Biotechnol Adv 21: 383-393.
3. Glick BR (2010) Using soil bacteria to facilitate phytoremediation. Biotechnol Adv 28: 367-374.
4. Al-Mailem DM, Sorkhoh NA, Al-Awadhi H, Eliyas M, Radwan SS (2010) Biodegradation of crude oil and pure hydrocarbons by extreme halophilic archaea from hypersaline coasts of the Arabian Gulf. Extremophiles 14: 321-328.
5. Margesin R, Schinner F (2001) Biodegradation and bioremediation of hydrocarbons in extreme environments. Appl Microbiol Biotechnol 56: 650-663.
6. Nievas ML, Commendatore MG, Esteves JL, Bucalá V (2008) Biodegradation pattern of hydrocarbons from a fuel oil-type complex residue by an emulsifier-producing microbial consortium. J Hazard Mater 154: 96-104.
7. Mulligan CN, Gibbs BF (2004) Types, production and applications of biosurfactants. Proc Indian Nat Sc Acad B70 1: 31-55.
8. Urum K, Pekdemir T (2004) Evaluation of biosurfactants for crude oil contaminated soil washing. Chemosphere 57: 1139-1150.
9. Lyman WJ, Reidy PJ, Levy B (1992) Mobility and degradation of organic contaminants in subsurface environments. CRC Press, USA.
10. Franzetti A, Gandolfi I, Bestetti G, Smyth TJP, Banat IM (2010) Production and applications of trehalose lipid biosurfactants. Eur J Lipid Sci Tech 112: 617-627.
11. Bento FM, Camargo FA, Okeke BC, Frankenberger WT (2005) Comparative bioremediation of soils contaminated with diesel oil by natural attenuation, biostimulation and bioaugmentation. Bioresour Technol 96: 1049-1055.
12. Glick BR (2012) Plant Growth-Promoting Bacteria: Mechanisms and Applications. Scientifica.
13. Lugtenberg BJJ, de Weger LA, Bennett JW (1991) Microbial stimulation of plant growth and protection from disease. Curr Opin Biotech 2: 457-464.
14. van Loon LC, Bakker PA, Pieterse CM (1998) Systemic resistance induced by rhizosphere bacteria. Annu Rev Phytopathol 36: 453-483.
15. Meyer JM (2000) Pyoverdines: pigments, siderophores and potential taxonomic markers of fluorescent Pseudomonas species. Arch Microbiol 174: 135-142.
16. Barrow GI, Feltham RKA (1993) Cowan and Steel's manual for the identification of medical bacteria, 3rd edition. Cambridge University Press, Cambridge, USA.
17. Schwyn B, Neilands JB (1987) Universal chemical assay for the detection and determination of siderophores. Anal Biochem 160: 47-56.
18. Karnwal A (2009) Production of Indole Acetic Acid by Fluorescent Pseudomonas In The Presence Of L-Tryptophan and Rice Root Exudates. Journal of Plant Pathology 91: 61-63.

19. Voisard C, Keel C, Haas D, Dèfago G (1989) Cyanide production by Pseudomonas fluorescens helps suppress black root rot of tobacco under gnotobiotic conditions. EMBO J 8: 351-358.

20. Nautiyal CS (1999) An efficient microbiological growth medium for screening phosphate solubilizing microorganisms. FEMS Microbiol Lett 170: 265-270.

21. Zajic JE, Seffens W, Panchal C (1983) Biosurfactants. CRC Cr Rev Biotechn 1: 87-107.

22. Cooper DG, Goldenberg BG (1987) Surface-active agents from two bacillus species. Appl Environ Microbiol 53: 224-229.

23. Meyer JM, Geoffroy VA, Baysse C, Cornelis P, Barelmann I, et al. (2002) Siderophore-mediated iron uptake in fluorescent Pseudomonas: characterization of the pyoverdine-receptor binding site of three cross-reacting pyoverdines. Arch Biochem Biophys 397: 179-183.

24. Patten CL, Glick BR (1996) Bacterial biosynthesis of indole-3-acetic acid. Can J Microbiol 42: 207-220.

25. Okolo JC, Amadi EN, Odu CTI (2005) Effects of soil treatments containing poultry manure on crude oil degradation in a sandy loam soil. Applied Ecology And Environmental Research 3: 47-53.

26. Akinola MO, Udo AS, Okwok N (2004) Effect of crude oil (Bonny Light) on germination, early seedling growth and pigment content in maize (Zea mays L.). Journal of Science Technology and Environment 4: 6-9.

27. Merkl N, Schultze-Kraft R, Infante C (2004) Phytoremediation in the Tropics—The Effect of Crude Oil on the Growth of Tropical Plants. Bioremediation Journal 8: 177-184.

28. Sathish-Kumar M, Binupriya AR, Baik SH, Yun SE (2008) Biodegradation of Crude Oil by Individual Bacterial Strains and a Mixed Bacterial Consortium Isolated from Hydrocarbon Contaminated Areas. CLEAN–Soil, Air, Water 36: 92-96.

29. Hamzah A, Phan CW, Bakar NFA, Wong KK (2013) Biodegradation of Crude Oil by Constructed Bacterial Consortia and the Constituent Single Bacteria Isolated From Malaysia. Bioremediation Journal 17: 1-10.

Permissions

The contributors of this book come from diverse backgrounds, making this book a truly international effort. This book will bring forth new frontiers with its revolutionizing research information and detailed analysis of the nascent developments around the world.

We would like to thank all the contributing authors for lending their expertise to make the book truly unique. They have played a crucial role in the development of this book. Without their invaluable contributions this book wouldn't have been possible. They have made vital efforts to compile up to date information on the varied aspects of this subject to make this book a valuable addition to the collection of many professionals and students.

This book was conceptualized with the vision of imparting up-to-date information and advanced data in this field. To ensure the same, a matchless editorial board was set up. Every individual on the board went through rigorous rounds of assessment to prove their worth. After which they invested a large part of their time researching and compiling the most relevant data for our readers. Conferences and sessions were held from time to time between the editorial board and the contributing authors to present the data in the most comprehensible form. The editorial team has worked tirelessly to provide valuable and valid information to help people across the globe.

Every chapter published in this book has been scrutinized by our experts. Their significance has been extensively debated. The topics covered herein carry significant findings which will fuel the growth of the discipline. They may even be implemented as practical applications or may be referred to as a beginning point for another development. Chapters in this book were first published by OMICS Group; hereby published with permission under the Creative Commons Attribution License or equivalent.

The editorial board has been involved in producing this book since its inception. They have spent rigorous hours researching and exploring the diverse topics which have resulted in the successful publishing of this book. They have passed on their knowledge of decades through this book. To expedite this challenging task, the publisher supported the team at every step. A small team of assistant editors was also appointed to further simplify the editing procedure and attain best results for the readers.

Our editorial team has been hand-picked from every corner of the world. Their multi-ethnicity adds dynamic inputs to the discussions which result in innovative outcomes. These outcomes are then further discussed with the researchers and contributors who give their valuable feedback and opinion regarding the same. The feedback is then collaborated with the researches and they are edited in a comprehensive manner to aid the understanding of the subject.

Apart from the editorial board, the designing team has also invested a significant amount of their time in understanding the subject and creating the most relevant covers. They scrutinized every image to scout for the most suitable representation of the subject and create an appropriate cover for the book.

The publishing team has been involved in this book since its early stages. They were actively engaged in every process, be it collecting the data, connecting with the contributors or procuring relevant information. The team has been an ardent support to the editorial, designing and production team. Their endless efforts to recruit the best for this project, has resulted in the accomplishment of this book. They are a veteran in the field of academics and their pool of knowledge is as vast as their experience in printing. Their expertise and guidance has proved useful at every step. Their uncompromising quality standards have made this book an exceptional effort. Their encouragement from time to time has been an inspiration for everyone.

The publisher and the editorial board hope that this book will prove to be a valuable piece of knowledge for researchers, students, practitioners and scholars across the globe.

List of Contributors

Anil Kumar Singh and Swaranjit Singh Cameotra
Institute of Microbial Technology, Sector 39 a, Chandigarh 160036, India

Evan Stephens, Ian L Ross and Ben Hankamer
The University of Queensland, Solar Biofuels Research Centre, Queensland, Australia
The University of Queensland, Institute for Molecular Bioscience, Queensland, Australia

Rocky de Nys
James Cook University, School of Marine and Tropical Biology, Queensland, Australia
James Cook University, Centre for Sustainable Tropical Fisheries and Aquaculture, Queensland, Australia

Javed Ahmad and Tauseef A Ansari
Department of Biochemical Engineering and Food Technology, Harcourt Butler Technological Institute, Nawabganj, Kanpur 208002, India

Hadi Belhaj and Hadil Abu Khalifeh
The Petroleum Institute, Abu Dhabi, UAE
Regent Group, Calgary, Canada

Nasser Al-Huraibi
Regent Group, Calgary, Canada

Abha Singh, Vinay Kumar and Srivastava JN
Microbiology Lab, Botany Department, Dayalbagh Educational Institute, Agra, Uttar Pradesh, India

Ghollami M and Roayaei M
Biology Department, Shahid Chamran University, Ahwaz, Iran

Ghavipanjeh F
Environment Group, Energy Department, Materials and Energy Research Center, Alborz, Iran

Rasekh B
Microbiology and Biotechnology Group, Research Institute of Petroleum Industry, Tehran, Iran

Garima Bhardwaj and Harish Kumar Chopra
Department of Chemistry, Sant Longowal Institute of Engineering and Technology, Longowal - 148106, Sangrur, Punjab, India

Swaranjit Singh Cameotra
Institute of Microbial Technology, Sector-39-A, Chandigarh-160036, India

Leandro Costa Lima Verde, Tiago Rodrigues e Silva, Bruna Martins Dellagnezze and Valéria Maia de Oliveira
Research Center for Chemistry, Biology and Agriculture (CPQBA), Campinas University - UNICAMP, CP 6171, CEP 13081-970, Campinas, SP, Brazil

Eugênio Vaz dos Santos Neto
Petrobras R&D Center, Cidade Universitária, Quadra 7, Rio de Janeiro, RJ, CEP: 21949-900, Brazil

He J, Arensman D and Nasr-El-Din HE
SPE, Texas A&M University, Texas, USA

Jones JC
School of Engineering, University of Aberdeen, UK

Wael Ismail, Noora A Alhamad, Ashraf M El Nayal and Riyad Y Hamzah
Biotechnology Program, College of Graduate Studies, Arabian Gulf University, Manama, Kingdom of Bahrain

Wael S El-Sayed
Taibah University, Faculty of Science, Biology Department, Al-Madinah Al-Monawarah, 344, Kingdom of Saudi Arabia
Microbiology Department, Faculty of Science, Ain Shams University, Cairo, Egypt

Yin-Ru Chiang
Biodiversity Research Center, Academia Sinica, Taipei 115, Taiwan

Rengathavasi Thavasi and Shilpy Sharma
Department of Chemical and Biological Sciences, Polytechnic Institute of New York University, 6 Metrotech Center, Brooklyn, New York 11201, USA

Singaram Jayalakshmi
CAS in Marine Biology, Annamalai University, Parangipettai 608502, TN, Indi

Mabkhot S Bin-Dahbag, Mohammed S Benzagouta and Mustafa M Kinawy
Department of Petroleum and Natural Gas Engineering, King Saud University, Saudi Arabia

Abdulrahman A Al Quraishi and Emad Al Mushaegeh
Oil and Gas Research Institute, King Abdulaziz City for Sciences and Technology, Saudi Arabia

Inas M Al Nashef
Department of Chemical Engineering, King Saud University, Saudi Arabia

Zhonghong Chen and Ming Zha
School of Geosciences, China University of Petroleum, Qingdao, Shandong, China

Jiqing Zhang and Li Wang
Petroleum Developing Center in Shengli Oilfield, SINOPEC, Dongying, Shandong, China

Ola A Olapade
Department of Biology and the Center for Sustainability and the Environment, Albion College, 611 East Porter Street, Albion, MI 49224, USA

Fazlurrahman Khan, Mohammad Sajid and Swaranjit Singh Cameotra
Environmental Biotechnology and Microbial Biochemistry Laboratory, Institute of Microbial Technology, Sector 39-A, Chandigarh-160036, India

Gupta Divya and Tiwari Archana
School of Biotechnology, Rajiv Gandhi Proudyogiki Vishwavidyalaya, Bhopal, M.P., India

Ramirez Alejandro Manzano
Department of Material Sciences CINVESTAV-IPN, Unidad Querétaro Libramiento Norponiente, Querétaro, México

Amandeep Girdhar, Mayuri Bhatia, Amol Kanampalliwar and Archana Tiwari
School of Biotechnology, Rajiv Gandhi Proudyogiki Vishwavidhyalaya, State Technological University, Airport Bypass Road, Bhopal 426033, Madhya Pradesh, India

Sunil Nagpal
Department of Biochemical Engineering, Indian Institute of Technology, Safdarganj Development Area, Hauz Khas, IIT Campus, New Delhi, Delhi 110016, India

Tukhvatullina AZ, Barskaya EE, Ganeeva YM, Yusupova TN and Romanov GV
AE Arbuzov Institute of Organic and Physical Chemistry of Kazan Scientific Center of Russian Academy of Sciences, Kazan, Russia

Kouryakov VN
Oil and Gas Research Institute of Russian Academy of Sciences, Moscow, Russia

Gisela Jakob, Juliane Wolf, Evan Stephens, Ian L Ross and Ben Hankamer
Institute for Molecular Bioscience, The University of Queensland, Queensland, Australia
Solar Biofuels Research Centre, The University of Queensland, Queensland, Australia

Tony VL Bui
Institute for Molecular Bioscience, The University of Queensland, Queensland, Australia

Clemens Posten
Institute of Life Science Engineering, Bioprocess Engineering, University of Karlsruhe, Karlsruhe, BW D-76131, Germany

Olaf Kruse
Department of Biology, Algae Biotechnology & Bioenergy, University of Bielefeld, Bielefeld, NRW D-33501, Germany

Khaled Abdalla Elraies and Shuaib Ahmed Kalwar
Faculty of Geoscience and Petroleum Engineering, Petroleum Engineering Department, Universiti Teknologi Petronas, Malaysia

Rajesh Kumar, Ram Naresh Bharagava and Manish Kumar
Department of Environmental Microbiology, School for Environmental Sciences, Babasaheb Bhimrao Ambedkar University, Vidya Vihar, Rae Bareli Road, Lucknow, Uttar Pradesh, India

Sanjeev K Singh
Division of Eco-Restoration, National Environmental Engineering Research Institute, Nehru Marg, Nagpur, India

Govind K
Department of Microbiology, College of Basic Science and Humanities, Pantnagar, Uttarakhand, India